Finite Elements

Theory, fast solvers, and applications in solid mechanics

Second Edition

Dietrich Braess
Ruhr-University, Bochum

Translated by Larry L. Schumaker

CAMBRIDGE
UNIVERSITY PRESS

CAMBRIDGE UNIVERSITY PRESS
Cambridge, New York, Melbourne, Madrid, Cape Town, Singapore, São Paulo

Cambridge University Press
The Edinburgh Building, Cambridge CB2 2RU, UK

Published in the United States of America by Cambridge University Press, New York

www.cambridge.org
Information on this title: www.cambridge.org/9780521011952

Originally published in German as *Finite Elemente*
by Springer Verlag, 1992

First published in English by Cambridge University Press 1997
as *Finite Elements*
Second edition 2001
Reprinted 2002
English translation © Cambridge University Press 1997, 2001

A catalogue record for this publication is available from the British Library

ISBN-13 978-0-521-01195-2 paperback
ISBN-10 0-521-01195-7 paperback

Transferred to digital printing 2005

1004851837

Contents

Chapter III
Nonconforming and Other Methods   105

Chapter V
Multigrid Methods 216

Chapter VI
Finite Elements in Solid Mechanics 269

Preface to the Second English Edition

The theory of finite elements and their applications is so vivid that the second printing gave rise to some additions. We have not only eliminated some misprints, but we have also added some new material that although being basic has turned out to be of interest in actual research or actual applications of finite elements during the last years. We will emphasize some of these extensions.

The introduction of finite element spaces in Chapter II, §5 is now focused such that all the ingredients of the formal definition at the end of that § are well motivated.

The general considerations of saddle point problems in Chapter III are augmented. The direct and converse theorems that are related to Fortin interpolation are presented now under a common aspect. Mixed methods are often connected with a softening of the energy functional that is wanted in some applications for good reasons. It is described in order to understand a different but equivalent variational formulation that has become popular in solid mechanics.

In Chapter IV only the standard proof of the Kantorowitch inequality has been replaced by a shorter one.

The multigrid theory requires less regularity assumptions if convergence with respect to the energy norm is considered. A quick introduction into that theory is now included, and multigrid algorithms are also considered in the framework of space decompositions.

Finite element computations in solid mechanics require often appropriate elements in order to avoid an effect called "locking" by engineers. From the mathematical point of view we have problems with a small parameter. Methods for treating nearly incompressible material serve as a model for positive results while negative results are easily described for a more general framework.

The author wants to thank numerous friends who have given valuable hints for improvements of the text. Finally thanks are going to Cambridge University Press for the continuation of the good cooperation.

Autumn, 2000   *Dietrich Braess*

Preface to the First English Edition

This book is based on lectures regularly presented to students in the third and fourth year at the Ruhr-University, Bochum. It was also used by the translater, Larry Schumaker, in a graduate course at Vanderbilt University in Nashville. I would like to thank him for agreeing to undertake the translation, and for the close cooperation in carrying it out. My thanks are also due to Larry and his students for raising a number of questions which led to improvements in the material itself.

Chapters I and II and selected sections of Chapters III and V provide material for a typical course. I have especially emphasized the differences with the numerical treatment of ordinary differential equations (for more details, see the preface to the German edition).

One may ask why I was not content with presenting only simple finite elements based on complete polynomials. My motivation for doing more was provided by problems in fluid mechanics and solid mechanics, which are treated to some extent in Chapter III and VI. I am not aware of other textbooks for mathematicians which give a mathematical treatment of finite elements in solid mechanics in this generality.

The English translation contains some additions as compared to the German edition from 1992. For example, I have added the theory for basic a posteriori error estimates since a posteriori estimates are often used in connection with local mesh refinements. This required a more general interpolation process which also applies to non-uniform grids. In addition, I have also included an analysis of locking phenomena in solid mechanics.

Finally, I would like to thank Cambridge University Press for their friendly cooperation, and also Springer-Verlag for agreeing to the publication of this English version.

Autumn, 1996   *Dietrich Braess*

Preface to the German Edition

The method of finite elements is one of the main tools for the numerical treatment of elliptic and parabolic partial differential equations. Because it is based on the variational formulation of the differential equation, it is much more flexible than finite difference methods and finite volume methods, and can thus be applied to more complicated problems. For a long time, the development of finite elements was carried out in parallel by both mathematicians and engineers, without either group acknowledging the other. By the end of the 60's and the beginning of the 70's, the material became sufficiently standardized to allow its presentation to students. This book is the result of a series of such lectures.

In contrast to the situation for ordinary differential equations, for elliptic partial differential equations, frequently no classical solution exists, and we often have to work with a so-called weak solution. This has consequences for both the theory and the numerical treatment. While it is true that classical solutions do exist under approriate regularity hypotheses, for numerical calculations we usually cannot set up our analisis in a framework in which the existence of classical solutions is guaranteed.

One way to get a suitable framework for solving elliptic boundary-value problems using finite elements is to pose them as variational problems. It is our goal in Chapter II to present the simplest possible introduction to this approach. In Sections 1 – 3 we discuss the existence of weak solutions in Sobolev spaces, and explain how the boundary conditions are incorporated into the variational calculation. To give the reader a feeling for the theory, we derive a number of properties of Sobolev spaces, or at least illustrate them. Sections 4 – 8 are devoted to the foundations of finite elements. The most difficult part of this chapter is §6 where approximation theorems are presented. To simplify matters, we first treat the special case of regular grids, which the reader may want to focus on in a first reading.

In Chapter III we come to the part of the theory of finite elements which requires deeper results from functional analysis. These are presented in §3. Among other things, the reader will learn about the famous Ladyshenskaja–Babuška–Brezzi condition, which is of great importance for the proper treatment of problems in fluid mechanics and for mixed methods in structural mechanics. In fact, without this knowledge and relying only on *common sense*, we would very likely find ourselves trying to solve problems in fluid mechanics using elements with an unstable behavior.

It was my aim to present this material with as little reliance on results from real analysis and functional analysis as possible. On the other hand, a certain basic knowledge is extremely useful. In Chapter I we briefly discuss the difference between the different types of partial differential equations. Students confronting the numerical solution of elliptic differential equations for the first time often find the finite difference method more accessible. However, the limits of the method usually become apparent only later. For completeness we present an elementary introduction to finite difference methods in Chapter I.

For fine discretizations, the finite element method leads to very large systems of equations. The operation count for solving them by direct methods grows like n^2. In the last two decades, very efficient solvers have been developed based on multigrid methods and on the method of conjugate gradients. We treat these subjects in detail in Chapters IV and V.

Structural mechanics provides a very important application area for finite elements. Since these kinds of problems usually involve systems of partial differential equations, often the elementary methods of Ch. II do not suffice, and we have to use the extra flexibility which the deeper results of Ch. III allow. I found it necessary to assemble a surprisingly wide set of building blocks in order to present a mathematically rigorous theory for the numerical treatment by finite elements of problems in linear elasticity theory.

Almost every section of the book includes a set of Problems, which are not only excercises in the strict sense, but also serve to further develop various formulae or results from a different viewpoint, or to follow a topic which would have disturbed the flow had it been included in the text itself. It is well-known that in the numerical treatment of partial differential equations, there are many opportunities to go down a false path, even if unintended, particularly if one is thinking in terms of classical solutions. Learning to avoid such pitfalls is one of the goals of this book.

This book is based on lectures regularly presented to students in the fifth through eighth semester at the Ruhr University, Bochum. Chapters I and II and parts of Chapters III and V were presented in one semester, while the method of conjugate gradients was left to another course. Chapter VI is the result of my collaboration with both mathematicians and engineers at the Ruhr University.

A text like this can only be written with the help of many others. I would like to thank F.-J. Barthold, C. Blömer, H. Blum, H. Cramer, W. Hackbusch, A. Kirmse, U. Langer, P. Peisker, E. Stein, R. Verfürth, G. Wittum and B. Worat for their corrections and suggestions for improvements. My thanks are also due to Frau L. Mischke, who typeset the text using TEX, and to Herr Schwarz for his help with technical problems relating to TEX. Finally, I would like to express my appreciation to Springer-Verlag for the publication of the German edition of this book, and for the always pleasant collaboration on its production.

Bochum, Autumn, 1991 *Dietrich Braess*

Notation

Notation for Differential Equations and Finite Elements

Ω	open set in $\mathbb{R}^n$
Γ	$= \partial\Omega$
Γ_D	part of the boundary on which Dirichlet conditions are prescribed
Γ_N	part of the boundary on which Neumann conditions are prescribed
Δ	Laplace operator
L	differential operator
a_{ik}, a_0	coefficient functions of the differential equation
$[\,\cdot\,]_*$	difference star, stencil
$L^2(\Omega)$	space of square-integrable functions over Ω
$H^m(\Omega)$	Sobolev space of L_2 functions with square-integrable derivatives up to order m
$H_0^m(\Omega)$	subspace of $H^m(\Omega)$ of functions with generalized zero bounary conditions
$C^k(\Omega)$	set of functions with continuous derivatives up to order k
$C_0^k(\Omega)$	subspace of $C^k(\Omega)$ of functions with compact support
γ	trace operator
$\|\cdot\|_m$	Sobolev norm of order m
$\|\cdot\|_m$	Sobolev semi-norm of order m
$\|\cdot\|_\infty$	supremum norm
$\|\cdot\|_{m,h}$	mesh-dependent norm
ℓ_2	space of square-summable sequences
H'	dual space of H
$\langle\cdot,\cdot\rangle$	dual pairing
$|\alpha|$	$= \sum \alpha_i$, order of multiindex α
∂_i	partial derivative $\frac{\partial}{\partial x_i}$
∂^α	partial derivative of order α
D	(Fréchet) derivative
α	ellipticity constant
ν	exterior normal
∂_ν	$\partial/\partial\nu$, derivative in the direction of the exterior normal
∇f	$(\partial f/\partial x_1, \partial f/\partial x_2, \ldots, \partial f/\partial x_n)$
div f	$\sum_{i=1}^n (\partial f_i/\partial x_i)$
S_h	finite element space
ψ_h	basis function in S_h
$\mathcal{T}_h$	partition of Ω
T	(triangular or quadrilateral) element in $\mathcal{T}_h$

T_{ref}	reference element
h_T, ρ_T	radii of circumscribed circle and incircle of T, respectively
κ	shape parameter of a partition
$\mu(T)$	area (volume) of T
$\mathcal{P}_t$	set of polynomials of degree $\leq t$
$\mathcal{Q}_t$	polynomial set (II.5.4) w.r.t. quadrilateral elements
$\mathcal{P}_{3,\mathrm{red}}$	cubic polynomial without bubble function term
Π_{ref}	set of polynomials which are formed by the restriction of S_h to a (reference) element
s	$= \dim \Pi_{\mathrm{ref}}$
Σ	set of linear functionals in the definition of affine families
$\mathcal{M}^k, \mathcal{M}_s^k, \mathcal{M}_{s,0}^k$	polynomial finite element spaces in L_2, H^{s+1} and H_0^{s+1}
$\mathcal{M}_{*,0}^1$	set of functions in $\mathcal{M}^1$ which are continuous at the midpoints of the sides and which satisfy zero boundary conditions in the same sense
RT_k	Raviart–Thomas element of degree k
I, I_h	interpolation operators on Π_{ref} and on S_h, respectively
A	stiffness or system matrix
$\delta_{..}$	Kronecker symbol
e	edge of an element
$\ker L$	kernel of the linear mapping L
$V^\perp$	orthogonal complement of V
V^0	polar of V
$\mathcal{L}$	Lagrange function
M	space of restrictions (for saddle point problems)
β	constant in the Brezzi condition
$H(\mathrm{div}, \Omega)$	$:= \{v \in L_2(\Omega)^d; \; \mathrm{div}\, v \in L_2(\Omega)\}, \; \Omega \in \mathbb{R}^d$
$L_{2,0}(\Omega)$	set of functions in $L_2(\Omega)$ with integral mean 0
B_3	cubic bubble functions
$\eta_{...}$	error estimator

Notation for the Method of Conjugate Gradients

∇f	gradient of f (column vector)
$\kappa(A)$	spectral condition number of the matrix A
$\sigma(A)$	spectrum of the matrix A
$\rho(A)$	spectral radius of the matrix A
$\lambda_{\min}(A)$	smallest eigenvalue of the matrix A
$\lambda_{\max}(A)$	largest eigenvalue of the matrix A
A^t	transpose of the matrix A
I	unit matrix
C	preconditioning matrix
g_k	gradient at the actual approximation x_k

d_k direction of the correction in step k

V_k $= \mathrm{span}[g_0, \ldots, g_{k-1}]$

$x'y$ Euclidean scalar product of the vectors x and y

$\|x\|_A$ $= \sqrt{x'Ax}$ (energy norm)

$\|x\|_\infty$ $= \max_i |x_i|$ (maximum norm)

T_k k-th Chebyshev polynomial

ω relaxation parameter

Notation for the Multigrid Method

$\mathcal{T}_\ell$ triangulation on the level ℓ

$S_\ell = S_{h_\ell}$ finite element space on the level ℓ

A_ℓ system matrix on the level ℓ

N_ℓ $= \dim S_\ell$

S smoothing operator

$r, \tilde{r}$ restrictions

p prolongation

$x^{\ell,k,m}, u^{\ell,k,m}$ variable on the level ℓ in the k-th iteration step and in the m-th substep

ν_1, ν_2 number of presmoothings or postsmoothings, respectively

ν $= \nu_1 + \nu_2$

μ $= 1$ for V-cycle, $= 2$ for W-cycle

q $= \ell_{\max}$

ψ_ℓ^j j-th basis function on the level ℓ

ρ_ℓ convergence rate of $\mathbf{MGM}_\ell$

ρ $= \sup_\ell \rho_\ell$

$\||\cdot\||_s$ discrete norm of order s ·

β measure of the smoothness of a function in S_h

$\mathcal{L}$ nonlinear operator

$\mathcal{L}_\ell$ nonlinear mapping on the level ℓ

$D\mathcal{L}$ derivative of $\mathcal{L}$

λ homotopy parameter for incremental methods

Notation for Solid Mechanics

u displacement

ϕ deformation

id identity mapping

C $= \nabla\phi^T \nabla\phi$ Cauchy–Green strain tensor

E strain

ε strain in a linear approximation

t Cauchy stress vector

T Cauchy stress tensor

T_R first Piola–Kirchhoff stress tensor

Σ_R second Piola–Kirchhoff stress tensor

$\hat{T}$	$= \hat{T}(F)$ response function for the Cauchy stress tensor
$\hat{\Sigma}$	$= \hat{\Sigma}(F)$ response function for the Piola–Kirchhoff stress tensor
$\tilde{\Sigma}$	$\tilde{\Sigma}(F^T F) = \hat{\Sigma}(F)$
$\bar{T}$	$\bar{T}(F F^T) = \hat{T}(F)$
σ	stress in linear approximation
S^2	unit sphere in $\mathbb{R}^3$
$\mathbb{M}^3$	set of 3×3 matrices
$\mathbb{M}^3_+$	set of matrices in $\mathbb{M}^3$ with positive determinants
$\mathbb{O}^3$	set of orthogonal 3×3 matrices
$\mathbb{O}^3_+$	$= \mathbb{O}^3 \cap \mathbb{M}^3_+$
$\mathbb{S}^3$	set of symmetric 3×3 matrices
$\mathbb{S}^3_>$	set of positive definite matrices in $\mathbb{S}^3$
ι_A	$= (\iota_1(A), \iota_2(A), \iota_3(A))$, invariants of A
$\wedge$	vector product in $\mathbb{R}^3$
$\text{diag}(d_1, \ldots, d_n)$	diagonal matrix with elements $d_1, \ldots, d_n$
λ, μ	Lamé constants
E	modulus of elasticity
ν	Poisson ratio
n	normal vector (different from Chs. II and III)
$\mathcal{C}$	$\sigma = \mathcal{C}\varepsilon$
$\hat{W}$	energy functional of hyperelastic materials
$\tilde{W}$	$\tilde{W}(F^T F) = \hat{W}(F)$
$\varepsilon : \sigma$	$= \sum_{ij} \varepsilon_{ij}\sigma_{ij}$
Γ_0, Γ_1	parts of the boundary on which u and $\sigma \cdot n$ are prescribed, respectively
Π	energy functional in the linear theory
$\nabla^{(s)}$	symmetric gradient
$as(\tau)$	skew-symmetric part of τ
$H^s(\Omega)^d$	$= [H^s(\Omega)]^d$
$H^1_\Gamma(\Omega)$	$:= \{v \in H^1(\Omega)R; \ v(x) = 0 \text{ for } x \in \Gamma_0\}$
$H(\text{div}, \Omega)$	$:= \{\tau \in L_2(\Omega); \ \text{div}\,\tau \in L_2(\Omega)\}$, τ is a vector or a tensor
$H(\text{rot}, \Omega)$	$:= \{\eta \in L_2(\Omega)^2; \ \text{rot}\,\eta \in L_2(\Omega)\}$, $\Omega \subset \mathbb{R}^2$
$H^{-1}(\text{div}, \Omega)$	$:= \{\tau \in H^{-1}(\Omega)^d; \ \text{div}\,\tau \in H^{-1}(\Omega)\}$, $\Omega \subset \mathbb{R}^d$
θ, γ, w	rotation, shear term, and transversal displacement of beams and plates
t	thickness of a beam, membrane, or plate
ℓ	length of a beam
$W_h, \Theta_h, \Gamma_h, Q_h$	finite element spaces in plate theory
π_h	L_2-projector onto Γ_h
R	restriction to Γ_h
P_h	L_2-projector onto Q_h

Chapter I
Introduction

In dealing with partial differential equations, it is useful to differentiate between several types. In particular, we classify partial differential equations of second order as *elliptic, hyperbolic,* and *parabolic.* Both the theoretical and numerical treatment differ considerably for the three types. For example, in contrast with the case of ordinary differential equations where either initial or boundary conditions can be specified, here the type of equation determines whether initial, boundary, or initial-boundary conditions should be imposed.

The most important application of the finite element method is to the numerical solution of elliptic partial differential equations. Nevertheless, it is important to understand the differences between the three types of equations. In addition, we present some elementary properties of the various types of equations. Our discussion will show that for differential equations of elliptic type, we need to specify boundary conditions and not initial conditions.

There are two main approaches to the numerical solution of elliptic problems: *finite difference methods* and *variational methods.* The finite element method belongs to the second category. Although finite element methods are particularly effective for problems with complicated geometry, finite difference methods are often employed for simple problems, primarily because they are simpler to use. We include a short and elementary discussion of them in this chapter.

§ 1. Examples and Classification of PDE's

Examples

We first consider some examples of second order partial differential equations which occur frequently in physics and engineering, and which provide the basic prototypes for elliptic, hyperbolic, and parabolic equations.

1.1 Potential Equation. Let Ω be a domain in $\mathbb{R}^2$. Find a function u on Ω with

$$u_{xx} + u_{yy} = 0. \tag{1.1}$$

This is a differential equation of second order. To determine a unique solution, we must also specify boundary conditions.

One way to get solutions of (1.1) is to identify $\mathbb{R}^2$ with the complex plane. It is known from function theory that if $w(z) = u(z) + i v(z)$ is a holomorphic function on Ω, then its real part u and imaginary part v satisfy the potential equation. Moreover, u and v are infinitely often differentiable in the interior of Ω, and attain their maximum and minimum values on the boundary.

For the case where $\Omega := \{(x, y) \in \mathbb{R}^2; x^2 + y^2 < 1\}$ is a disk, there is a simple formula for the solution. Since $z^k = (re^{i\phi})^k$ is holomorphic, it follows that

$$r^k \cos k\phi, \quad r^k \sin k\phi, \quad \text{for } k = 0, 1, 2, \ldots,$$

satisfy the potential equation. If we expand these functions on the boundary in Fourier series,

$$u(\cos \phi, \sin \phi) = a_0 + \sum_{k=1}^{\infty} (a_k \cos k\phi + b_k \sin k\phi),$$

we can represent the solution in the interior as

$$u(x, y) = a_0 + \sum_{k=1}^{\infty} r^k (a_k \cos k\phi + b_k \sin k\phi). \tag{1.2}$$

The differential operator in (1.1) is the *two-dimensional Laplace operator*. For functions of d variables, it is

$$\Delta u := \sum_{i=1}^{d} \frac{\partial^2 u}{\partial x_i^2}.$$

The potential equation is a special case of the Poisson equation.

1.2 Poisson Equation. Let Ω be a domain in $\mathbb{R}^d$, $d = 2$ or 3. Here $f : \Omega \to \mathbb{R}$ is a prescribed charge density in Ω, and the solution u of the Poisson equation

$$-\Delta u = f \quad \text{in } \Omega \tag{1.3}$$

describes the potential throughout Ω. As with the potential equation, this type of problem should be posed with boundary conditions.

1.3 The Plateau Problem as a Prototype of a Variational Problem. Suppose we stretch an ideal elastic membrane over a wire frame to create a drum. Suppose the wire frame is described by a closed, rectifiable curve in $\mathbb{R}^3$, and suppose that its parallel projection onto the (x, y)-plane is a curve with no double points. Then the position of the membrane can be described as the graph of a function $u(x, y)$. Because of the elasticity, it must assume a position such that its surface area

$$\int_\Omega \sqrt{1 + u_x^2 + u_y^2}\, dx dy$$

is minimal.

In order to solve this nonlinear variational problem approximately, we introduce a simplification. Since $\sqrt{1 + z} = 1 + \frac{z}{2} + \mathcal{O}(z^2)$, for small values of u_x and u_y we can replace the integrand by a quadratic expression. This leads to the problem

$$\frac{1}{2} \int_\Omega (u_x^2 + u_y^2)\, dx dy \to \text{min!} \tag{1.4}$$

The values of u on the boundary $\partial\Omega$ are prescribed by the given curve. We now show that the minimum is characterized by the associated Euler equation

$$\Delta u = 0. \tag{1.5}$$

Since such variational problems will be dealt with in more detail in Chapter II, here we establish (1.5) only on the assumption that a minimal solution u exists in $C^2(\Omega) \cap C^0(\bar\Omega)$. If a solution belongs to $C^2(\Omega) \cap C^0(\bar\Omega)$, it is called a *classical solution*. Let

$$D(u, v) := \int_\Omega (u_x v_x + u_y v_y)\, dx dy$$

and $D(v) := D(v, v)$. The quadratic form D satisfies the binomial formula

$$D(u + \alpha v) = D(u) + 2\alpha D(u, v) + \alpha^2 D(v).$$

Let $v \in C^1(\Omega)$ and $v|_{\partial\Omega} = 0$. Since $u + \alpha v$ for $\alpha \in \mathbb{R}$ is an admissible function for the minimum problem (1.4), we have $\frac{\partial}{\partial\alpha} D(u + \alpha v) = 0$ for $\alpha = 0$. Using

4

I. Introduction

the above binomial formula, we get $D(u, v) = 0$. Now applying Green's integral formula, we have

$$0 = D(u, v) = \int_\Omega (u_x v_x + u_y v_y)\, dx dy$$

$$= -\int_\Omega v(u_{xx} + u_{yy})\, dx dy + \int_{\partial\Omega} v(u_x dy - u_y\, dx).$$

The contour integral vanishes because of the boundary condition for v. The first integral vanishes for all $v \in C^1(\Omega)$ if and only if $\Delta u = u_{xx} + u_{yy} = 0$. This proves that (1.5) characterizes the solution of the (linearized) Plateau problem.

1.4 The Wave Equation as a Prototype of a Hyperbolic Differential Equation. The motion of particles in an ideal gas is subject to the following three laws, where as usual, we denote the velocity by v, the density by ρ, and the pressure by p:

1. *Continuity Equation.*
$$\frac{\partial \rho}{\partial t} = -\rho_0 \operatorname{div} v.$$

Because of conservation of mass, the change in the mass contained in a (partial) volume V must be equal to the flow through its surface, i.e., it must be equal to $\int_{\partial V} \rho v \cdot n\, dO$. Applying Gauss' integral theorem, we get the above equation. Here ρ is approximated by the fixed density ρ_0.

2. *Newton's Law.*
$$\rho_0 \frac{\partial v}{\partial t} = -\operatorname{grad} p.$$

The gradient in pressure induces a force field which causes the acceleration of the particles.

3. *State Equation.*
$$p = c^2 \rho.$$

In ideal gases, the pressure is proportional to the density for constant temperature.

Using these three laws, we conclude that

$$\frac{\partial^2}{\partial t^2} p = c^2 \frac{\partial^2 \rho}{\partial t^2} = -c^2 \frac{\partial}{\partial t} \rho_0 \operatorname{div} v = -c^2 \operatorname{div}(\rho_0 \frac{\partial v}{\partial t})$$

$$= c^2 \operatorname{div} \operatorname{grad} p = c^2 \Delta p.$$

Other examples of the *wave equation*

$$u_{tt} = c^2 \Delta u$$

arise in two space dimensions for vibrating membranes, and in the one-dimension-al case for a vibrating string. In one space dimension, the equation simplifies when c is normalized to 1:

$$u_{tt} = u_{xx}. \tag{1.6}$$

The wave equation leads to a well-posed problem (see Definition 1.8 below) when combined with initial conditions of the form

$$\begin{aligned} u(x,0) &= f(x), \\ u_t(x,0) &= g(x). \end{aligned} \tag{1.7}$$

1.5 Solution of the One-dimensional Wave Equation. To solve the wave equation (1.6)–(1.7), we apply the transformation of variables

$$\begin{aligned} \xi &= x+t, \\ \eta &= x-t. \end{aligned} \tag{1.8}$$

Applying the chain rule $u_x = u_\xi \frac{\partial \xi}{\partial x} + u_\eta \frac{\partial \eta}{\partial x}$, etc., we easily get

$$\begin{aligned} u_x &= u_\xi + u_\eta, & u_{xx} &= u_{\xi\xi} + 2u_{\xi\eta} + u_{\eta\eta}, \\ u_t &= u_\xi - u_\eta, & u_{tt} &= u_{\xi\xi} - 2u_{\xi\eta} + u_{\eta\eta}. \end{aligned} \tag{1.9}$$

Substituting the formulae (1.9) in (1.6) gives

$$4u_{\xi\eta} = 0.$$

The general solution is

$$\begin{aligned} u &= \phi(\xi) + \psi(\eta) \\ &= \phi(x+t) + \psi(x-t), \end{aligned} \tag{1.10}$$

where ϕ and ψ are functions which can be determined from the initial conditions (1.7):

$$\begin{aligned} f(x) &= \phi(x) + \psi(x), \\ g(x) &= \phi'(x) - \psi'(x). \end{aligned}$$

After differentiating the first equation, we have two equations for ϕ' and ψ' which are easily solved:

$$\begin{aligned} \phi' &= \frac{1}{2}(f'+g), & \phi(\xi) &= \frac{1}{2}f(\xi) + \frac{1}{2}\int_{x_0}^{\xi} g(s)\, ds, \\ \psi' &= \frac{1}{2}(f'-g), & \psi(\eta) &= \frac{1}{2}f(\eta) - \frac{1}{2}\int_{x_0}^{\eta} g(s)\, ds. \end{aligned}$$

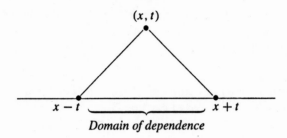

(x, t)

$x - t$ $x + t$

Domain of dependence

Fig. 1. Domain of dependence for the wave equation

Finally, using (1.10) we get

$$u(x, t) = \frac{1}{2}[f(x + t) + f(x - t)] + \frac{1}{2} \int_{x-t}^{x+t} g(s)\, ds. \qquad (1.11)$$

We emphasize that the solution $u(x, t)$ depends only on the initial values between the points $x - t$ and $x + t$ (see Fig. 1). [If the constant c is not normalized to be 1, the dependence is on all points between $x - ct$ and $x + ct$]. This corresponds to the fact that in the underlying physical system, any change of data can only propagate with a finite velocity.

The solution u in (1.11) was derived on the assumption that it is twice differentiable. If the initial functions f and g are not differentiable, then neither are ϕ, ψ and u. However, the formula (1.11) remains correct and makes sense even in the nondifferentiable case.

1.6 The Heat Equation as a Prototype of a Parabolic Equation. Let $T(x, t)$ be the distribution of temperature in an object. Then the heat flow is given by

$$F = -\kappa \operatorname{grad} T,$$

where κ is the diffusion constant which depends on the material. Because of conservation of energy, the change in energy in a volume element is the sum of the heat flow through the surface and the amount of heat injection Q. Using the same arguments as for conservation of mass in Example 1.4, we have

$$\frac{\partial E}{\partial t} = -\operatorname{div} F + Q$$
$$= \operatorname{div} \kappa \operatorname{grad} T + Q$$
$$= \kappa \Delta T + Q,$$

where κ is assumed to be constant. Introducing the constant $a = \partial E / \partial T$ for the specific heat (which also depends on the material), we get

$$\frac{\partial T}{\partial t} = \frac{\kappa}{a} \Delta T + \frac{1}{a} Q.$$

For a one-dimensional rod and $Q = 0$, with $u = T$ this simplifies to

$$u_t = \sigma u_{xx}, \tag{1.12}$$

where $\sigma = \kappa/a$. As before, we may assume the normalization $\sigma = 1$ by an appropriate choice of units.

Parabolic problems typically lead to *initial-boundary-value problems*.

We first consider the heat distribution on a rod of finite length ℓ. Then, in addition to the initial values, we also have to specify the temperature or the heat fluxes on the boundaries. For simplicity, we restrict ourselves to the case where the temperature is constant at both ends of the rod as a function of time. Then, without loss of generality, we can assume that

$$\sigma = 1, \quad \ell = \pi \quad \text{and} \quad u(0, t) = u(\pi, t) = 0;$$

cf. Problem 1.10. Suppose the initial values are given by the Fourier series expansion

$$u(x, 0) = \sum_{k=1}^{\infty} a_k \sin kx, \quad 0 < x < \pi.$$

Obviously, the functions $e^{-k^2 t} \sin kx$ satisfy the heat equation $u_t = u_{xx}$, and thus

$$u(x, t) = \sum_{k=1}^{\infty} a_k e^{-k^2 t} \sin kx, \quad t \geq 0 \tag{1.13}$$

is a solution of the given initial-value problem.

For an infinitely long rod, the boundary conditions drop out. Now we need to know something about the decay of the initial values at infinity, which we ignore here. In this case we can write the solution using Fourier integrals instead of Fourier series. This leads to the representation

$$u(x, t) = \frac{1}{2\sqrt{\pi t}} \int_{-\infty}^{+\infty} e^{-\xi^2/4t} f(x - \xi) \, d\xi, \tag{1.14}$$

where the initial value $f(x) := u(x, 0)$ appears explicitly. Note that the solution at a point (x, t) depends on the initial values on the entire domain, and the propagation of the data occurs with infinite speed.

Classification of PDE's

Problems involving ordinary differential equations can be posed with either initial or boundary conditions. This is no longer the case for partial differential equations. Here the question of whether initial or boundary conditions should be applied depends on the *type of the differential equation*.

The general linear partial differential equation of second order in n variables $x = (x_1, \ldots, x_n)$ has the form

$$-\sum_{i,k=1}^{n} a_{ik}(x)u_{x_i x_k} + \sum_{i=1}^{n} b_i(x)u_{x_i} + c(x)u = f(x). \qquad (1.15)$$

If the functions a_{ik}, b_i and c do not depend on x, then the partial differential equation has *constant coefficients*. Since $u_{x_i x_k} = u_{x_k x_i}$ for any function which is twice continuously differentiable, without loss of generality we can assume the symmetry $a_{ik}(x) = a_{ki}(x)$. Then the corresponding $n \times n$ matrix

$$A(x) := (a_{ik}(x))$$

is symmetric.

1.7 Definition. (1) The equation (1.15) is called *elliptic at the point* x provided $A(x)$ is positive definite.

(2) The equation (1.15) is called *hyperbolic at the point* x provided $A(x)$ has one negative and $n - 1$ positive eigenvalues.

(3) The equation (1.15) is called *parabolic at the point* x provided $A(x)$ is positive semidefinite, but is not positive definite, and the rank of $(A(x), b(x))$ equals n.

(4) An equation is called *elliptic, hyperbolic* or *parabolic* provided it has the corresponding property for all points of the domain. $\square$

In the elliptic case, the equation (1.15) is usually written in the compact form

$$L u = f, \qquad (1.16)$$

where L is an *elliptic differential operator of order* $\overset{.}{2}$. The part with the derivatives of highest order, i.e., $-\sum a_{ik}(x)u_{x_i x_k}$, is called the *principal part* of L. For hyperbolic and parabolic problems there is a special variable which is usually time. Thus, hyperbolic differential equations can often be written in the form

$$u_{tt} + Lu = f, \qquad (1.17)$$

while parabolic ones can often be written in the form

$$u_t + Lu = f, \qquad (1.18)$$

where L is an elliptic differential operator.

If a differential equation is invariant under *isometric mappings* (i.e., under translation and rotation), then the elliptic operator has the form

$$Lu = -a_0 \Delta u + c_0 u.$$

The above examples all display this invariance.

Well-posed Problems

What happens if we consider a partial differential equation in a framework which is meant for a different type?

To answer this question, we first turn to the wave equation (1.6), and attempt to solve the *boundary-value problem* in the domain

$$\Omega = \{(x,t) \in \mathbb{R}^2; \ a_1 < x + t < a_2, \ b_1 < x - t < b_2\}.$$

Here Ω is a rotated rectangle, and its sides are parallel to the coordinate axes ξ, η defined in (1.8). In view of $u(\xi, \eta) = \phi(\xi) + \psi(\eta)$, the values of u on opposite sides of Ω can differ only by a constant. Thus, the boundary-value problem with general data is not solvable. This also follows for differently shaped domains by similar but somewhat more involved considerations.

Next we study the potential equation (1.1) in the domain $\{(x, y) \in \mathbb{R}^2; \ y \geq 0\}$ as an *initial-value problem*, where y plays the role of time. Let $n > 0$. Assuming

$$u(x, 0) = \frac{1}{n} \sin nx,$$
$$u_y(x, 0) = 0,$$

we clearly get the *formal solution*

$$u(x, y) = \frac{1}{n} \cosh ny \ \sin nx,$$

which grows like e^{ny}. Since n can be arbitrarily large, we draw the following conclusion: there exist *arbitrarily small initial values* for which the corresponding solution at $y = 1$ is *arbitrarily large*. This means that solutions of this problem, when they exist, are not stable with respect to perturbations of the initial values.

Using the same arguments, it is immediately clear from (1.13) that a solution of a parabolic equation is well-behaved for $t > t_0$, but not for $t < t_0$. However, sometimes we want to solve the heat equation in the backwards direction, e.g., in order to find out what initial temperature distribution is needed in order to get a prescribed distribution at a later time $t_1 > 0$. This is a well-known improperly posed problem. By (1.13), we can prescribe at most the low frequency terms of the temperature at time t_1, but by no means the high frequency ones.

Considerations of this type led Hadamard [1932] to consider the solvability of differential equations (and similarly structured problems) together with the stability of the solution.

1.8 Definition. A problem is called *well posed* provided it has a unique solution which depends continuously on the given data. Otherwise it is called *improperly posed.*

In principle, the question of whether a problem is well posed can depend on the choice of the norm used for the corresponding function spaces. For example, from (1.11) we see that problem (1.6)–(1.7) is well posed. The mapping

$$C(\mathbb{R}) \times C(\mathbb{R}) \longrightarrow C(\mathbb{R} \times \mathbb{R}_+),$$
$$f, g \longmapsto u$$

defined by (1.11) is continuous provided $C(\mathbb{R})$ is endowed with the usual maximum norm, and $C(\mathbb{R} \times \mathbb{R}_+)$ is endowed with the weighted norm

$$\|u\| := \max_{x,t} \left\{ \frac{|u(x, t)|}{1 + |t|} \right\}.$$

The maximum principle to be discussed in the next section is a useful tool for showing that elliptic and parabolic differential equations are well posed.

Problems

1.9 Consider the potential equation in the disk $\Omega := \{(x, y) \in \mathbb{R}^2; \; x^2 + y^2 < 1\}$, with the boundary condition

$$\frac{\partial}{\partial r} u(x) = g(x) \quad \text{for } x \in \partial\Omega$$

on the derivative in the normal direction. Find the solution when g is given by the Fourier series

$$g(\cos\phi, \sin\phi) = \sum_{k=1}^{\infty} (a_k \cos k\phi + b_k \sin k\phi)$$

without a constant term. (The reason for the lack of a constant term will be explained in Ch. II, §3.)

1.10   Consider the heat equation (1.12) for a rod with $\sigma \neq 1$, $\ell \neq \pi$ and $u(0, t) = u(\ell, t) = T_0 \neq 0$. How should the scalars, i.e., the constants in the transformations $t \longmapsto \alpha t$, $x \longmapsto \beta x$, $u \longmapsto u + \gamma$, be chosen so that the problem reduces to the normalized one?

1.11 Solve the heat equation for a rod with the temperature fixed only at the left end. Suppose that at the right end, the rod is isolated, so that the heat flow, and thus $\partial T/\partial x$, vanishes there.

Hint: For k odd, the functions $\phi_k(x) = \sin kx$ satisfy the boundary conditions $\phi_k(0) = 0$, $\varphi'(\frac{\pi}{2}) = 0$.

1.12 Suppose u is a solution of the wave equation, and that at time $t = 0$, u is zero outside of a bounded set. Show that the energy

$$\int_{\mathbf{R}^d} [u_t^2 + c^2(\operatorname{grad} u)^2]\, dx \qquad (1.19)$$

is constant.

Hint: Write the wave equation in the symmetric form

$$u_t = c \operatorname{div} v,$$
$$v_t = c \operatorname{grad} u,$$

and represent the time derivative of the integrand in (1.19) as the divergence of an appropriate expression.

§ 2. The Maximum Principle

An important tool for the analysis of finite difference methods is the discrete analog of the so-called maximum principle. Before turning to the discrete case, we examine a simple continuous version.

In the following, Ω denotes a bounded domain in $\mathbb{R}^d$. Let

$$Lu := -\sum_{i,k=1}^{d} a_{ik}(x)u_{x_i x_k} \tag{2.1}$$

be a linear elliptic differential operator L. This means that the matrix $A = (a_{ik})$ is symmetric and positive definite on Ω. For our purposes we need a quantitative measure of ellipticity.

For convenience, the reader may assume that the coefficients a_{ik} are continuous functions, although the results remain true under less restrictive hypotheses.

2.1 Maximum Principle. *For $u \in C^2(\Omega) \cap C^0(\bar{\Omega})$, let*

$$Lu = f \leq 0 \quad in \ \Omega.$$

Then u attains its maximum over $\bar{\Omega}$ on the boundary of Ω. Moreover, if u attains a maximum at an interior point of a connected set Ω, then u must be constant on Ω.

Here we prove the first assertion. For a proof of the second one, see Gilbarg and Trudinger [1983].

(1) We first carry out the proof under the stronger assumption that $f < 0$. Suppose that for some $x_0 \in \Omega$,

$$u(x_0) = \sup_{x \in \Omega} u(x) > \sup_{x \in \partial\Omega} u(x).$$

Applying the linear coordinate transformation $x \longmapsto \xi = Ux$, the differential operator becomes

$$Lu = -\sum_{i,k}(U^t A(x)U)_{ik}u_{\xi_i \xi_k}$$

in the new coordinates. In view of the symmetry, we can find an orthogonal matrix U so that $U^T A(x_0)U$ is diagonal. By the definiteness of $A(x_0)$, we deduce that these diagonal elements are positive. Since x_0 is a maximal point,

$$u_{\xi_i} = 0, \quad u_{\xi_i \xi_i} \leq 0$$

at $x = x_0$. This means that

$$Lu(x_0) = -\sum_i (U^T A(x_0) U)_{ii} u_{\xi_i \xi_i} \geq 0,$$

in contradiction with $Lu(x_0) = f(x_0) < 0$.

(2) Now suppose that $f(x) \leq 0$ and that there exists $x = \bar{x} \in \Omega$ with $u(\bar{x}) > \sup_{x \in \partial\Omega} u(x)$. The auxiliary function $h(x) := (x_1 - \bar{x}_1)^2 + (x_2 - \bar{x}_2)^2 + \cdots + (x_d - \bar{x}_d)^2$ is bounded on $\partial\Omega$. Now if $\delta > 0$ is chosen sufficiently small, then the function

$$w := u + \delta h$$

attains its maximum at a point x_0 in the interior. Since $h_{x_i x_k} = 2\delta_{ik}$, we have

$$
\begin{aligned}
Lw(x_0) &= Lu(x_0) + \delta Lh(x_0) \\
&= f(x_0) - 2\delta \sum_i a_{ii}(x_0) < 0.
\end{aligned}
$$

This leads to a contradiction just as in the first part of the proof. □

Examples

The maximum principle has interesting interpretations for the equations (1.1)–(1.3). If the charge density vanishes in a domain Ω, then the potential is determined by the potential equation. Without any charge, the potential in the interior cannot be larger than its maximum on the boundary. The same holds if there are only negative charges.

Next we consider the variational problem 1.3. Let $c := \max_{x \in \partial\Omega} u(x)$. If the solution u does not attain its maximum on the boundary, then

$$w(x) := \min\{u(x), c\}$$

defines an admissible function which is different from u. Now the integral $D(w, w)$ exists in the sense of Lebesgue, and

$$D(w, w) = \int_{\Omega_1} (u_x^2 + u_y^2) \, dx \, dy < \int_{\Omega} (u_x^2 + u_y^2) \, dx \, dy,$$

where $\Omega_1 := \{(x, y) \in \Omega;\ u(x) < c\}$. Thus, w leads to a smaller (generalized) surface than u. We can smooth w to get a differentiable function which also provides a smaller surface. This means that the minimal solution must satisfy the maximum principle.

Corollaries

A number of simple consequences of the maximum principle can be easily derived by elementary means, such as taking the difference of two functions, or by replacing u by $-u$.

2.2 Definition. An elliptic operator of the form (2.1) is called *uniformly elliptic* provided there exists a constant $\alpha > 0$ such that

$$\xi' A(x)\xi \geq \alpha \|\xi\|^2 \quad \text{for } \xi \in \mathbb{R}^d, x \in \Omega. \tag{2.2}$$

The largest such constant α is called the *constant of ellipticity*.

2.3 Corollary. Suppose L is a linear elliptic differential operator.

(1) *Minimum Principle.* If $Lu = f \geq 0$ on Ω, then u attains its minimum on the boundary of Ω.

(2) *Comparison Principle.* Suppose $u, v \in C^2(\Omega) \cap C^0(\bar{\Omega})$ and

$$\begin{aligned} Lu &\leq Lv \quad \text{in } \Omega, \\ u &\leq v \quad \text{on } \partial\Omega. \end{aligned}$$

Then $u \leq v$ in Ω.

(3) *Continuous Dependence on the Boundary Data.* The solution of the linear equation $Lu = f$ with Dirichlet boundary conditions depends continuously on the boundary values. Suppose u_1 and u_2 are solutions of the linear equation $Lu = f$ with two different boundary values. Then

$$\sup_{x \in \Omega} |u_1(x) - u_2(x)| = \sup_{z \in \partial\Omega} |u_1(z) - u_2(z)|.$$

(4) *Continuous Dependence on the Right-Hand Side.* Let L be uniformly elliptic in Ω. Then there exists a constant c which depends only on Ω and the ellipticity constant α such that

$$|u(x)| \leq \sup_{z \in \partial\Omega} |u(z)| + c \sup_{z \in \Omega} |Lu(z)| \tag{2.3}$$

for every $u \in C^2(\Omega) \cap C^0(\bar{\Omega})$.

(5) *Elliptic Operators with Helmholtz Terms.* There is a weak form of the maximum principle for the general differential operator

$$Lu := -\sum_{i,k=1}^{d} a_{ik}(x) u_{x_i x_k} + c(x)u \quad \text{with } c(x) \geq 0. \tag{2.4}$$

In particular, $Lu \leq 0$ implies

$$\max_{x \in \Omega} u(x) \leq \max\{0, \max_{x \in \partial\Omega} u(x)\}. \tag{2.5}$$

Proof. (1) Apply the maximum principle to $v := -u$.

(2) By construction, $Lw = Lv - Lu \geq 0$ and $w \geq 0$ on $\partial\Omega$, where $w := v - u$. It follows from the minimum principle that inf $w \geq 0$, and thus $w(x) \geq 0$ in Ω.

(3) $Lw = 0$ for $w := u_1 - u_2$. It follows from the maximum principle that $w(x) \leq \sup_{z \in \partial\Omega} w(z) \leq \sup_{z \in \partial\Omega} |w(z)|$. Similarly, the minimum principle implies $w(x) \geq -\sup_{z \in \partial\Omega} |w(z)|$.

(4) Suppose Ω is contained in a circle of radius R. Since we are free to choose the coordinate system, we may assume without loss of generality that the center of this circle is at the origin. Let

$$w(x) = R^2 - \sum_i x_i^2.$$

Since $w_{x_i x_k} = -2\delta_{ik}$, clearly $Lw \geq 2n\alpha$ and $0 \leq w \leq R^2$ in Ω, where α is the ellipticity constant appearing in Definition 2.2. Let

$$v(x) := \sup_{z \in \partial\Omega} |u(z)| + w(x) \cdot \frac{1}{2n\alpha} \sup_{z \in \partial\Omega} |Lu(z)|.$$

Then by construction, $Lv \geq |Lu|$ in Ω, and $v \geq |u|$ on $\partial\Omega$. The comparison principle in (2) implies $-v(x) \leq u(x) \leq +v(x)$ in Ω. Since $w \leq R^2$, we get (2.3) with $c = R^2/2n\alpha$.

(5) It suffices to give a proof for $x_0 \in \Omega$ and $u(x_0) = \sup_{z \in \Omega} u(z) > 0$. Then $Lu(x_0) - c(x_0)u(x_0) \leq Lu(x_0) \leq 0$, and moreover, the principal part $Lu - cu$ defines an elliptic operator. Now the proof proceeds as for Theorem 2.1. $\quad\square$

Problem

2.4 For a uniformly elliptic differential operator of the form (2.4), show that the solution depends continuously on the data.

§ 3. Finite Difference Methods

The finite difference method for the numerical solution of an elliptic partial differential equation involves computing approximate values for the solution at points on a rectangular grid. To compute these values, derivatives are replaced by divided differences. The stability of the method follows from a discrete analog of the maximum principle, which we will call the *discrete maximum principle*. For simplicity, we assume that Ω is a domain in $\mathbb{R}^2$.

Discretization

The first step in the discretization is to put a two-dimensional grid over the domain Ω. For simplicity, we restrict ourselves to a grid with constant mesh size h in both variables; see Fig. 2:

$$\Omega_h := \{(x, y) \in \Omega; \; x = kh, \; y = \ell h \quad \text{with } k, \ell \in \mathbb{Z}\},$$
$$\partial\Omega_h := \{(x, y) \in \partial\Omega; \; x = kh \; \text{ or } \; y = \ell h \quad \text{with } k, \ell \in \mathbb{Z}\}.$$

We want to compute approximations to the values of u on Ω_h. These approximate values define a function U on $\Omega_h \cup \partial\Omega_h$. We can think of U as a vector of dimension equal to the number of grid points.

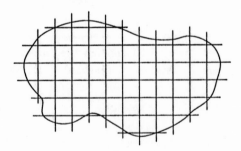

Fig. 2. A grid on a domain Ω

We get an equation at each point $z_i = (x_i, y_i)$ of Ω_h by evaluating the differential equation $Lu = f$, after replacing the derivatives in the representation (2.4) by divided differences. We choose the center of the divided difference to be the grid point of interest, and mark the neighboring points with subscripts indicating their direction relative to the center (see Fig. 3).

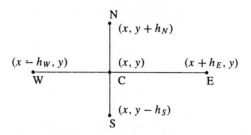

Fig. 3. Coordinates of the neighboring points of C for nonuniform step sizes. The labels of the neighbors refer to the directions east, south, west, and north.

If (x, y) is a point on a square grid whose distance to the boundary is greater than h, we can choose $h_N = h_W = h_S = h_E$ (see Fig. 2). However, for points in the neighborhood of the boundary, we have to choose $h_E \neq h_W$ or $h_N \neq h_S$. Using the Taylor formula, we see that for $u \in C^3(\Omega)$,

$$u_{xx} = \frac{2}{h_E(h_E + h_W)} u_E - \frac{2}{h_E h_W} u_C + \frac{2}{h_W(h_E + h_W)} u_W + \mathcal{O}(h), \quad (3.1)$$

where h is the maximum of h_E and h_W. In the special case where the step sizes are the same, i.e., $h_E = h_W = h$, we get the simpler formula

$$u_{xx} = \frac{1}{h^2}(u_E - 2u_C + u_W) + \mathcal{O}(h^2) \quad \text{for } u \in C^4(\Omega), \quad (3.2)$$

with an error term of second order. Analogous formulae hold for approximating u_{yy} in terms of the values u_C, u_S and u_N. To approximate the mixed derivative u_{xy} by a divided difference, we also need either the values at the NW and SE positions, or those at the NE and SW positions.

Discretization of the Poisson equation $-\Delta u = f$ leads to a system of the form

$$\alpha_C u_C + \alpha_E u_E + \alpha_S u_S + \alpha_W u_W + \alpha_N u_N = h^2 f(x_C) \quad \text{for } x_C \in \Omega_h, \quad (3.3)$$

where for each $z_C \in \Omega_h$, u_C is the associated function value. The variables with a subscript indicating a compass direction are values of u at points which are neighbors of x_C. If the differential equation has constant coefficients and we use a uniform grid, then the coefficients α_* appearing in (3.3) for a point x_C not near the boundary do not depend on C. We can write them in a matrix which we call the *difference star* or *stencil:*

$$\begin{bmatrix} \alpha_{NW} & \alpha_N & \alpha_{NE} \\ \alpha_W & \alpha_C & \alpha_E \\ \alpha_{SW} & \alpha_S & \alpha_{SE} \end{bmatrix}_*. \quad (3.4)$$

For example, for the Laplace operator, (3.2) yields the so-called *standard five-point stencil*

$$\frac{1}{h^2}\begin{bmatrix} & -1 & \\ -1 & +4 & -1 \\ & -1 & \end{bmatrix}_* .$$

To get a higher order discretization error for the Laplace operator, we can use the *nine-point stencil*

$$\frac{1}{6h^2}\begin{bmatrix} -1 & -4 & -1 \\ -4 & 20 & -4 \\ -1 & -4 & -1 \end{bmatrix}_* .$$

These stencils are also called *multipoint formulae* in the literature.

3.1 An Algorithm for the Discretization of the Dirichlet Problem.
1. Choose a step size $h > 0$, and construct Ω_h and $\partial\Omega_h$.
2. Let n and m be the numbers of points in Ω_h and $\partial\Omega_h$, respectively. Number the points of Ω_h from 1 to n. Usually this is done so that the coordinates (x_i, y_i) appear in lexicographical order. Number the boundary points as $n+1$ to $n+m$.
3. Insert the given values at the boundary points:
$$U_i = u(z_i) \quad \text{for } i = n+1, \ldots, n+m.$$
4. For every interior point $z_i \in \Omega_h$, write the difference equation with z_i as center point which gives the discrete analog of $Lu(z_i) = f(z_i)$:
$$\sum_{\ell=C,E,S,W,N} \alpha_\ell U_\ell = f(z_i). \tag{3.5}$$
If a neighboring point z_ℓ belongs to the boundary $\partial\Omega_h$, move the associated term $\alpha_\ell U_\ell$ in (3.5) to the right-hand side.
5. Step 4 leads to a system
$$A_h U = f$$
of n equations in n unknowns U_i. Solve this system and identify the solution U as an approximation to u on the grid Ω_h. (Usually U is called a *numerical solution* of the PDE.)

3.2 Examples. (1) Let Ω be an isosceles right triangle whose nondiagonal sides are of length 7; see Fig. 4. Suppose we want to solve the Laplace equation $\Delta u = 0$ with Dirichlet boundary conditions. For $h = 2$, Ω_h contains three points. We get the following system of equations for U_1, U_2 and U_3:

$$U_1 - \frac{1}{4}U_2 - \frac{1}{4}U_3 = \frac{1}{4}U_4 + \frac{1}{4}U_{11},$$
$$-\frac{1}{6}U_1 + U_2 \qquad = \frac{1}{6}U_5 + \frac{1}{3}U_6 + \frac{1}{3}U_7,$$
$$-\frac{1}{6}U_1 \qquad + U_3 = \frac{1}{3}U_8 + \frac{1}{3}U_9 + \frac{1}{6}U_{10}.$$

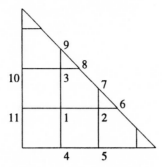

Fig. 4. Grid for Example 3.2

(2) Suppose we want to solve the Poisson equation in the unit square:

$$-\Delta u = f \quad \text{in } \Omega = [0, 1]^2,$$
$$u = 0 \quad \text{on } \partial\Omega.$$

Choose a grid on Ω with mesh size $h = 1/m$. For convenience, we use the double indexing system $U_{ij} \approx u(\frac{i}{m}, \frac{j}{m})$, $1 \le i, j \le m - 1$. This leads to the system

$$4U_{i,j} - U_{i-1,j} - U_{i+1,j} - U_{i,j-1} - U_{i,j+1} = f_{i,j}, \quad 1 \le i, j \le m - 1, \quad (3.6)$$

where $f_{i,j} = h^2 f(\frac{i}{m}, \frac{j}{m})$. Here terms with indices 0 or m are taken to be zero.

Discrete Maximum Principle

When using the standard five-point stencil (and also in Example 3.2) every value U_i is a weighted average of neighboring values. This clearly implies that no value can be larger than the maximum of its neighbors, and is a special case of the following more general result.

3.3 Star Lemma. *Let $k \ge 1$. Suppose α_i and p_i, $0 \le i \le k$, are such that*

$$\alpha_i < 0 \quad \text{for } i = 1, 2, \ldots, k,$$

$$\sum_{i=0}^{k} \alpha_i \ge 0, \quad \sum_{i=0}^{k} \alpha_i p_i \le 0.$$

In addition, let $p_0 \ge 0$ or $\sum_{i=0}^{k} \alpha_i = 0$. Then $p_0 \ge \max_{1 \le i \le k} p_i$ implies

$$p_0 = p_1 = \cdots = p_k. \quad (3.7)$$

Proof. The hypotheses imply that

$$\sum_{i=1}^{k} \alpha_i (p_i - p_0) = \sum_{i=0}^{k} \alpha_i (p_i - p_0) = \sum_{i=0}^{k} \alpha_i p_i - p_0 \sum_{i=0}^{k} \alpha_i \le 0.$$

Since $\alpha_i < 0$ for $i = 1, \ldots, k$ and $p_i - p_0 \le 0$, all summands appearing in the sums on the left-hand side are nonnegative. Hence, every summand equals 0. Now $\alpha_i \ne 0$ implies (3.7).   □

In the following, it is important to note that the discretization can change the topological structure of Ω. If Ω is connected, it does not follow that Ω_h is connected (with an appropriate definition). The situation shown in Fig. 5 leads to a system with a reducible matrix. To guarantee that the matrix is irreducible, we have to use a sufficiently small mesh size.

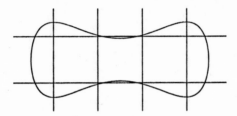

Fig. 5. Connected domain Ω for which Ω_h is not connected

3.4 Definition. Ω_h is said to be *(discretely) connected* provided that between every pair of points in Ω_h, there exists a path of grid lines which remains inside of Ω.

Clearly, using a finite difference method to solve the Poisson equation, we get a system with an irreducible matrix if and only if Ω_h is discretely connected.

We are now in a position to formulate the discrete maximum principle. Note that the hypotheses for the standard five-point stencil for the Laplace operator are satisfied.

3.5 Discrete Maximum Principle. *Let U be a solution of the linear system which arises from the discretization of*

$$Lu = f \quad in \ \Omega \quad with \ f \le 0$$

using a difference star which satisfies the following three conditions at every grid point in Ω_h:
(i) All of the coefficients except for the one at the center are nonpositive.
(ii) The coefficient in one of the directions is negative, say $\alpha_E < 0$.
(iii) The sum of all of the coefficients is nonnegative.
Then

$$\max_{z_i \in \Omega_h} U_i \le \max_{z_j \in \partial \Omega_h} U_j. \tag{3.8}$$

Furthermore, suppose the maximum over all the grid points is attained in the interior, the coefficients $\alpha_E, \alpha_S, \alpha_W$ and α_N in all four cardinal directions are negative, and Ω_h is discretely connected. Then U is constant.

Proof. (1) If the maximum value over $\bar{\Omega}_h$ is attained at $z_i \in \Omega_h$, then U_i must have the same values at all of the neighboring points which appear in the difference star of z_i. This follows from the star lemma when U_C is identified with p_0 and $U_E, U_S, \ldots$ are identified with $p_1, p_2, \ldots$.

(2) The assertion now follows using the technique of *marching to the boundary*. Consider all points of Ω_h and $\partial\Omega_h$ which lie on the same horizontal grid line as the point z_i. It follows from (1) by induction that the maximum is attained at all points on this line lying between z_i and the first encountered boundary point.

(3) If Ω_h is connected, by Definition 3.4 we can choose a polygonal path between z_i and any point z_k in Ω_h. Repeating the argument of (2), we get $U_i = U_k$, and thus U is constant. $\qquad\Box$

The discrete maximum principle implies that the discrete solution U has properties which correspond exactly to those in Corollary 2.3. In particular, we have both the comparison principle and the continuous dependence on f and on the boundary data. In addition, we have

3.6 Corollary. *If the hypotheses of the first part of the discrete maximum principle 3.5 are satisfied, then the system $A_h U = f$ in 3.1(5) has a unique solution.*

Proof. The corresponding homogeneous system $A_h U = 0$ is associated with the discretization of the homogeneous differential equation with zero boundary condition. By 3.5, $\max U_i = \min U_i = 0$. Thus the homogeneous system has only the trivial solution, and the matrix A_h is nonsingular. $\qquad\Box$

Problem

3.7 Obviously there is always the danger of a misprint in formulas as (3.1). Verify the formula by applying it to the functions 1, x, and x^2 and the points $-h_W$, 0, and h_E. Why is this test sufficient?

§ 4. A Convergence Theory for Difference Methods

It is relatively easy to establish the convergence of finite difference methods, provided that the solution u of the differential equation is sufficiently smooth up to the boundary, and its second derivatives are bounded. Although these assumptions are quite restrictive, it is useful to carry out the analysis in this framework to provide a first impression of the more general convergence theory. Under weaker assumptions, the analysis is much more complicated; see e.g. Hackbusch [1986].

Consistency

In the following we shall write L_h for the difference operator (which also specifies the method). Then given $u \in C(\Omega)$, $L_h u$ is a function defined at all points in Ω_h. The symbol A_h will denote the resulting matrix.

4.1 Definition. A finite difference method L_h is called *consistent* with the elliptic equation $Lu = f$ provided

$$Lu - L_h u = o(1) \quad \text{on } \Omega_h \text{ as } h \to 0,$$

for every function $u \in C^2(\bar{\Omega})$. A method has *consistency order* m provided that for every $u \in C^{m+2}(\bar{\Omega})$,

$$Lu - L_h u = \mathcal{O}(h^m) \quad \text{on } \Omega_h \text{ as } h \to 0.$$

The five-point formula (3.1) for the Laplace operator derived by Taylor expansions has order 1 for an arbitrary grid, and order 2 when the four neighbors of the center point are at the same distance from it.

Local and Global Error

The definition of consistency relates to the *local error* $Lu - L_h u$. However, the convergence of a method depends on the *global error*

$$\eta(z_i) := u(z_i) - U_i$$

as z_i runs over Ω_h. The two errors are connected by

4.2 A Difference Equation for the Global Error. Let

$$Lu = f \quad \text{in } \Omega,$$

and suppose

$$A_h U = F$$

is the associated linear system over Ω_h with $F_i = f(z_i)$. In addition, suppose that for the points on the boundary,

$$U(z_j) = u(z_j) \quad \text{for } z_j \in \partial\Omega_h.$$

In view of the linearity of the difference operator, it follows that the global error η satisfies

$$
\begin{aligned}
(L_h\eta)_i &= (L_h u)(z_i) - (A_h U)_i \\
&= (L_h u)(z_i) - f(z_i) = (L_h u)(z_i) - (Lu)(z_i) \\
&= -r_i,
\end{aligned}
\tag{4.1}
$$

where $r := Lu - L_h u$ is the local error on Ω_h. Thus, η can be interpreted as the solution of the discrete boundary-value problem

$$
\begin{aligned}
L_h\eta &= -r \quad \text{in } \Omega_h, \\
\eta &= 0 \quad \text{on } \partial\Omega_h.
\end{aligned}
\tag{4.2}
$$

4.3 Remark. If we eliminate those variables in (4.2) which belong to $\partial\Omega_h$, we get a system of the form

$$A_h \tilde{\eta} = -r.$$

Here $\tilde{\eta}$ is the vector with components $\tilde{\eta}_i = \eta(z_i)$ for $z_i \in \Omega_h$. This shows that convergence is assured provided r tends to 0 and the inverses A_h^{-1} remain bounded as $h \to 0$. This last condition is called *stability*. Thus, *consistency and stability imply convergence.*

In order to illustrate the error calculation by the perturbation method of (4.1), we will turn our attention for a moment to a more formal argument. We investigate the differences between the solutions of the two linear systems of equations

$$Ax = b,$$

$$(A + F)y = b,$$

where F is regarded as a small perturbation. Obviously, $(A + F)(x - y) = Fx$. Thus, the error $x - y = (A + F)^{-1} Fx$ is small provided F is small and $(A + F)^{-1}$ is bounded. – In estimating the global error by the above perturbation calculation, it is important to note that the given elliptic operator and the difference operator operate on different spaces.

We will estimate the size of the solution of (4.2) by considering the difference equation rather than via the norm of the inverse matrices $\| A_h^{-1} \|$.

4.4 Lemma. *Suppose Ω is contained in the disk $B_R(0) := \{(x, y) \in \mathbb{R}^2; \ x^2 + y^2 < R^2\}$. Let V be the solution of the equation*

$$L_h V = 1 \quad in \ \Omega_h,$$
$$V = 0 \quad on \ \partial\Omega_h, \qquad\qquad (4.3)$$

where L_h is the standard five-point stencil. Then

$$0 \le V(x_i, y_i) \le \frac{1}{4}(R^2 - x_i^2 - y_i^2). \qquad\qquad (4.4)$$

Proof. Consider the function $w(x, y) := \frac{1}{4}(R^2 - x^2 - y^2)$, and set $W_i = w(x_i, y_i)$. Since w is a polynomial of second degree, the derivatives of higher order which were dropped when forming the difference star vanish. Hence we have $(L_h W)_i = Lw(x_i, y_i) = 1$. Moreover, $W \ge 0$ on $\partial\Omega$. The discrete comparison principle implies that $V \le W$, while the minimum principle implies $V \ge 0$, and (4.4) is proved. □

The essential fact about (4.4) is that it provides a bound which is independent of h. – This lemma can be extended to any elliptic differential equation for which the finite difference approximation is exact for polynomials of degree 2. In this case the factor $\frac{1}{4}$ in (4.4) is replaced by a number which depends on the constant of ellipticity.

4.5 Convergence Theorem. *Suppose the solution of the Poisson equation is a C^2 function, and that the derivatives u_{xx} and u_{yy} are uniformly continuous in Ω. Then the approximations obtained using the five-point stencil converge to the solution. In particular*

$$\max_{z \in \Omega_h} |U_h(z) - u(z)| \to 0 \ as \ h \to 0. \qquad\qquad (4.5)$$

Proof. By the Taylor expansion at the point (x_i, y_i),

$$L_h u(x_i, y_i) = -u_{xx}(\xi_i, y_i) - u_{yy}(x_i, \eta_i),$$

where ξ_i and η_i are certain numbers. Because of the uniform continuity, the local discretization error $\max_i |r_i|$ tends to 0. It now follows from (4.2) and Lemma 4.4 that

$$\max |\eta_i| \le \frac{R^2}{4} \max |r_i|, \qquad\qquad (4.6)$$

which gives the convergence assertion. □

Analogously, using (4.6), we can get $\mathcal{O}(h)$ or $\mathcal{O}(h^2)$ estimates for the global error, provided u is in $C^3(\bar{\Omega})$ or in $C^4(\bar{\Omega})$, respectively.

Limits of the Convergence Theory

The hypotheses on the derivatives required for the above convergence theorem are often too restrictive.

4.6 Example. Suppose we want to find the solution of the potential equation in the unit disk satisfying the (Dirichlet) boundary condition

$$u(\cos\varphi, \sin\varphi) = \sum_{k=2}^{\infty} \frac{1}{k(k-1)} \cos k\varphi.$$

Since it is absolutely and uniformly convergent, the series represents a continuous function. By (1.2), the solution of the boundary-value problem in polar coordinates is

$$u(x, y) = \sum_{k=2}^{\infty} \frac{r^k}{k(k-1)} \cos k\varphi. \tag{4.7}$$

Now on the x-axis, the second derivative

$$u_{xx}(x, 0) = \sum_{k=2}^{\infty} x^{k-2} = \frac{1}{1-x}$$

is unbounded in a neighborhood of the boundary point $(1, 0)$, and thus Theorem 4.5 is not directly applicable.

A complete convergence theory can be found e.g. in Hackbusch [1986]. It uses the stability of the differential operator in the sense of the L_2-norm, while here the maximum norm was used (but see Problem 4.8). Since the main topic of this book is the finite element method, we restrict ourselves here to a simple generalization. Using an approximation-theoretical argument, we can extend the convergence theorem at least to a disk with arbitrary continuous boundary values.

By the Weierstrass approximation theorem, every periodic continuous function can be approximated arbitrarily well by a trigonometric polynomial. Thus, for given $\varepsilon > 0$, there exists a trigonometric polynomial

$$v(\cos\varphi, \sin\varphi) = a_0 + \sum_{k=1}^{m} (a_k \cos k\varphi + b_k \sin k\varphi)$$

with $|v - u| < \frac{\varepsilon}{4}$ on $\partial\Omega$. Let

$$v(x, y) = a_0 + \sum_{k=1}^{m} r^k (a_k \cos k\varphi + b_k \sin k\varphi)$$

and let V be the numerical solution obtained by the finite difference method. By the maximum principle and the discrete maximum principle, it follows that

$$|u - v| < \frac{\varepsilon}{4} \quad \text{in } \Omega, \quad |U - V| < \frac{\varepsilon}{4} \quad \text{in } \Omega_h. \tag{4.8}$$

Note that the second estimate in (4.8) is uniform for all h. Moreover, since the derivatives of v up to order 4 are bounded in Ω, by the convergence theorem, $|V - v| < \frac{\varepsilon}{2}$ in Ω_h for sufficiently small h. Then by the triangle inequality,

$$|u - U| \le |u - v| + |v - V| + |V - U| < \varepsilon \quad \text{in } \Omega_h. \qquad \square$$

Here we have used an explicit representation for the solution of the Poisson equation on the disk, but we needed only the fact that the boundary values which produce nice solutions are dense. We obtain a generalization if we put this property into an abstract hypothesis.

4.7 Theorem. *Suppose the set of solutions of the Poisson equation whose derivatives u_{xx} and u_{yy} are uniformly continuous in Ω is dense in the set*

$$\{u \in C(\bar{\Omega}); \quad Lu = f\}.$$

Then the numerical solution obtained using the five-point stencil converges, i.e., (4.5) holds.

Problems

4.8 Let L_h be the difference operator obtained from the Laplace operator using (3.1), and let $\Omega_{h,0}$ be the set of (interior) points of Ω_h for which all four neighbors also belong to Ω_h. In order to take into account the fact that the consistency error on the boundary may be larger, in analogy with (4.3) we need to find the solution of

$$L_h V = 1 \text{ in } \Omega_h \backslash \Omega_{h,0},$$
$$L_h V = 0 \text{ in } \Omega_{h,0},$$
$$V = 0 \text{ on } \partial\Omega_h.$$

Show (for simplicity, on a square) that

$$0 \le V \le h^2 \text{ in } \Omega_h.$$

4.9 Consider the eigenvalue problem for the Laplacian on the unit square:

$$-\Delta u = \lambda u \text{ in } \Omega = (0, 1)^2,$$
$$u = 0 \quad \text{on } \partial\Omega.$$

Then

$$u_{k\ell}(x, y) = \sin k\pi x \sin \ell\pi y, \quad k, \ell = 1, 2, \ldots, \tag{4.9}$$

are the eigenfunctions with the eigenvalues $(k^2 + \ell^2)\pi^2$. Show that if $h = 1/n$, the restrictions of these functions to the grid are the eigenfunctions of the difference operator corresponding to the five-point stencil. Which eigenvalues are better approximated, the small ones or the large ones?

Chapter II
Conforming Finite Elements

The mathematical treatment of the finite element method is based on the variational formulation of elliptic differential equations. Solutions of the most important differential equations can be characterized by minimal properties, and the corresponding variational problems have solutions in certain function spaces called Sobolev spaces. The numerical treatment involves minimization in appropriate finite-dimensional linear subspaces. A suitable choice for these subspaces, both from a practical and from a theoretical point of view, are the so-called *finite element spaces*.

For linear differential equations, it suffices to work with Hilbert space methods. In this framework, we immediately get the existence of so-called *weak solutions*. Regularity results, to the extent they are needed for the finite element theory, will be presented without proof.

This chapter contains a theory of the simple methods which suffice for the treatment of scalar elliptic differential equations of second order. The aim of this chapter are the error estimates in §7 for the finite element solutions. They refer to the L_2-norm and to the Sobolev norm $\| \cdot \|_1$. Some of the more general results presented here will also be used later in our discussion in Chapter III of other elliptic problems whose treatment requires additional techniques.

The paper of Courant [1943] is generally considered to be the first mathematical contribution to a finite element theory, although a paper of Schellbach [1851] had appeared already a century earlier. If we don't take too narrow a view, finite elements also appear in some work of Euler. The method first became popular at the end of the sixties, when engineers independently developed and named the method. The long survey article of Babuška and Aziz [1972] laid a broad foundation for many of the deeper functional analytic tools, and the first textbook on the subject was written by Strang and Fix [1973].

Independently, the method of finite elements became an established technique in engineering sciences for computations in structural mechanics. The developments there began around 1956, e.g. with the paper of Turner, Clough, Martin, and Topp [1956] who also created the name *finite elements* and the paper by Argyris [1957]. The book by Zienkiewicz [1971] also had great impact. An interesting review of the history was presented by Oden [1991].

§ 1. Sobolev Spaces

In the following, let Ω be an open subset of $\mathbb{R}^n$ with piecewise smooth boundary.

The *Sobolev spaces* which will play an important role in this book are built on the function space $L_2(\Omega)$. $L_2(\Omega)$ consists of all functions u which are square-integrable over Ω in the sense of Lebesgue. We identify two functions u and v whenever $u(x) = v(x)$ for $x \in \Omega$, except on a set of measure zero. $L_2(\Omega)$ becomes a Hilbert space with the scalar product

$$(u, v)_0 := (u, v)_{L_2} = \int_\Omega u(x)v(x)dx \qquad (1.1)$$

and the corresponding norm

$$\|u\|_0 = \sqrt{(u, u)_0}. \qquad (1.2)$$

1.1 Definition. $u \in L_2(\Omega)$ possesses the *(weak) derivative* $v = \partial^\alpha u$ in $L_2(\Omega)$ provided that $v \in L_2(\Omega)$ and

$$(\phi, v)_0 = (-1)^{|\alpha|}(\partial^\alpha \phi, u)_0 \quad \text{for all } \phi \in C_0^\infty(\Omega). \qquad (1.3)$$

Here $C^\infty(\Omega)$ denotes the space of infinitely differentiable functions, and $C_0^\infty(\Omega)$ denotes the subspace of such functions which are nonzero only on a compact subset of Ω.

If a function is differentiable in the classical sense, then its weak derivative also exists, and the two derivatives coincide. In this case (1.3) becomes Green's formula for integration by parts.

The concept of the weak derivative carries over to other differential operators. For example, let $u \in L_2(\Omega)^n$ be a vector field. Then $v \in L_2(\Omega)$ is the divergence of u in the weak sense, $v = \operatorname{div} u$ for short, provided $(\phi, v)_0 = -(\operatorname{grad} \phi, u)_0$ for all $\phi \in C_0^\infty(\Omega)$.

Introduction to Sobolev Spaces

1.2 Definition. Given an integer $m \geq 0$, let $H^m(\Omega)$ be the set of all functions u in $L_2(\Omega)$ which possess weak derivatives $\partial^\alpha u$ for all $|\alpha| \leq m$. We can define a scalar product on $H^m(\Omega)$ by

$$(u, v)_m := \sum_{|\alpha| \leq m} (\partial^\alpha u, \partial^\alpha v)_0$$

with the associated norm

$$\|u\|_m := \sqrt{(u, u)_m} = \sqrt{\sum_{|\alpha| \leq m} \|\partial^\alpha u\|^2_{L_2(\Omega)}}. \tag{1.4}$$

The corresponding semi-norm

$$|u|_m := \sqrt{\sum_{|\alpha| = m} \|\partial^\alpha u\|^2_{L_2(\Omega)}} \tag{1.5}$$

is also of interest.

We shall often write H^m instead of $H^m(\Omega)$. Conversely, we will write $\| \cdot \|_{m,\Omega}$ instead of $\| \cdot \|_m$ whenever it is important to distinguish the domain.

The letter H is used in honor of David Hilbert.

$H^m(\Omega)$ is complete with respect to the norm $\| \cdot \|_m$, and is thus a Hilbert space. We shall make use of the following result which is often used to introduce the Sobolev spaces without recourse to the concept of weak derivative.

1.3 Theorem. *Let $\Omega \subset \mathbb{R}^n$ be an open set with piecewise smooth boundary, and let $m \geq 0$. Then $C^\infty(\Omega) \cap H^m(\Omega)$ is dense in $H^m(\Omega)$.*

By Theorem 1.3, $H^m(\Omega)$ is the completion of $C^\infty(\Omega) \cap H^m(\Omega)$, provided that Ω is bounded. This suggests a corresponding generalization for functions satisfying zero boundary conditions.

1.4 Definition. We denote the completion of $C_0^\infty(\Omega)$ w.r.t. the Sobolev norm $\| \cdot \|_m$ by $H_0^m(\Omega)$.

Obviously, the Hilbert space $H_0^m(\Omega)$ is a closed subspace of $H^m(\Omega)$. Moreover, $H_0^0(\Omega) = L_2(\Omega)$, and we have the following inclusions:

$$L_2(\Omega) = H^0(\Omega) \supset H^1(\Omega) \supset H^2(\Omega) \supset \cdots$$
$$\| \quad\quad\quad \cup \quad\quad\quad \cup$$
$$H_0^0(\Omega) \supset H_0^1(\Omega) \supset H_0^2(\Omega) \supset \cdots$$

The above Sobolev spaces are based on $L_2(\Omega)$ and the L_2-norm. Analogous Sobolev spaces can be defined for arbitrary L_p-norms with $p \neq 2$. They are useful in the study of *nonlinear* elliptic problems. We denote the spaces analogous to H^m and H_0^m by $W^{m,p}$ and $W_0^{m,p}$, respectively.

Friedrichs' Inequality

In spaces with generalized homogeneous boundary conditions, i.e. in H_0^m, the semi-norm (1.5) is equivalent to the norm (1.4).

1.5 Poincaré–Friedrichs Inequality. *Suppose Ω is contained in an n-dimensional cube with side length s. Then*

$$\|v\|_0 \leq s|v|_1 \quad \text{for all } v \in H_0^1(\Omega). \tag{1.6}$$

Proof. Since $C_0^\infty(\Omega)$ is dense in $H_0^1(\Omega)$, it suffices to establish the inequality for $v \in C_0^\infty(\Omega)$. We may assume that $\Omega \subset W := \{(x_1, x_2, \ldots, x_n);\ 0 < x_i < s\}$, and set $v = 0$ for $x \in W \backslash \Omega$. Then

$$v(x_1, x_2, \ldots, x_n) = v(0, x_2, \ldots, x_n) + \int_0^{x_1} \partial_1 v(t, x_2, \ldots, x_n)dt.$$

The boundary term vanishes, and using the Cauchy–Schwarz inequality gives

$$|v(x)|^2 \leq \int_0^{x_1} 1^2 dt \int_0^{x_1} |\partial_1 v(t, x_2, \ldots, x_n)|^2 dt$$

$$\leq s \int_0^s |\partial_1 v(t, x_2, \ldots, x_n)|^2 dt.$$

Since the right-hand side is independent of x_1, it follows that

$$\int_0^s |v(x)|^2 dx_1 \leq s^2 \int_0^s |\partial_1 v(x)|^2 dx_1.$$

To complete the proof, we integrate over the other coordinates to obtain

$$\int_W |v|^2 dx \leq s^2 \int_W |\partial_1 v|^2 dx \leq s^2 |v|_1^2.$$

$\square$

The Poincaré–Friedrichs inequality is often called Friedrichs' inequality or the Poincaré inequality for short.

1.6 Remark. The proof of the Poincaré–Friedrichs inequality only requires zero boundary conditions on a part of the boundary. If $\Gamma = \partial\Omega$ is piecewise smooth, it suffices that the function vanishes on a part of the boundary Γ_D, where Γ_D is a set with positive $(n-1)$-dimensional measure. – Moreover, if zero Dirichlet boundary conditions are prescribed on the whole boundary, then it is sufficient that Ω is located between two hyperplanes whose distance apart is s. $\square$

Applying Friedrichs' inequality to derivatives, we see that

$$|\partial^\alpha v|_0 \leq s|\partial_1 \partial^\alpha v|_0 \quad \text{for } |\alpha| \leq m-1,\ v \in H_0^m(\Omega).$$

Now induction implies

1.7 Theorem. *If Ω is bounded, then $|\cdot|_m$ is a norm on $H_0^m(\Omega)$ which is equivalent to $\|\cdot\|_m$. If Ω is contained in a cube with side length s, then*

$$|v|_m \leq \|v\|_m \leq (1+s)^m |v|_m \quad \text{for all } v \in H_0^m(\Omega). \tag{1.7}$$

Possible Singularities of H^1 functions

It is well known that $L_2(\Omega)$ also contains unbounded functions. Whether such functions also belong to higher order Sobolev spaces depends on the dimension of the domain. We illustrate this with the most important space $H^1(\Omega)$.

1.8 Remark. If $\Omega = [a, b]$ is a real interval, then $H^1[a, b] \subset C[a, b]$, i.e., each element in $H^1[a, b]$ has a representer which lies in $C[a, b]$.

Proof. Let $v \in C^\infty[a, b]$ or more generally in $C^1[a, b]$. Then for $|x - y| \leq \delta$, the Cauchy–Schwarz inequality gives

$$|v(x) - v(y)| = \left| \int_x^y Dv(t)dt \right| \leq \left| \int_x^y 1^2 dt \right|^{1/2} \cdot \left| \int_x^y [Dv(t)]^2 dt \right|^{1/2} \leq \sqrt{\delta}\, \|v\|_1.$$

Thus, every Cauchy sequence in $H^1[a, b] \cap C^\infty[a, b]$ is equicontinuous and bounded. The theorem of Arzelà-Ascoli implies that the limiting function is continuous. $\qquad\square$

The analogous assertion already fails for a two-dimensional domain Ω. The function

$$u(x, y) = \log\log\frac{2}{r}, \tag{1.8}$$

where $r^2 = x^2 + y^2$, is an unbounded H^1 function on the unit disk $D := \{(x, y) \in \mathbb{R}^2;\ x^2 + y^2 < 1\}$. The fact that u lies in $H^1(D)$ follows from

$$\int_0^{1/2} \frac{dr}{r \log^2 r} < \infty.$$

For an n-dimensional domain with $n \geq 3$,

$$u(x) = r^{-\alpha}, \quad \alpha < (n-2)/2, \tag{1.9}$$

is an H^1 function with a singularity at the origin. Clearly, the singularity in (1.9) becomes stronger with increasing n.

The fact that functions in H^2 over a domain in $\mathbb{R}^2$ are continuous will be established in §3 in connection with an imbedding and a trace theorem.

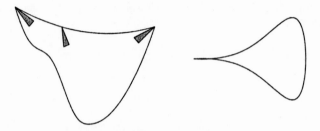

Fig. 6. Domains which satisfy and fail to satisfy the cone condition, respectively

Compact Imbeddings

A continuous linear mapping $L : U \to V$ between normed linear spaces U and V is called *compact* provided that the image of the unit ball in U is a relatively compact set in V. In particular, if $U \subset V$ and the imbedding $j : U \hookrightarrow V$ is compact, we call it a *compact imbedding*.

By the theorem of Arzelà-Ascoli, the C^1 functions v for which

$$\sup_{\Omega} |v(x)| + \sup_{\Omega} |\nabla v(x)| \qquad (1.10)$$

is bounded by a given number form a relatively compact subset of $C^0(\Omega)$. The quantity (1.10) is a norm on C^1. In this sense, $C^1(\Omega)$ is compactly imbedded in $C^0(\Omega)$. The analogous assertion also holds for Sobolev spaces, although as we have seen, H^1 functions can exhibit singularities.

1.9 Rellich Selection Theorem. *Given $m \geq 0$, let Ω be a Lipschitz domain,[1] and suppose that it satisfies a cone condition (see Fig. 6), i.e., the interior angles at each vertex are positive, and so a nontrivial cone can be positioned in Ω with its tip at the vertex. Then the imbedding $H^{m+1}(\Omega) \hookrightarrow H^m(\Omega)$ is compact.*

---

[1] A function $f : \mathbb{R}^n \supset D \to \mathbb{R}^m$ is called *Lipschitz continuous* provided that for some number c, $\|f(x) - f(y)\| \leq c\|x - y\|$ for all $x, y \in D$. A hypersurface in $\mathbb{R}^n$ is a *graph* whenever it can be represented in the form $x_k = f(x_1, \ldots, x_{k-1}, x_{k+1}, \ldots, x_n)$, with $1 \leq k \leq n$ and some suitable domain in $\mathbb{R}^{n-1}$. A domain $\Omega \subset \mathbb{R}^n$ is called a *Lipschitz domain* provided that for every $x \in \partial\Omega$, there exists a neighborhood of $\partial\Omega$ which can be represented as the graph of a Lipschitz continuous function.

Problems

1.10 Let Ω be a bounded domain. With the help of Friedrichs' inequality, show that the constant function $u = 1$ is not contained in $H_0^1(\Omega)$, and thus $H_0^1(\Omega)$ is a proper subspace of $H^1(\Omega)$.

1.11 Let $\Omega \subset \mathbb{R}^n$ be a sphere with center at the origin. Show that $u(x) = \|x\|^s$ possesses a weak derivative in $L_2(\Omega)$ if $2s > 2 - n$ or if $s = 0$ (the trivial case).

1.12 A variant of Friedrichs' inequality. Let Ω be a domain which satisfies the hypothesis of Theorem 1.9. Then there is a constant $c = c(\Omega)$ such that

$$\|v\|_0 \le c\big(|\bar{v}| + |v|_1\big) \quad \text{for all } v \in H^1(\Omega) \tag{1.11}$$

$$\text{with} \quad \bar{v} = \frac{1}{\mu(\Omega)} \int_\Omega v(x)dx.$$

Hint: This variant of Friedrichs' inequality can be established using the technique from the proof of the inequality 1.5 only under restrictive conditions on the domain. Use the compactness of $H^1(\Omega) \hookrightarrow L_2(\Omega)$ in the same way as in the proof of Lemma 6.2 below.

1.13 Let $\Omega_1, \Omega_2 \subset \mathbb{R}^n$ be bounded, and suppose that for the bijective continuously differentiable mapping $F : \Omega_1 \to \Omega_2$, $\|DF(x)\|$ and $\|(DF(x))^{-1}\|$ are bounded for $x \in \Omega$. Verify that $v \in H^1(\Omega_2)$ implies $v \circ F \in H^1(\Omega_1)$.

1.14 Exhibit a function in $C[0, 1]$ which is not contained in $H^1[0, 1]$. – To illustrate that $H_0^0(\Omega) = H^0(\Omega)$, exhibit a sequence in $C_0^\infty(0, 1)$ which converges to the constant function $v = 1$ in the $L_2[0, 1]$ sense.

1.15 Let ℓ_p denote the space of infinite sequences $(x_1, x_2, \ldots)$ satisfying the condition $\sum_k |x_k|^p < \infty$. It is a Banach space with the norm

$$\|x\|_p := \|x\|_{\ell_p} := \left(\sum_k |x_k|^p\right)^{1/p}, \quad 1 \le p < \infty.$$

Since $\| \cdot \|_2 \le \| \cdot \|_1$, the imbedding $\ell_1 \hookrightarrow \ell_2$ is continuous. Is it also compact?

1.16 Consider
  (a) the Fourier series $\sum_{k=-\infty}^{+\infty} c_k e^{ikx}$ on $[0, 2\pi]$,
  (b) the Fourier series $\sum_{k,\ell=-\infty}^{+\infty} c_{k\ell} e^{ikx+i\ell y}$ on $[0, 2\pi]^2$.
Express the condition $u \in H^m$ in terms of the coefficients. In particular, show the equivalence of the assertions $u \in L_2$ and $c \in \ell_2$.

Show that in case (b), $u_{xx} + u_{yy} \in L^2$ implies $u_{xy} \in L^2$.

§ 2. Variational Formulation of Elliptic Boundary-Value Problems of Second Order

A function which satisfies a given partial differential equation of second order and assumes prescribed boundary values is called a *classical solution* provided it lies in $C^2(\Omega) \cap C^0(\bar{\Omega})$ in the case of Dirichlet boundary conditions, and in $C^2(\Omega) \cap C^1(\bar{\Omega})$ in the case of Neumann boundary conditions, respectively. Classical solutions exist if the boundary of the underlying domain is sufficiently smooth, and if certain additional conditions are satisfied in the case where Neumann boundary conditions are specified on part of the boundary. In general, higher derivatives of a classical solution need not be bounded (see Example 2.1), and thus the simple convergence theory presented in Ch. I for the finite difference method may not be applicable.

In this section we discuss the variational formulation of boundary-value problems. It provides a natural approach to their numerical treatment using finite elements, and also furnishes a simple way to establish the existence of so-called *weak solutions*.

Fig. 7. Domain with reentrant corner (cf. Example 2.1)

2.1 Example. Consider the two-dimensional domain

$$\Omega = \{(x, y) \in \mathbb{R}^2; \ x^2 + y^2 < 1, \ x < 0 \text{ or } y > 0\} \tag{2.1}$$

with reentrant corner (see Fig. 7) and identify $\mathbb{R}^2$ with $\mathbb{C}$. Then $w(z) := z^{2/3}$ is analytic in Ω, and its imaginary part $u(z) := \operatorname{Im} w(z)$ is a harmonic function solving the boundary-value problem

$$\Delta u = 0 \qquad \text{in } \Omega,$$

$$u(e^{i\varphi}) = \sin(\frac{2}{3}\varphi) \quad \text{for } 0 \le \varphi \le \frac{3\pi}{2},$$

$$u = 0 \qquad \text{elsewhere on } \partial\Omega.$$

Since $w'(z) = \frac{2}{3}z^{-1/3}$, even the first derivatives of u are not bounded as $z \to 0$. — The singularity will be no problem when we look for a solution in the right Sobolev space.

Variational Formulation

Before formulating linear elliptic problems as variational problems, we first present the following abstract result.

2.2 Characterization Theorem. *Let V be a linear space, and suppose*

$$a : V \times V \to \mathbb{R}$$

is a symmetric positive bilinear form, i.e., $a(v, v) > 0$ for all $v \in V$, $v \neq 0$. In addition, let

$$\ell : V \to \mathbb{R}$$

be a linear functional. Then the quantity

$$J(v) := \frac{1}{2}a(v, v) - \langle \ell, v \rangle$$

attains its minimum over V at u if and only if

$$a(u, v) = \langle \ell, v \rangle \quad \text{for all } v \in V. \tag{2.2}$$

Moreover, there is at most one solution of (2.2).

Remark. The set of linear functionals ℓ is a linear space. Instead of $\ell(v)$, we prefer to write $\langle \ell, v \rangle$ in order to emphasize the symmetry with respect to ℓ and v.

Proof. For $u, v \in V$ and $t \in \mathbb{R}$, we have

$$J(u + tv) = \frac{1}{2}a(u + tv, u + tv) - \langle \ell, u + tv \rangle$$

$$= J(u) + t[a(u, v) - \langle \ell, v \rangle] + \frac{1}{2}t^2 a(v, v). \tag{2.3}$$

If $u \in V$ satifies (2.2), then (2.3) with $t = 1$ implies

$$J(u + v) = J(u) + \frac{1}{2}a(v, v) \quad \text{for all } v \in V \tag{2.4}$$

$$> J(u), \quad \text{if } v \neq 0.$$

Thus, u is a unique minimal point. Conversely, if J has a minimum at u, then for every $v \in V$, the derivative of the function $t \mapsto J(u + tv)$ must vanish at $t = 0$. By (2.3) the derivative is $a(u, v) - \langle \ell, v \rangle$, and (2.2) follows. □

The relation (2.4) which describes the size of J at a distance v from a minimal point u will be used frequently below.

Reduction to Homogeneous Boundary Conditions

In the following, let L be a second order elliptic partial differential operator with divergence structure

$$Lu := -\sum_{i,k=1}^{n} \partial_i(a_{ik}\partial_k u) + a_0 u, \qquad (2.5)$$

where

$$a_0(x) \geq 0 \quad \text{for } x \in \Omega.$$

We begin by transforming the associated Dirichlet problem

$$\begin{aligned} Lu &= f \quad \text{in } \Omega, \\ u &= g \quad \text{on } \partial\Omega \end{aligned} \qquad (2.6)$$

into one with homogeneous boundary conditions. To this end, we assume that there is a function u_0 which coincides with g on the boundary and for which Lu_0 exists. Then

$$\begin{aligned} Lw &= f_1 \quad \text{in } \Omega, \\ w &= 0 \quad \text{on } \partial\Omega, \end{aligned} \qquad (2.7)$$

where $w := u - u_0$ and $f_1 := f - Lu_0$. For simplicity, we usually assume that the boundary condition in (2.6) is already homogeneous.

We now show that the boundary-value problem (2.7) characterizes the solution of the variational problem. A similar analysis was carried out already by L. Euler, and thus the differential equation $Lu = f$ is called the *Euler equation* or the *Euler–Lagrange* equation for the variational problem.

2.3 Minimal Property. *Every classical solution of the boundary-value problem*

$$-\sum_{i,k} \partial_i(a_{ik}\partial_k u) + a_0 u = f \quad \text{in } \Omega,$$
$$u = 0 \quad \text{on } \partial\Omega$$

is a solution of the variational problem

$$J(v) := \int_\Omega \left[\frac{1}{2}\sum_{i,k} a_{ik}\partial_i v \partial_k v + \frac{1}{2}a_0 v^2 - fv \right] dx \longrightarrow \min ! \qquad (2.8)$$

among all functions in $C^2(\Omega) \cap C^0(\bar\Omega)$ with zero boundary values.

Proof. The proof proceeds with the help of Green's formula

$$\int_\Omega v\partial_i w\, dx = -\int_\Omega w\partial_i v\, dx + \int_{\partial\Omega} vw\, \nu_i\, ds. \qquad (2.9)$$

Here v and w are assumed to be C^1 functions, and ν_i is the i-th component of the outward-pointing normal ν. Inserting $w := a_{ik}\partial_k u$ in (2.9), we have

$$\int_\Omega v\partial_i(a_{ik}\partial_k u)\, dx = -\int_\Omega a_{ik}\partial_i v\partial_k u\, dx, \qquad (2.10)$$

provided $v = 0$ on $\partial\Omega$. Let[2]

$$a(u, v) := \int_\Omega \left[\sum_{i,k} a_{ik}\partial_i u\, \partial_k v + a_0 uv\right] dx, \qquad (2.11)$$

$$\langle \ell, v\rangle := \int_\Omega fv\, dx.$$

Summing (2.10) over i and k gives that for every $v \in C^1(\Omega) \cap C(\bar{\Omega})$ with $v = 0$ on $\partial\Omega$,

$$a(u, v) - \langle \ell, v\rangle = \int_\Omega v\left[-\sum_{i,k}\partial_i(a_{ik}\partial_k u) + a_0 u - f\right] dx$$

$$= \int_\Omega v[Lu - f]\, dx = 0,$$

provided $Lu = f$. This is true if u is a classical solution. Now the characterization theorem implies the minimal property. $\qquad\square$

The same method of proof shows that every solution of the variational problem which lies in the space $C^2(\Omega) \cap C^0(\bar{\Omega})$ is a classical solution of the boundary-value problem.

The above connection was observed by Thomson in 1847, and later by Dirichlet for the Laplace equation. Dirichlet asserted that the boundedness of $J(u)$ from below implies that J attains its minimum for some function u. This argument is now called the *Dirichlet principle*. However, in 1870 Weierstrass showed that it does not hold in general. In particular, the integral

$$J(u) = \int_0^1 u^2(t)dt \qquad (2.12)$$

has infimum 0 in the set $\{v \in C^0[0, 1];\ v(0) = v(1) = 1\}$, but the value 0 is never assumed for any function in $C[0, 1]$ with the given boundary values.

---

[2] The use of the letter a for the bilinear form and also in the expressions a_{ik} and a_0 for the coefficient functions should be no cause for confusion.

Existence of Solutions

The difficulty with the nonexistence of solutions vanishes if we solve the variational problem (2.8) in a suitable Hilbert space. This is why we don't work in the function space $C^2(\Omega)$, although to get classical solutions this would be desirable. – In Theorem 2.2 only the linear structure was used for the characterization of a solution. But, for existence, the choice of the topology is crucial.

2.4 Definition. Let H be a Hilbert space. A bilinear form $a : H \times H \to \mathbb{R}$ is called *continuous* provided there exists $C > 0$ such that

$$|a(u, v)| \leq C \|u\| \|v\| \quad \text{for all } u, v \in H.$$

A symmetric continuous bilinear form a is called *H-elliptic*, or for short *elliptic* or *coercive*, provided for some $\alpha > 0$,

$$a(v, v) \geq \alpha \|v\|^2 \quad \text{for all } v \in H. \tag{2.13}$$

Clearly, every H-elliptic bilinear form a induces a norm via

$$\|v\|_a := \sqrt{a(v, v)}. \tag{2.14}$$

This is equivalent to the norm of the Hilbert space H. The norm (2.14) is called the *energy norm*.

As usual, the space of continuous linear functionals on a normed linear space V will be denoted by V'.

2.5 The Lax–Milgram Theorem (*for Convex Sets*). *Let V be a closed convex set in a Hilbert space H, and let $a : H \times H \to \mathbb{R}$ be an elliptic bilinear form. Then, for every $\ell \in H'$, the variational problem*

$$J(v) := \frac{1}{2} a(v, v) - \langle \ell, v \rangle \longrightarrow \min !$$

has a unique solution in V.

Proof. J is bounded from below since

$$J(v) \geq \frac{1}{2} \alpha \|v\|^2 - \|\ell\| \|v\|$$

$$= \frac{1}{2\alpha} (\alpha \|v\| - \|\ell\|)^2 - \frac{\|\ell\|^2}{2\alpha} \geq -\frac{\|\ell\|^2}{2\alpha}.$$

Let $c_1 := \inf\{J(v); v \in V\}$, and let (v_n) be a minimizing sequence. Then

$$\alpha \|v_n - v_m\|^2 \le a(v_n - v_m, v_n - v_m)$$
$$= 2a(v_n, v_n) + 2a(v_m, v_m) - a(v_n + v_m, v_n + v_m)$$
$$= 4J(v_n) + 4J(v_m) - 8J\left(\frac{v_m + v_n}{2}\right)$$
$$\le 4J(v_n) + 4J(v_m) - 8c_1,$$

since V is convex and thus $\frac{1}{2}(v_n + v_m) \in V$. Now $J(v_n), J(v_m) \to c_1$ implies $\|v_n - v_m\| \to 0$ for $n, m \to \infty$. Thus, (v_n) is a Cauchy sequence in H, and $u = \lim_{n \to \infty} v_n$ exists. Since V is closed, we also have $u \in V$. The continuity of J implies $J(u) = \lim_{n \to \infty} J(v_n) = \inf_{v \in V} J(v)$.

We now show that the solution is unique. Suppose u_1 and u_2 are both solutions. Clearly, $u_1, u_2, u_1, u_2, \ldots$ is a minimizing sequence. As we saw above, every minimizing sequence is a Cauchy sequence. This is only possible if $u_1 = u_2$. ☐

2.6 Remarks. (1) The above proof makes use of the following *parallelogram law*: the sum of the squares of the lengths of the diagonals in any parallelogram is equal to the sum of the squares of the lengths of the sides.

(2) In the special case $V = H$, Theorem 2.5 implies that given $\ell \in H'$, there exists an element $u \in H$ with

$$a(u, v) = \langle \ell, v \rangle \quad \text{for all } v \in H.$$

(3) If we further specialize to the case $a(u, v) := (u, v)$, where (u, v) is the defining scalar product on H, then we obtain the *Riesz representation theorem*: given $\ell \in H'$, there exists an element $u \in H$ with

$$(u, v) = \langle \ell, v \rangle \quad \text{for all } v \in H.$$

This defines a mapping $H' \to H$, $\ell \mapsto u$ which is called the *canonical imbedding of H' in H*.

(4) The Characterization Theorem 2.2 can be generalized to convex sets as follows. The function u is the minimal solution in a convex set V if and only if the so-called *variational inequality*

$$a(u, v - u) \ge \langle \ell, v - u \rangle \quad \text{for all } v \in V \tag{2.15}$$

holds. We leave the proof to the reader.

If the underlying space has finite dimension, i.e., is the Euclidean space $\mathbb{R}^N$, then instead of (2.13) we only need to require that

$$a(v, v) > 0 \quad \text{for all } v \in H, v \ne 0. \tag{2.16}$$

Then the compactness of the unit ball implies (2.13) for some $\alpha > 0$. The fact that (2.16) does not suffice in the infinite-dimensional case can already be seen in the example (2.12). To make this point even clearer, we consider another simple example.

2.7 Example. Let $H = \ell_2$ be the space of infinite sequences $(x_1, x_2, \ldots)$, equipped with the norm $\|x\|^2 := \sum_m x_m^2$. The form

$$a(x, y) := \sum_{m=1}^{\infty} 2^{-m} x_m y_m$$

is positive and continuous but not coercive, and $\langle \ell, x \rangle := \sum_{m=1}^{\infty} 2^{-m} x_m$ defines a continuous linear functional. However, $J(x) = \frac{1}{2} a(x, x) - \langle \ell, x \rangle$ does not attain a minimum in ℓ_2. Indeed, a necessary condition for a minimal solution in this case is that $x_m = 1$ for $m = 1, 2, \ldots$, and this contradicts $\sum_m x_m^2 < \infty$. □

With the above preparations, we can now make the concept of a solution of the boundary-value problem more precise.

2.8 Definition. A function $u \in H_0^1(\Omega)$ is called a *weak solution* of the second order elliptic boundary-value problem

$$\begin{aligned} Lu &= f \quad \text{in } \Omega, \\ u &= 0 \quad \text{on } \partial\Omega, \end{aligned} \tag{2.17}$$

with homogeneous Dirichlet boundary conditions, provided that

$$a(u, v) = (f, v)_0 \quad \text{for all } v \in H_0^1(\Omega), \tag{2.18}$$

where a is the associated bilinear form defined in (2.11).

In other cases we shall also refer to a function as a weak solution of an elliptic boundary-value problem provided it is a solution of an associated variational problem. – Throughout the above, we have implicitly assumed that the coefficient functions are sufficiently smooth. For the following theorem, $a_{ij} \in L_\infty(\Omega)$ and $f \in L_2(\Omega)$ suffice.

2.9 Existence Theorem. *Let L be a second order uniformly elliptic partial differential operator. Then the Dirichlet problem (2.17) always has a weak solution in $H_0^1(\Omega)$. It is a minimum of the variational problem*

$$\frac{1}{2} a(v, v) - (f, v)_0 \longrightarrow \min !$$

over $H_0^1(\Omega)$.

Proof. Let[3] $c := \sup\{|a_{ik}(x)|;\ x \in \Omega, 1 \le i, k \le n\}$. Then the Cauchy–Schwarz inequality implies

$$\left| \sum_{i,k} \int a_{ik} \partial_i u \partial_k v\, dx \right| \le c \sum_{i,k} \int |\partial_i u \partial_k v|\, dx$$

$$\le c \sum_{i,k} \left[\int (\partial_i u)^2 dx \int (\partial_k v)^2 dx \right]^{1/2}$$

$$\le C |u|_1\, |v|_1,$$

where $C = cn^2$. If we also assume that $C \ge \sup\{|a_0(x)|;\ x \in \Omega\}$, then we get

$$\left| \int a_0 u v\, dx \right| \le C \int |uv|\, dx \le C \cdot \|u\|_0 \cdot \|v\|_0$$

in an analogous way. Combining these, we have

$$a(u, v) \le C \|u\|_1 \|v\|_1.$$

Next, the uniform ellipticity implies the pointwise estimate

$$\sum_{i,k} a_{ik} \partial_i v \partial_k v \ge \alpha \sum_i (\partial_i v)^2,$$

for C^1 functions. Integrating both sides and using $a_0 \ge 0$ leads to

$$a(v, v) \ge \alpha \sum_i \int_\Omega (\partial_i v)^2 dx = \alpha |v|_1^2 \quad \text{for all } v \in H^1(\Omega). \qquad (2.19)$$

By Friedrichs' inequality, $|\cdot|_1$ and $\|\cdot\|_1$ are equivalent norms on H_0^1. Thus, a is an H^1-elliptic bilinear form on $H_0^1(\Omega)$. By the Lax–Milgram Theorem, there exists a unique weak solution which is also a solution of the variational problem. $\qquad \square$

2.10 Example. In the model problem

$$-\Delta u = f \quad \text{in } \Omega,$$

$$u = 0 \quad \text{on } \partial\Omega,$$

the associated bilinear form is $a(u, v) = \int \nabla u \cdot \nabla v\, dx$. We will also write $(\nabla u, \nabla v)_0$ for $\int \nabla u \cdot \nabla v\, dx$. Thus, the solution is determined by

$$(\nabla u, \nabla v)_0 = (f, v)_0 \quad \text{for all } v \in H_0^1(\Omega). \qquad (2.20)$$

We see that the divergence of ∇u in the sense of Definition 1.1 exists, and $-\Delta u = -\operatorname{div} \operatorname{grad} u = f$.

---

[3] $c, c_1, c_2, \ldots$ are generic constants, i.e. they can change from line to line. In general, we reserve C for the value of the norm of a in the sense of Definition 2.4.

Inhomogeneous Boundary Conditions

We now return to equation (2.6) with inhomogeneous boundary conditions. Let $u_0 \in C^2(\Omega) \cap C^0(\bar{\Omega}) \cap H^1(\Omega)$ be a function which coincides with g on the boundary of Ω. The weak formulation of (2.7) is now

Find $w \in H_0^1(\Omega)$ with

$$a(w, v) = (f - Lu_0, v)_0 \quad for\ all\ v \in H_0^1(\Omega).$$

Since $(Lu_0, v) = a(u_0, v)$, this can now be written in the following form:

Find $u \in H^1(\Omega)$ with

$$a(u, v) = (f, v)_0 \quad for\ all\ v \in H_0^1(\Omega),$$
$$u - u_0 \in H_0^1(\Omega). \tag{2.21}$$

The second part of (2.21) can be considered as a weak formulation of the boundary condition.

It follows from density considerations that it suffices to assume that $u_0 \in H^1(\Omega)$. On the other hand, it is not always possible to satisfy this requirement. In fact, it is not even satisfied in some cases for which a classical solution is known.

Example (Hadamard [1932]). Let r and φ be the polar coordinates in the unit disk $\Omega = B_1 := \{x \in \mathbb{R}^2;\ \|x\| < 1\}$. The function $u(r, \varphi) := \sum_{k=1}^{\infty} k^{-2} r^{k!} \sin(k!\varphi)$ is harmonic in Ω. If we identify $\mathbb{R}^2$ with $\mathbb{C}$, then $u(z) = \text{Im} \sum_{k=1}^{\infty} k^{-2} z^{k!}$. This shows that $\int |\nabla u|^2 dx = \infty$, and thus $u \notin H^1$. There does not exist any function in H^1 with the same boundary value as u, since for a given boundary value, the harmonic function is always the one with the smallest value of the H^1-semi-norm.

Problems

2.11 Let Ω be bounded with $\Gamma := \partial\Omega$, and let $g : \Gamma \to \mathbb{R}$ be a given function. Find the function $u \in H^1(\Omega)$ with minimal H^1-norm which coincides with g on Γ. Under what conditions on g can this problem be handled in the framework of this section?

2.12 Consider the elliptic, but not uniformly elliptic, bilinear form

$$a(u, v) := \int_0^1 x^2 u'v'\, dx$$

on the interval $[0, 1]$. Show that the problem $\frac{1}{2}a(u, u) - \int_0^1 u\,dx \to$ min ! does not have a solution in $H_0^1(0, 1)$. – What is the associated (ordinary) differential equation?

2.13 Prove that in a convex set, the solution to the variational problem is characterized by (2.15).

2.14 In connection with Example 2.7, consider the continuous linear mapping

$$L : \ell_2 \to \ell_2,$$
$$(Lx)_k = 2^{-k} x_k.$$

Show that the range of L is not closed.
Hint: The closure contains the point $y \in \ell_2$ with $y_k = 2^{-k/2}$, $k = 1, 2, \dots$.

2.15 Show that

$$\int_\Omega \phi \operatorname{div} v \, dx = - \int_\Omega \operatorname{grad} \phi \cdot v \, dx + \int_{\partial\Omega} \phi v \cdot v \, ds \qquad (2.22)$$

for all sufficiently smooth functions v and ϕ with values in $\mathbb{R}^n$ and $\mathbb{R}$, respectively.
Here

$$\operatorname{div} v := \sum_{i=1}^n \frac{\partial v}{\partial x_i} .$$

§ 3. The Neumann Boundary-Value Problem.
A Trace Theorem

In passing from a partial differential equation to an associated variational problem, Dirichlet boundary conditions are explicitly built into the function space. This kind of boundary condition is therefore called *essential*. In contrast, Neumann boundary conditions, which are conditions on derivatives on the boundary, are implicitly forced, and thus are called *natural boundary conditions*.

Ellipticity in H^1

Suppose L is the uniformly elliptic differential operator in (2.5), and that a is the corresponding bilinear form (2.11). We now require that $a_0(x)$ be bounded from below by a positive number. After possibly reducing the number α in (2.19), we can assume

$$a_0(x) \geq \alpha > 0 \quad \text{for all } x \in \Omega.$$

Now we get the bound

$$a(v, v) \geq \alpha |v|_1^2 + \alpha \|v\|_0^2 = \alpha \|v\|_1^2 \quad \text{for all } v \in H^1(\Omega), \tag{3.1}$$

which has one more term $\int a_0(x)v^2 dx \geq \alpha \|v\|_0^2$ than the bound in (2.19). Thus, the quadratic form $a(v, v)$ is elliptic on the entire space $H^1(\Omega)$, and not just on the subspace $H_0^1(\Omega)$. In addition, for $f \in L_2(\Omega)$ and $g \in L_2(\partial\Omega)$ we can define a linear functional by

$$\langle \ell, v \rangle := \int_\Omega f v \, dx + \int_\Gamma g v \, ds, \tag{3.2}$$

where as usual, $\Gamma := \partial\Omega$. The following theorem shows that $\langle \ell, v \rangle$ is well defined for all $v \in H^1(\Omega)$, and that ℓ is a bounded linear functional.[4]

3.1 Trace Theorem. *Let Ω be bounded, and suppose Ω has a piecewise smooth boundary. In addition, suppose Ω satisfies the cone condition. Then there exists a bounded linear mapping*

$$\gamma : H^1(\Omega) \to L_2(\Gamma), \quad \|\gamma(v)\|_{0,\Gamma} \leq c\|v\|_{1,\Omega}, \tag{3.3}$$

such that $\gamma v = v_{|\Gamma}$ for all $v \in C^1(\bar\Omega)$.

---

[4] There are sharper results for Sobolev spaces with non-integer indices.

Clearly, γv is the *trace* of v on the boundary, i.e., the restriction of v to the boundary. We know that the evaluation of an H^1 function at a single point does not always make sense. Theorem 3.1 asserts that the restriction of v to the boundary is at least an L_2 function.

We delay the proof of the trace theorem until the end this section.

Boundary-Value Problems with Natural Boundary Conditions

3.2 Theorem. *Suppose the domain Ω satisfies the hypotheses of the trace theorem. Then the variational problem*

$$J(v) := \frac{1}{2} a(v, v) - (f, v)_{0,\Omega} - (g, v)_{0,\Gamma} \longrightarrow \text{min}!$$

has exactly one solution $u \in H^1(\Omega)$. The solution of the variational problem lies in $C^2(\Omega) \cap C^1(\bar{\Omega})$ if and only if there exists a classical solution of the boundary-value problem

$$Lu = f \quad \text{in } \Omega,$$
$$\sum_{i,k} v_i a_{ik} \partial_k u = g \quad \text{on } \Gamma, \tag{3.4}$$

in which case the two solutions are identical. Here $v := v(x)$ is the outward-pointing normal defined almost everywhere on Γ.

Proof. Since a is an H^1-elliptic bilinear form, the existence of a unique minimum $u \in H^1(\Omega)$ follows from the Lax–Milgram Theorem. In particular, u is characterized by

$$a(u, v) = (f, v)_{0,\Omega} + (g, v)_{0,\Gamma} \quad \text{for all } v \in H^1(\Omega). \tag{3.5}$$

Now suppose (3.5) is satisfied for $u \in C^2(\Omega) \cap C^1(\bar{\Omega})$. For $v \in H_0^1(\Omega)$, $\gamma v = 0$, and we deduce from (3.5) that

$$a(u, v) = (f, v)_0 \quad \text{for all } v \in H_0^1(\Omega).$$

By (2.21), u is also a solution of the Dirichlet problem, where we define the boundary condition using u. Thus, in the interior we have

$$Lu = f \quad \text{in } \Omega. \tag{3.6}$$

For $v \in H^1(\Omega)$, Green's formula (2.9) yields

$$\int_\Omega v \partial_i (a_{ik} \partial_k u) \, dx = - \int_\Omega \partial_i v a_{ik} \partial_k u \, dx + \int_\Gamma v a_{ik} \partial_k u \, v_i \, ds.$$

Hence,

$$a(u, v) - (f, v)_0 - (g, v)_{0,\Gamma} = \int_\Omega v[Lu - f] \, dx + \int_\Gamma [\sum_{i,k} \nu_i a_{ik} \partial_k u - g] v \, ds. \quad (3.7)$$

Now it follows from (3.5) and (3.6) that the second integral on the right-hand side of (3.7) vanishes. Suppose the function $v_0 := \nu_i a_{ik} \partial_k u - g$ does not vanish. Then $\int_\Gamma v_0^2 ds > 0$. Since $C^1(\bar\Omega)$ is dense in $C^0(\bar\Omega)$, there exists $v \in C^1(\bar\Omega)$ with $\int_\Gamma v_0 \cdot v \, ds > 0$. This is a contradiction, and the boundary condition must be satisfied.

On the other hand, from (3.7) we can immediately see that every classical solution of (3.4) satisfies (3.5). $\square$

Neumann Boundary Conditions

For the Helmholtz equation

$$-\Delta u + a_0(x) u = f \quad \text{in } \Omega,$$

the natural boundary condition is

$$\frac{\partial u}{\partial \nu} := \nu \cdot \nabla u = g \quad \text{on } \partial\Omega.$$

We call it *the Neumann boundary condition*. Here $\partial u / \partial \nu$ is the normal derivative, i.e., the direction perpendicular to the tangent plane (if the boundary is smooth). [In the general case, the boundary condition in (3.4) also involves the normal direction if we define orthogonality w.r.t. the metric induced by the quadratic form with the matrix $a_{ik} = a_{ik}(x)$.]

Clearly, the Poisson equation with Neumann boundary conditions,

$$\begin{aligned}
-\Delta u &= f \quad \text{in } \Omega, \\
\frac{\partial u}{\partial \nu} &= g \quad \text{on } \partial\Omega,
\end{aligned} \quad (3.8)$$

only determines a function up to an additive constant. This suggests that in formulating the weak version of this problem we should restrict ourselves to the subspace $V := \{v \in H^1(\Omega); \int_\Omega v \, dx = 0\}$. The bilinear form $a(u, v) = \int_\Omega \nabla u \cdot \nabla v \, dx$ is not H^1-elliptic, but in view of the variant (1.11) of Friedrichs' inequality, it is V-elliptic.

We claim that any solution of (3.8) must satisfy a certain *compatibility condition* relating the functions f and g to each other. With $w := \nabla u$, equation (3.8) becomes

$$-\operatorname{div} w = f \text{ in } \Omega, \quad \nu' w = g \text{ on } \Gamma.$$

By the Gauss Integral Theorem, $\int_\Omega \operatorname{div} w \, dx = \int_{\partial\Omega} w\nu ds$, and thus

$$\int_\Omega f \, dx + \int_\Gamma g \, ds = 0.$$

This condition is not only necessary, but also sufficient. By the Lax–Milgram Theorem, we get $u \in V$ with

$$a(u, v) = (f, v)_{0,\Omega} + (g, v)_{0,\Gamma} \qquad (3.9)$$

for all $v \in V$. Because of the compatibility condition, (3.9) also holds for $v = \text{const}$, and thus for all $v \in H^1(\Omega)$. As in Theorem 3.2, we now deduce that every classical solution of the variational problem satisfies the equation (3.8).

Another method to deal with the pure Neumann problem (3.8) will be discussed in Problem III.4.21.

Mixed Boundary Conditions

In physical problems, we often encounter Neumann or natural boundary conditions whenever the flow over the boundary is prescribed. Sometimes a Neumann condition is prescribed on only part of the boundary.

3.3 Example. Suppose we want to determine the stationary temperature distribution in an isotropic body $\Omega \subset \mathbb{R}^3$. On the part of the boundary where the body is mechanically clamped, the temperature is prescribed. We denote this part of the boundary by Γ_D. On the rest of the boundary $\Gamma_N = \Gamma \setminus \Gamma_D$, we assume that the heat flux is so small that it can be considered to be 0. If there are no heat sources in Ω, then we have to solve the elliptic boundary-value problem

$$\Delta u = 0 \text{ in } \Omega,$$
$$u = g \text{ on } \Gamma_D,$$
$$\frac{\partial u}{\partial \nu} = 0 \text{ on } \Gamma_N.$$

This problem leads in a natural way to a Hilbert space which lies between $H^1(\Omega)$ and $H^1_0(\Omega)$. Consider functions of the form

$$u \in C^\infty(\Omega) \cap H^1(\Omega), \quad u \text{ vanishes in a neighborhood of } \Gamma_D.$$

Then the closure of this set w.r.t. the H^1-norm leads to the desired space. This is a subspace of $H^1(\Omega)$, and by Remark 1.6, under very general hypotheses $|\cdot|_1$ is a norm which is equivalent to $\|\cdot\|_1$.

Proof of the Trace Theorem

We now present the proof of the trace theorem. For the sake of clarity, we restrict ourselves to domains in $\mathbb{R}^2$. The generalization to domains in $\mathbb{R}^n$ is straightforward, and can be left to the reader as an exercise.

Suppose the boundary is piecewise smooth. In addition, suppose a cone condition is satisfied at the (finitely many) points where the boundary is not smooth. Then we can divide the boundary into finitely many boundary pieces $\Gamma_1, \Gamma_2, \ldots, \Gamma_m$ so that for every piece Γ_i, after a rotation of the coordinate system, we have

1. For some function $\phi = \phi_i \in C^1[y_1, y_2]$,
$$\Gamma_i = \{(x, y) \in \mathbb{R}^2; x = \phi(y), \ y_1 \le y \le y_2\}.$$

2. The domain $\Omega_i = \{(x, y) \subset \mathbb{R}^2; \ \phi(y) < x < \phi(y) + r, \ y_1 < y < y_2\}$ is contained in Ω, where $r > 0$.

We now apply an argument used earlier for the Poincaré–Friedrichs inequality. For $v \in C^1(\bar{\Omega})$ and $(x, y) \in \Gamma$,

$$v(\phi(y), y) = v(\phi(y) + t, y) - \int_0^t \partial_1 v(\phi(y) + s, y)ds,$$

where $0 \le t \le r$. Integrating over t from 0 to r gives

$$rv(\phi(y), y) = \int_0^r v(\phi(y) + t, y)dt - \int_0^r \partial_1 v(\phi(y) + t, y)(r - t)dt.$$

We take the square of this equation, and use *Young's inequality* $(a+b)^2 \le 2a^2 + 2b^2$. Applying the Cauchy–Schwarz inequality to the squares of the integrals gives

$$r^2 v^2(\phi(y), y) \le 2 \int_0^r 1dt \int_0^r v^2(\phi(y) + t, y)dt$$
$$+ 2 \int_0^r t^2 dt \int_0^r |\partial_1 v(\phi(y) + t, y)|^2 dt.$$

We now insert the values $\int 1dt = r$ and $\int t^2 dt = r^3/3$. Dividing by r^2 and integrating over y, we get

$$\int_{y_1}^{y_2} v^2(\phi(y), y)dy \le 2r^{-1} \int_{\Omega_i} v^2 dx dy + r \int_{\Omega_i} |\partial_1 v|^2 dx dy.$$

The arc length differential on Γ is given by $ds = \sqrt{1 + \phi'^2} dy$. Thus, we have

$$\int_{\Gamma_i} v^2 ds \le c_i [2r^{-1} \|v\|_0^2 + r|v|_1^2],$$

where $c_i = \max\{\sqrt{1 + \phi'^2}; \ y_1 \le y \le y_2\}$. Setting $c = (r + 2r^{-1}) \sum_{i=1}^m c_i$, we finally get

$$\|v\|_{0,\Gamma} \le c\|v\|_{1,\Omega}.$$

Thus, the restriction $\gamma : H^1(\Omega) \cap C^1(\bar{\Omega}) \to L_2(\Gamma)$ is a bounded mapping on a dense set. Because of the completeness of $L_2(\Gamma)$, it can be extended to all of $H^1(\Omega)$ without enlarging the bound. $\square$

Note that the cone condition excludes *cusps* in the domain. The domain

$$\Omega := \{(x, y) \in \mathbb{R}^2;\ 0 < y < x^5 < 1\}$$

has a cusp at the origin, and $H^1(\Omega)$ contains the function

$$u(x, y) = x^{-1},$$

whose trace is not square-integrable over Γ. $\qquad\qquad\square$

We would like to point out that Green's formula (2.9) also holds for functions $u, w \in H^1(\Omega)$, provided that Ω satisfies the hypotheses of the trace theorem.

The space $H^1(\Omega)$ is isomorphic to a direct sum

$$H^1(\Omega) \sim H_0^1(\Omega) \oplus \gamma(H^1(\Omega)).$$

Specifically, every $u \in H^1(\Omega)$ can be decomposed as

$$u = v + w,$$

according to the following rule. Let w be the solution of the variational problem $|w|_1^2 \to$ min ! More exactly, suppose

$$(\nabla w, \nabla v)_{0,\Omega} = 0 \qquad \text{for all } v \in H_0^1(\Omega),$$
$$w - u \in H_0^1(\Omega).$$

Let $v := u - w \in H_0^1(\Omega)$. Here γ is an injective mapping on the set of functions w which appear in the decomposition.

We now consider the connection with continuous functions. As usual, the norm $\|u\|_\infty = \|u\|_{\infty,\Omega}$ is based on the essential supremum of $|u|$ over Ω. [It is not a Sobolev norm.]

3.4 Remarks. (1) Let $\Omega \subset \mathbb{R}^2$ be a convex polygonal domain, or a domain with Lipschitz continuous boundary. Then $H^2(\Omega)$ is compactly imbedded in $C(\bar{\Omega})$, and

$$\|v\|_\infty \le c\|v\|_2 \quad \text{for all } v \in H^2(\Omega), \tag{3.10}$$

for some number $c = c(\Omega)$.

(2) For every open connected domain $\Omega \subset \mathbb{R}^2$, $H^2(\Omega)$ is compactly imbedded in $C(\Omega)$.

The above results are not the sharpest possible in this framework. Because of their importance, and because they follow simply from the trace theorem, we now give their proofs.

Choose an angle φ and a radius r with the following property: for every two points $x, y \in \bar{\Omega}$ with $\|x - y\| < r$, there exists a cone K with angle φ, diameter r, and tip at x such that $y \in \partial K$. We now rotate the coordinate system so that x and y differ in only the first coordinate. By the trace theorem, we deduce that

$$\|v\|_{0,\partial K} \leq c(r, \varphi)\|v\|_{1,K}.$$

Then for $v \in H^2(\Omega)$, $\partial_1 v \in H^1(\Omega)$, and thus $\|\partial_1 v\|_{0,\partial K} \leq c(r, \varphi)\|v\|_{2,\Omega}$. By Remark 1.8, $|v(x) - v(y)| \leq c(r, \varphi) \sqrt{\|x - y\|} \cdot \|v\|_{2,\Omega}$. Thus, v is Hölder continuous with exponent $1/2$. Using $\|v\|_{0,K} \leq \|v\|_{2,\Omega}$, we get a bound for $|v|$ in K, and (3.10) follows. The Arzelà-Ascoli Theorem now establishes the compactness assertion.

For every $m \geq 1$, we can find a polygonal domain Ω_m which contains all of the points $x \in \Omega$ whose distance from 0 is at most m, and whose distance from $\partial \Omega$ is at least $1/m$. Since $\Omega = \bigcup_{m>0} \Omega_m$, (2) follows from (1).   □

Practical Consequences of the Trace Theorem

For practical applications, it is tempting to believe that only classical solutions are of importance, and that weak solutions with singularities are nothing more than interesting mathematical objects. However, this is far from the truth, as the following example shows.

3.5 Example. Suppose we erect a *tent* over a disk with radius R such that its height at the center is 1. Find the shape of the tent which has the minimal surface area. Suppose $u(x)$ is the height of the tent at the point x. Then it is well known that the surface area is given by

$$\int_{B_R} \sqrt{1 + (\nabla u)^2}\, dx.$$

Here B_R is the disk with radius R and center at 0. Now $\sqrt{1 + (\nabla u)^2} \leq 1 + \frac{1}{2}(\nabla u)^2$, and for small gradients the difference between the two sides of the inequality is small. Thus, we are led to the variational problem

$$\frac{1}{2} \int_{B_R} (\nabla v)^2 dx \longrightarrow \text{min}! \tag{3.11}$$

subject to the constraints

$$v(0) = 1,$$
$$v = 0 \quad \text{on } \partial B_R.$$

We now show that the constraint $v(0) = 1$ will be ignored by the solution of the variational problem. The singular function

$$w_0(x) = \log\log\frac{eR}{r} \quad \text{with } r = r(x) = \|x\|$$

has a finite Dirichlet integral (3.11). We smooth it to get

$$w_\varepsilon(x) = \begin{cases} w_0(x) & \text{for } r(x) \geq \varepsilon, \\ \log\log\dfrac{eR}{\varepsilon} & \text{for } 0 \leq r(x) < \varepsilon. \end{cases}$$

Now $|w_\varepsilon|_{1,B_R} \leq |w_0|_{1,B_R}$, and $w_\varepsilon(0)$ tends to ∞ as $\varepsilon \to 0$. Thus,

$$u_\varepsilon = \frac{w_\varepsilon(x)}{w_\varepsilon(0)}$$

for $\varepsilon = 1, 1/2, 1/3, \ldots$ provides a minimizing sequence for $J(v)$ which converges almost everywhere to the zero function. This means that the requirement $u(0) = 1$ was ignored.

The situation is different if we require that $u = 1$ on a curve segment. This would be the case if the tent were attached to a ring on the tent pole or if the tent is put over a rope so that it assumes the shape of a roof. While the evaluation of an H^1 function at a point does not make any sense, its evaluation on a line in the L_2 sense is possible. A condition on a function which is defined on a curve segment will be respected almost everywhere.

Fig. 8. Tent attached to a loop of a rope to prevent an extreme force concentration at the tip

For most larger tents, the boundary of the tent at the tip is a ring instead of a single point. Or (see Fig. 8) the tent may be attached to a loop of rope. This avoids very high forces, since the force applies to the ring or loop, rather than at a single point. The trace theorem explains why the point is to be replaced by a one-dimensional curve.

Problems

3.6 Show that every classical solution of the equations and inequalities

$$-\Delta u + a_0 u = f \quad \text{in } \Omega,$$

$$\left.\begin{array}{l} u \geq 0, \quad \dfrac{\partial u}{\partial \nu} \geq 0, \\[2mm] \quad u \cdot \dfrac{\partial u}{\partial \nu} = 0 \end{array}\right\} \text{on } \Gamma,$$

is a solution of a variational problem in the convex set

$$V^+ := \{v \in H^1(\Omega); \ \gamma v \geq 0 \text{ almost everywhere on } \Gamma\}.$$

– It is known from integration theory that the subset $\{\phi \in L_2(\Gamma); \ \phi \geq 0 \text{ almost}$ everywhere on $\Gamma\}$ is closed in $L_2(\Gamma)$.

3.7 Suppose the domain Ω has a piecewise smooth boundary, and let $u \in H^1(\Omega) \cap C(\bar{\Omega})$. Show that $u \in H_0^1(\Omega)$ is equivalent to $u = 0$ on $\partial\Omega$.

3.8 Suppose the domain Ω is divided into two subdomains Ω_1 and Ω_2 by a piecewise smooth curve Γ_0. Let $\alpha_1 \gg \alpha_2 > 0$ and $a(x) = \alpha_i$ for $x \in \Omega_i$, $i = 1, 2$. Show that for every classical solution of the variational problem

$$\int_\Omega [\frac{1}{2} a(x)(\nabla v(x))^2 - f(x)v(x)]dx \longrightarrow \min!$$

in $H_0^1(\Omega)$, the quantity $a(x)\frac{\partial u}{\partial n}$ is continuous on the curve Γ_0. [The discontinuity of $a(x)$ now implies that $\frac{\partial u}{\partial n}$ is not continuous there.]

§ 4. The Ritz–Galerkin Method
and Some Finite Elements

There is a simple natural approach to the numerical solution of elliptic boundary-value problems. Instead of minimizing the functional J definining the corresponding variational problem over all of $H^m(\Omega)$ or $H_0^m(\Omega)$, respectively, we minimize it over some suitable finite-dimensional subspace [Ritz 1908]. The standard notation for the subspace is S_h. Here h stands for a discretization parameter, and the notation suggests that the approximate solution will converge to the true solution of the given (continuous) problem as $h \to 0$.

We first consider approximation in general subspaces, and later show how to apply it to a model problem.

The solution of the variational problem

$$J(v) := \frac{1}{2}a(v, v) - \langle \ell, v \rangle \longrightarrow \min_{S_h}! \tag{4.1}$$

in the subspace S_h can be computed using the Characterization Theorem 2.2. In particular, u_h is a solution provided

$$a(u_h, v) = \langle \ell, v \rangle \quad \text{for all } v \in S_h. \tag{4.2}$$

Suppose $\{\psi_1, \psi_2, \ldots, \psi_N\}$ is a basis for S_h. Then (4.2) is equivalent to

$$a(u_h, \psi_i) = \langle \ell, \psi_i \rangle, \quad i = 1, 2, \ldots, N.$$

Assuming u_h has the form

$$u_h = \sum_{k=1}^{N} z_k \psi_k, \tag{4.3}$$

we are led to the system of equations

$$\sum_{k=1}^{N} a(\psi_k, \psi_i) z_k = \langle \ell, \psi_i \rangle, \quad i = 1, 2, \ldots, N, \tag{4.4}$$

which we can write in matrix-vector form as

$$Az = b, \tag{4.5}$$

where $A_{ik} := a(\psi_k, \psi_i)$ and $b_i := \langle \ell, \psi_i \rangle$. Whenever a is an H^m-elliptic bilinear form, the matrix A is positive definite:

$$
\begin{aligned}
z'Az &= \sum_{i,k} z_i A_{ik} z_k \\
&= a\left(\sum_k z_k \psi_k, \sum_i z_i \psi_i \right) = a(u_h, u_h) \\
&\geq \alpha \|u_h\|_m^2,
\end{aligned}
$$

and so $z'Az > 0$ for $z \neq 0$. Here we have made use of the bijective mapping $\mathbb{R}^N \longrightarrow S_h$ which is defined by (4.3). Without explicitly referring to this canonical mapping, in the sequel we will identify the function space S_h with $\mathbb{R}^N$.

In engineering sciences, and in particular if the problem comes from continuum mechanics, the matrix A is called the *stiffness matrix* or *system matrix*.

Methods. There are several related methods:

Rayleigh–Ritz Method: Here the minimum of J is sought in the space S_h. Instead of the basis-free derivation via (4.2), usually one finds u_h as in (4.3) by solving the equation $(\partial/\partial z_i) J(\sum_k z_k \psi_k) = 0$.

Galerkin Method: The weak equation (4.2) is solved for problems where the bilinear form is not necessarily symmetric. If the weak equations arise from a variational problem with a positive quadratic form, then often the term *Ritz–Galerkin Method* is used.

Petrov–Galerkin Method: Here we seek $u_h \in S_h$ with

$$
a(u_h, v) = \langle \ell, v \rangle \quad \text{for all } v \in T_h,
$$

where the two N-dimensional spaces S_h and T_h need not be the same. The choice of a space of test functions which is different from S_h is particularly useful for problems with singularities.

As we saw in §§2 and 3, the boundary conditions determine whether a problem should be formulated in $H^m(\Omega)$ or in $H_0^m(\Omega)$. For the purposes of a unified notation, in the following we always suppose $V \subset H^m(\Omega)$, and that the bilinear form a is always V-elliptic, i.e.,

$$
a(v, v) \geq \alpha \|v\|_m^2 \quad \text{and} \quad |a(u, v)| \leq C \|u\|_m \|v\|_m \quad \text{for all } u, v \in V,
$$

where $0 < \alpha \leq C$. The norm $\| \cdot \|_m$ is thus equivalent to the energy norm (2.14), which we use to get our first error bounds. – In addition, let $\ell \in V'$ with $|\langle \ell, v \rangle| \leq \|\ell\| \cdot \|v\|_m$ for $v \in V$. Here $\|\ell\|$ is the (dual) norm of ℓ.

4.1 Remark. (Stability) Independent of the choice of the subspace S_h of V, the solution of (4.2) always satisfies

$$\|u_h\|_m \leq \alpha^{-1}\|\ell\|.$$

Proof. Let u_h be a solution of (4.2). Substituting $v = u_h$, we get

$$\alpha\|u_h\|_m^2 \leq a(u_h, u_h) = \langle \ell, u_h \rangle \leq \|\ell\|\,\|u_h\|_m.$$

Dividing by $\|u_h\|_m$, we get the assertion. $\qquad\square$

The following lemma is of fundamental importance in establishing error bounds for finite element approximations. The line of proof is typical, and we will make frequent use of variants of the technique. The relation (4.7) below is often denoted as *Galerkin orthogonality*.

4.2 Céa's Lemma. *Suppose the bilinear form a is V-elliptic with $H_0^m(\Omega) \subset V \subset H^m(\Omega)$. In addition, suppose u and u_h are the solutions of the variational problem in V and $S_h \subset V$, respectively. Then*

$$\|u - u_h\|_m \leq \frac{C}{\alpha} \inf_{v_h \in S_h} \|u - v_h\|_m. \tag{4.6}$$

Proof. By the definition of u and u_h,

$$a(u, v) = \langle \ell, v \rangle \quad \text{for all } v \in V,$$
$$a(u_h, v) = \langle \ell, v \rangle \quad \text{for all } v \in S_h.$$

Since $S_h \subset V$, it follows by subtraction that

$$a(u - u_h, v) = 0 \quad \text{for all } v \in S_h. \tag{4.7}$$

Let $v_h \in S_h$. With $v = v_h - u_h \in S_h$, it now follows immediately from (4.7) that $a(u - u_h, v_h - u_h) = 0$, and

$$\begin{aligned}
\alpha\|u - u_h\|_m^2 &\leq a(u - u_h, u - u_h) \\
&= a(u - u_h, u - v_h) + a(u - u_h, v_h - u_h) \\
&\leq C\|u - u_h\|_m\|u - v_h\|_m.
\end{aligned}$$

After dividing by $\|u - u_h\|_m$, we get $\alpha\|u - u_h\|_m \leq C\|u - v_h\|_m$, and the assertion is established. $\qquad\square$

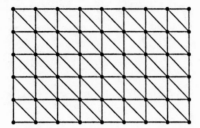

Fig. 9. A uniform triangulation of a rectangle

According to Céa's lemma, the accuracy of a numerical solution depends essentially on choosing function spaces which are capable of approximating the solution u well. For polynomials, the order of approximation is determined by the smoothness of the solution. However, for boundary-value problems, the smoothness of the solution typically decreases as we approach the boundary. Thus, it doesn't make much sense to use polynomials that are defined on the whole domain and to insist on a high accuracy by forcing the degree of the polynomials to be high. As we shall see in §§6 and 7, it makes more sense to use piecewise polynomials, and to achieve the desired accuracy by making the associated partition of Ω sufficiently fine. The so-called *h-p-methods* combine refinements of the partitions and an increase of the degree of the polynomials; see Schwab [1998].

Model Problem

4.3 Example (Courant [1943]). Suppose we want to solve the Poisson equation in the unit square (or in a general domain which can be triangulated with congruent triangles):

$$-\Delta u = f \quad \text{in } \Omega = (0, 1)^2,$$
$$u = 0 \quad \text{on } \partial\Omega.$$

Suppose we partition $\bar{\Omega}$ with a uniform triangulation of mesh size h, as shown in Fig. 9. Choose

$$S_h := \{v \in C(\bar{\Omega}); v \text{ is linear in every triangle and } v = 0 \text{ on } \partial\Omega\}. \qquad (4.8)$$

In every triangle, $v \in S_h$ has the form $v(x, y) = a + bx + cy$, and is uniquely defined by its values at the three vertices of the triangle. Thus, $\dim S_h = N =$ number of interior mesh points. Globally, v is determined by its values at the N grid points (x_j, y_j). Now choose a basis $\{\psi_i\}_{i=1}^N$ with

$$\psi_i(x_j, y_j) = \delta_{ij}.$$

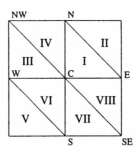

Fig. 10. Numbering of the elements in a neighborhood of the center C and the neighboring points in the compass directions: E, S, W, N, NW and SE

Table 1. Derivatives of the basis functions ψ_C shown in Fig. 10 (ψ_C has the value 1 at C and is 0 at other nodes)

	I	II	III	IV	V	VI	VII	VIII
$\partial_1 \psi_C$	$-h^{-1}$	0	h^{-1}	0	0	h^{-1}	0	$-h^{-1}$
$\partial_2 \psi_C$	$-h^{-1}$	0	0	$-h^{-1}$	0	h^{-1}	h^{-1}	0

We compute the elements of the system matrix A_{ij}, where again we choose local indices and exploit the symmetry, (see Fig. 11 and Table 1)

$$
\begin{aligned}
a(\psi_C, \psi_C) &= \int_{I-VIII} (\nabla \psi_C)^2 dx\,dy \\
&= 2 \int_{I+III+IV} [(\partial_1 \psi_C)^2 + (\partial_2 \psi_C)^2] dx\,dy \\
&= 2 \int_{I+III} (\partial_1 \psi_C)^2 dx\,dy + 2 \int_{I+IV} (\partial_2 \psi_C)^2 dx\,dy \\
&= 2h^{-2} \int_{I+III} dx\,dy + 2h^{-2} \int_{I+IV} dx\,dy \\
&= 4,
\end{aligned}
$$

$$
\begin{aligned}
a(\psi_C, \psi_N) &= \int_{I+IV} \nabla \psi_C \cdot \nabla \psi_N dx\,dy \\
&= \int_{I+IV} \partial_2 \psi_C \partial_2 \psi_N dx\,dy = \int_{I+IV} (-h^{-1})h^{-1} dx\,dy \\
&= -1.
\end{aligned}
$$

Here ψ_N is the nodal function for the point north of C. By symmetry, a similar computation gives

$$
a(\psi_C, \psi_E) = a(\psi_C, \psi_S) = a(\psi_C, \psi_W) = a(\psi_C, \psi_N) = -1.
$$

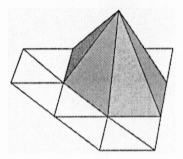

Fig. 11. Nodal basis function

Finally, we find that

$$a(\psi_C, \psi_{NW}) = \int_{III+IV} [\partial_1 \psi_C \partial_1 \psi_{NW} + \partial_2 \psi_C \partial_2 \psi_{NW}] dx dy = 0.$$

In evaluating $a(\psi_C, \psi_{SE})$, note that all products in the integrals vanish. Thus we get a system of linear equations with exactly the same matrix as in the finite difference method based on the standard five-point stencil

$$\begin{bmatrix} & -1 & \\ -1 & 4 & -1 \\ & -1 & \end{bmatrix}_* . \tag{4.9}$$

We should emphasize that this connection with difference methods does not hold in general. The finite element method provides the user with a great deal of freedom, and for most other finite element approximations and other equations, there is no equivalent finite difference star. In general, the finite element approximation does not even satisfy the discrete maximum principle. – The same holds, by the way, for the method of finite volumes. Once again, we get the same matrix only in the above simple case [Hackbusch 1989].

Problems

4.4 As usual, let u and u_h be the functions which minimize J over V and S_h, respectively. Show that u_h is also a solution of the minimum problem

$$a(u - v, u - v) \longrightarrow \min_{v \in S_h} !$$

Because of this, the mapping

$$R_h : V \longrightarrow S_h$$

$$u \longmapsto u_h$$

is called the *Ritz projector*.

4.5 Consider the potential equation with inhomogeneous boundary conditions

$$-\Delta u = 0 \quad \text{in } \Omega = (0, 1)^2,$$
$$u = u_0 \quad \text{on } \partial\Omega,$$

and suppose we select the same regular triangulation as in Example 4.3. In addition, let u_0 be piecewise linear on the boundary. Then u_0 can be extended continuously to Ω so that u_0 is linear in every triangle and vanishes at the interior nodes. Show that in this situation, i.e. for inhomogeneous boundary conditions and with S_h as in (4.8), we get the same linear system as in Chapter I.

4.6 Suppose in Example 4.3 that on the bottom side of the square we replace the Dirichlet boundary condition by the natural boundary condition $\partial u/\partial v = 0$. Verify that this leads to the difference star

$$\begin{bmatrix} & -1 & \\ -1/2 & 2 & -1/2 \end{bmatrix}_*$$

at these boundary points.

4.7 In Example 4.3, does the part $-u_{xx}$ of the Laplace operator $-\Delta$ lead to a stencil which has only nonzero terms in one horizontal line, as is the case for the finite difference method?

4.8 Given the variational problem

$$\int_\Omega [a_1(\partial_1 v)^2 + a_2(\partial_2 v)^2 + a_3(\partial_1 v - \partial_2 v)^2 - 2fv]\,dx \longrightarrow \text{min!}$$

with $a_1, a_2, a_3 > 0$, find the associated Euler differential equation and the difference star, using the same form of approximating function as in Example 4.3.

§ 5. Some Standard Finite Elements

In practice, the spaces over which we solve the variational problems associated with boundary-value problems are called *finite element spaces*. We partition the given domain Ω into (finitely many) subdomains, and consider functions which reduce to a polynomial on each subdomain. The subdomains are called *elements*. For planar problems, they can be triangles or quadrilaterals. For three-dimensional problems, we can use tetrahedra, cubes, rectangular parallelepipeds, etc. For simplicity, we restrict our discussion primarily to the two-dimensional case.

Here is a list of some of the important properties characterizing different finite element spaces:

1. The kind of partition used on the domain: triangles or quadrilaterals. If all elements are congruent, we say that the partition is *regular*.

2. In two variables, we refer to

$$\mathcal{P}_t := \{u(x, y) = \sum_{\substack{i + k \le t \\ i, k \ge 0}} c_{ik} x^i y^k\} \tag{5.1}$$

as the set of *polynomials of degree $\le t$*. If all polynomials of degree $\le t$ are used, we call them finite elements with *complete polynomials*.

The restrictions of the polynomials to the edges of the triangles or quadrilaterals are polynomials in one variable. Sometimes we will require that their degree be smaller than t (e.g. at most $t - 1$). Such a condition will be part of the specification of the elements.

The admissible polynomial degrees in the elements or on their edges are a local property.

3. Continuity and differentiability properties: A finite element is said to be a C^k *element* provided it is contained in $C^k(\Omega)$.[5] This property is of a global character and is often concealed in interpolation conditions.

We remark that according to this scheme, the Courant triangles in Example 4.3 would be classified as linear triangular elements in $C^0(\Omega)$.

We use the terminology *conforming finite element* if the functions lie in the Sobolev space in which the variational problem is posed. Nonconforming elements will be studied in Chapter III.

---

[5] The use of the terminology *element* may be somewhat confusing. We decompose the domain into *elements* which are geometric objects, while the *finite elements* are actually functions. However, we will deviate from this convention when discussing e.g. C^k elements or linear elements, where the meaning is clear from the context.

   The formal definition of finite elements will be given in the definitions 5.8 and
5.12. The reader will recognize the three properties above in the triple (T, Π, Σ)
after the significance of the differentiability conditions will be clear.

Requirements on the Meshes

For simplicity, in the following let Ω be a polygonal domain which can be partitioned into triangles or quadrilaterals. The partition is by no means required to be
as regular as the one shown in the model problem in §4.

5.1 Definition. (1) A partition $\mathcal{T} = \{T_1, T_2, \ldots, T_M\}$ of Ω into triangular or
quadrilateral elements is called *admissible* provided the following properties hold
(see Fig. 12):

  i. $\bar{\Omega} = \bigcup_{i=1}^{M} T_i$.
 ii. If $T_i \cap T_j$ consists of exactly one point, then it is a common vertex of T_i and
     T_j.
iii. If for $i \neq j$, $T_i \cap T_j$ consists of more than one point, then $T_i \cap T_j$ is a common
     edge of T_i and T_j.

   (2) We will write $\mathcal{T}_h$ instead of $\mathcal{T}$ when every element has diameter at most
$2h$.

   (3) A family of partitions $\{\mathcal{T}_h\}$ is called *shape regular* provided that there
exists a number $\kappa > 0$ such that every T in $\mathcal{T}_h$ contains a circle of radius ρ_T with

$$\rho_T \geq h_T/\kappa,$$

where h_T is half the diameter of T.

   (4) A family of partitions $\{\mathcal{T}_h\}$ is called *uniform* provided that there exists
a number $\kappa > 0$ such that every element T in $\mathcal{T}_h$ contains a circle with radius
$\rho_T \geq h/\kappa$. We will often use the terminology κ-*regular*.

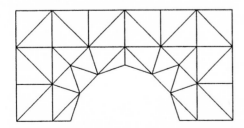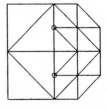

Fig. 12. An admissible triangulation (left), and one which is not because of two
hanging nodes marked by o (right).

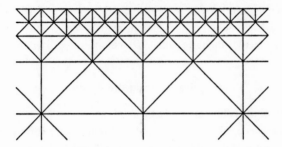

Fig. 13. A triangulation which is shape regular but not uniform

Since $h = \max_{T \in \mathcal{T}} h_T$, uniformity is a stronger requirement than shape regularity. Clearly, the triangulations shown in Figs. 13 and 14 are shape regular, independent of how many steps of the refinement in the neighborhood of the boundary or of a reentrant corner are carried out. However, if the number of steps depends on h, the partitions are no longer uniform.

In practice, we almost always use shape-regular meshes, and very frequently even uniform ones.

Significance of the Differentiability Properties

In the conforming treatment of second order elliptic problems, we choose finite elements which lie in H^1. We shall show that it is possible to use functions which are continuous but not necessarily continuously differentiable. Thus, the functions are much less smooth than required for a classical solution of the boundary-value problem.

In the following, we will always assume unless otherwise indicated that the partitions satisfy the requirements of 5.1. We say that a function u on Ω satisfies a given property *piecewise* provided that its restriction to every element has that property.

5.2 Theorem. *Let $k \geq 1$ and suppose Ω is bounded. Then a piecewise infinitely differentiable function $v : \bar{\Omega} \to \mathbb{R}$ belongs to $H^k(\Omega)$ if and only if $v \in C^{k-1}(\bar{\Omega})$.*

Proof. It suffices to give the proof for $k = 1$. For $k > 1$ the assertion then follows immediately from a consideration of the derivatives of order $k - 1$. In addition, for simplicity we restrict ourselves to domains in $\mathbb{R}^2$.

(1) Let $v \in C(\bar{\Omega})$, and suppose $\mathcal{T} = \{T_j\}_{j=1}^M$ is a partition of Ω. For $i = 1, 2$, define $w_i : \Omega \to \mathbb{R}$ piecewise by $w_i(x) := \partial_i v(x)$ for $x \in \Omega$, where on the edges we can take either of the two limiting values. Let $\phi \in C_0^\infty(\Omega)$. Green's formula

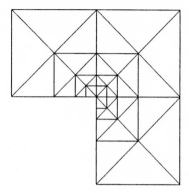

Fig. 14. Nonuniform triangulations with a reentrant vertex

can be applied in every element T_j to give

$$\int_\Omega \phi w_i \, dx dy = \sum_j \int_{T_j} \phi \partial_i v \, dx dy$$

$$= \sum_j \left\{ -\int_{T_j} \partial_i \phi v \, dx dy + \int_{\partial T_j} \phi v \, \nu_i \, ds \right\}. \qquad (5.2)$$

Since v was assumed to be continuous, the integrals over the interior edges cancel. Moreover, ϕ vanishes on $\partial\Omega$, and we are left with the integral over the domain

$$- \int_\Omega \partial_i \phi v \, dx dy.$$

By Definition 1.1, w_i is the weak derivative of v.

(2) Let $v \in H^1(\Omega)$. We do not establish the continuity of v by working backwards through the formulae (although this would be possible), but instead employ an approximation-theoretical argument. Consider v in the neighborhood of an edge, and rotate the edge so that it lies on the y-axis. Suppose the edge becomes the interval $[y_1 - \delta, y_2 + \delta]$ on the y-axis with $y_1 < y_2$ and $\delta > 0$. We now investigate the auxiliary function

$$\psi(x) := \int_{y_1}^{y_2} v(x, y) dy.$$

First, suppose $v \in C^\infty(\Omega)$. It now follows from the Cauchy–Schwarz inequality that

$$|\psi(x_2) - \psi(x_1)|^2 = \left| \int_{x_1}^{x_2} \int_{y_1}^{y_2} \partial_1 v \, dx dy \right|^2$$

$$\leq \left| \int_{x_1}^{x_2} \int_{y_1}^{y_2} 1 \, dx dy \right| \cdot |v|_{1,\Omega}^2$$

$$\leq |x_2 - x_1| \cdot |y_2 - y_1| \cdot |v|_{1,\Omega}^2.$$

Because of the density of $C^\infty(\Omega)$ in $H^1(\Omega)$, this assertion also holds for $v \in H^1(\Omega)$. Thus the function $x \mapsto \psi(x)$ is continuous, and in particular at $x = 0$. Since y_1 and y_2 are arbitrary except for $y_1 < y_2$, this can only happen if the piecewise continuous function v is continuous on the edge. $\qquad\square$

If no other additional conditions are required, continuous finite elements are easily constructed. In view of Theorem 5.2, this is of great advantage for the solution of second order boundary-value problems using conforming finite elements. The construction of C^1 elements, which according to Theorem 5.2 are required for the conforming treatment of problems of fourth order, is more difficult.

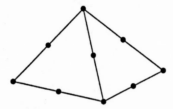

Fig. 15. Piecewise quadratic polynomials that interpolate at the points (•) are continuous at the interface

Triangular Elements with Complete Polynomials

The simplest triangular elements to construct are C^0 elements made up of complete polynomials.

5.3 Remark. Let u be a polynomial of degree t. If we apply an affine linear transformation and express u in the new coordinates, we again get a polynomial of degree t. Thus, the set of polynomials $\mathcal{P}_t$ is invariant under affine linear transformations.

5.4 Remark. Let $t \geq 0$. Given a triangle T, suppose $z_1, z_2, \ldots, z_s$ are the $s = 1 + 2 + \cdots + (t + 1)$ points in T which lie on $t + 1$ lines, as in Fig. 16. Then for every $f \in C(T)$, there is a unique polynomial p of degree $\leq t$ satisfying the interpolation conditions

$$p(z_i) = f(z_i), \quad i = 1, 2, \ldots, s. \tag{5.3}$$

Proof. The result is trivial for $t = 0$. We now assume it has been established for $t - 1$, and prove it for t. In view of the invariance under affine transformations, we can assume that one of the edges of T lies on the x-axis. Suppose it is the one containing the points $z_1, z_2, \ldots, z_{t+1}$. There exists a univariate polynomial $p_0 = p_0(x)$ with

$$p_0(z_i) = f(z_i), \quad i = 1, 2 \ldots, t + 1.$$

By the induction hypothesis, there also exists a polynomial $q = q(x, y)$ of degree $t - 1$ with

$$q(z_i) = \frac{1}{y_i}[f(z_i) - p_0(z_i)], \quad i = t + 2, \ldots, s.$$

Clearly, $p(x, y) = p_0(x) + yq(x, y)$ satisfies (5.3).   □

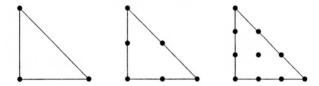

Fig. 16. Nodes of the nodal basis for linear, quadratic, and cubic triangular elements $\mathcal{M}_0^1$, $\mathcal{M}_0^2$, and $\mathcal{M}_0^3$

5.5 Definition. Suppose that for a given finite element space, there is a set of points which uniquely determines any function in the space by its values at the points. Then the set of functions in the space which take on a nonzero value at precisely one of the points form a basis for the space, called the *nodal basis*.

The following construction, which assures continuity by using sufficiently many points on the edges of the triangle, is typical for the construction of C^0 elements.

5.6 A Nodal Basis for C^0 Elements. Let $t \geq 1$, and suppose we are given a triangulation of Ω. In each triangle, we place $s := (t + 1)(t + 2)/2$ points as indicated in Fig. 16, so that there are $t + 1$ points on each edge. By Remark 5.4, in each triangle a polynomial of degree $\leq t$ is determined by choosing values at these points. The restriction of any such polynomial to an edge is a polynomial of degree $\leq t$ in one variable. Now given an edge, the two polynomials on either side interpolate the same values at the $t + 1$ points on that edge, and thus must reduce to the same one-dimensional polynomial. This ensures that our elements are globally continuous.

In dealing with finite elements with complete polynomials, we make use of the following notation from the literature:

$$\begin{aligned}
\mathcal{M}^k &:= \mathcal{M}_k(\mathcal{T}) := \{v \in L_2(\Omega); \ v|_T \in \mathcal{P}_k \text{ for every } T \in \mathcal{T}\}, \\
\mathcal{M}_0^k &:= \mathcal{M}^k \cap C^0(\Omega) = \mathcal{M}^k \cap H^1(\Omega), \\
\mathcal{M}_{0,0}^k &:= \mathcal{M}^k \cap H_0^1(\Omega).
\end{aligned} \quad (5.4)$$

$\mathcal{M}_0^1$ is also called the *conforming P_1 element* or *Courant triangle*.

Table 2. Interpolation with some standard finite elements
- •   Function value prescribed
- ⊙   Function value and 1st derivative prescribed
- ◎   Function value and 1st and 2nd derivatives prescribed
- ⊥   Normal derivative prescribed

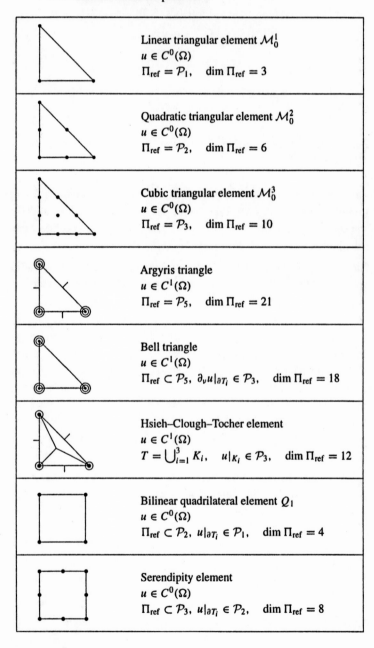

Linear triangular element $\mathcal{M}_0^1$
$u \in C^0(\Omega)$
$\Pi_{\mathrm{ref}} = \mathcal{P}_1, \quad \dim \Pi_{\mathrm{ref}} = 3$

Quadratic triangular element $\mathcal{M}_0^2$
$u \in C^0(\Omega)$
$\Pi_{\mathrm{ref}} = \mathcal{P}_2, \quad \dim \Pi_{\mathrm{ref}} = 6$

Cubic triangular element $\mathcal{M}_0^3$
$u \in C^0(\Omega)$
$\Pi_{\mathrm{ref}} = \mathcal{P}_3, \quad \dim \Pi_{\mathrm{ref}} = 10$

Argyris triangle
$u \in C^1(\Omega)$
$\Pi_{\mathrm{ref}} = \mathcal{P}_5, \quad \dim \Pi_{\mathrm{ref}} = 21$

Bell triangle
$u \in C^1(\Omega)$
$\Pi_{\mathrm{ref}} \subset \mathcal{P}_5, \ \partial_\nu u|_{\partial T_i} \in \mathcal{P}_3, \quad \dim \Pi_{\mathrm{ref}} = 18$

Hsieh–Clough–Tocher element
$u \in C^1(\Omega)$
$T = \bigcup_{i=1}^3 K_i, \quad u|_{K_i} \in \mathcal{P}_3, \quad \dim \Pi_{\mathrm{ref}} = 12$

Bilinear quadrilateral element Q_1
$u \in C^0(\Omega)$
$\Pi_{\mathrm{ref}} \subset \mathcal{P}_2, \ u|_{\partial T_i} \in \mathcal{P}_1, \quad \dim \Pi_{\mathrm{ref}} = 4$

Serendipity element
$u \in C^0(\Omega)$
$\Pi_{\mathrm{ref}} \subset \mathcal{P}_3, \ u|_{\partial T_i} \in \mathcal{P}_2, \quad \dim \Pi_{\mathrm{ref}} = 8$

Remarks on C^1 Elements

The construction of C^1 elements is considerably more difficult. There are two well-known constructions of triangular elements based on polynomials of degree 5. Recall that dim $\mathcal{P}_5 = 21$. We now assume that we are given values for derivatives up to order 2 at each of the vertices of the triangle. This uses $3 \times 6 = 18$ degrees of freedom.

To construct the *Argyris element*, we use the remaining three degrees of freedom to specify the values of the normal derivatives at the midpoint of each side of the triangle. The following argument shows that this leads to a global C^1 function.

Consider the univariate polynomials which are the restrictions of two neighboring polynomials to a common edge of the triangle. At the ends of the edge, these two polynomials must both interpolate the values of the given derivatives up to order 2. Since this interpolation problem has a unique solution, we get the desired continuity of the function and its tangential derivative. The normal derivatives along the edge are polynomials of degree 4. They both interpolate the given derivatives up to order 1 at the ends of the edge, along with the given value at the center of the edge. Since these five pieces of data uniquely determine a polynomial of degree 4, we have established the continuity of the normal derivative.

To construct the *triangular element of Bell*, we use the same data at the vertices as for the Argyris element. But now we restrict ourselves to the class of polynomials of degree 5 whose normal derivatives on the sides of the triangle are polynomials of degree 3 rather than 4. Again the normal derivative along an edge is uniquely determined by the derivative information at the vertices, and we get a continuous derivative. The number of degrees of freedom for this element is 3 less than for the Argyris element (see Table 2).

The *Hsieh–Clough–Tocher element* is constructed by a completely different process. First we subdivide the triangle T into three subtriangles by connecting its vertices to its center of gravity. We now build a C^1 function consisting piecewise of cubic polynomials. At each of the vertices of the original triangle, we specify the function value and the first derivatives. In addition, we specify the normal derivative at the midpoint of each of the three sides of T. It can be shown that the three cubic polynomials join together to form a C^1 function on T. The fact that two adjoining macro-elements join with C^1 continuity can be established in the same way as for the other C^1 elements. This element has exactly 12 degrees of freedom.

The *reduced Hsieh–Clough–Tocher element* is constructed in a similar way, except that now we insist that the normal derivatives along the edges of T be linear rather than quadratic. The analysis now proceeds as in our construction of the Bell element from the Argyris element (cf. Problem 6.15).

Because it involves a subpartition, the Hsieh–Clough–Tocher element is called a *macro-element*.

It should be noted that the continuity of derivatives along the element boundaries is easy to handle in terms of the Bernstein–Bézier representation of polynomials.

Bilinear Elements

The polynomial families $\mathcal{P}_t$ are not used on rectangular partitions of a domain. We can see why by looking at the simplest example, the bilinear element. Instead of using $\mathcal{P}_t$ as we did for triangles, on rectangular elements we use the polynomial family which contains *tensor products*:

$$\mathcal{Q}_t := \{u(x, y) = \sum_{0 \leq i,k \leq t} c_{ik} x^i y^k\}. \tag{5.5}$$

If more general quadrilateral elements are involved, we can use appropriately transformed families.

We consider first a rectangular grid whose grid lines run parallel to the coordinate axes. On each rectangle we use

$$u(x, y) = a + bx + cy + dxy, \tag{5.6}$$

where the four parameters are uniquely determined by the values of u at the four vertices of the rectangle. Although u is a polynomial of degree 2, its restriction to each edge is a linear function. Because of this, we automatically get global continuity of the elements since neighboring bilinear pieces share the same node information.

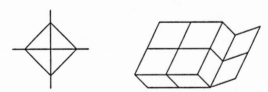

Fig. 17. A rectangle rotated by 45°, and a parallelogram element

The polynomial form (5.6) is not usable on a grid which has been rotated by 45° as in Fig. 17. Indeed, the term dxy in (5.6) vanishes at all of the vertices of the rotated square.

We can get the correct polynomial form for general parallelograms (and thus for the rotated elements shown in Fig. 17) by means of a linear transformation. However, the treatment of general quadrilaterals requires the more general class of so-called *isoparametric* mappings, which are discussed in Chapter III. – It is possible to combine parallelograms with triangles, in order to make the partition of Ω more flexible.

Suppose the edges of a parallelogram element lie on lines of the form

$$\alpha_1 x + \beta_1 y = \gamma_1,$$
$$\alpha_2 x + \beta_2 y = \gamma_2 \tag{5.7}$$

(where the coefficients vary from element to element). Then with the transformation

$$\xi = \alpha_1 x + \beta_1 y,$$
$$\eta = \alpha_2 x + \beta_2 y,$$

we get the *bilinear* function

$$u(x, y) = a + b\xi + c\eta + d\xi\eta \tag{5.8}$$

which is linear along the edges of the parallelogram.

5.7 Remarks. (1) We can also characterize the Q_1 elements obtained by the above construction in a coordinate-free way:

$$S = \{v \in C^0(\bar{\Omega}); \text{ for every element } T, v|_T \in \mathcal{P}_2,$$
$$\text{and the restriction to each edge belongs to } \mathcal{P}_1\}.$$

(2) If every edge of an element contains $t_k + 1$ nodes of a nodal basis, and the restrictions to the edges are all polynomials of degree t_k at most, then we automatically get globally continuous elements. – Note that the maximum degree of a polynomial restricted to an edge does not increase under a linear transformation.

Quadratic Rectangular Elements

One of the most popular elements on rectangles (or on more general parallelograms) consists of piecewise polynomials of degree 3 whose restrictions to the edges are quadratic polynomials. Using the coordinates shown in Fig. 18, we can write such a polynomial in the form

$$\begin{aligned}
u(x, y) = {}& a + bx + cy + dxy \\
& + e(x^2 - 1)(y - 1) + f(x^2 - 1)(y + 1) \\
& + g(x - 1)(y^2 - 1) + h(x + 1)(y^2 - 1)
\end{aligned}$$

(cf. Problem 5.16). There are eight degrees of freedom. The first four are determined by the values at the vertices. The remaining parameters e, f, g and h can be computed directly from the values at the midpoints of the sides. This element is called the *eight node element* or the *serendipity element*. If we add the term

$$k(x^2 - 1)(y^2 - 1),$$

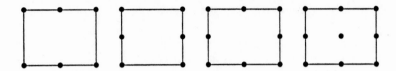

Fig. 18. Rectangular elements with 6, 8, or 9 nodes for a rectangle with edges
on the lines $|x| = 1$, $|y| = 1$.

we get one more degree of freedom, and can then interpolate a value at the center
of the rectangle. By dropping some degrees of freedom, we can also get useful six
node elements (with $e = f = 0$ or $g = h = 0$, respectively), as shown in Fig. 18.

Affine Families

In the above discussion of special finite element spaces, we have implicitly made
use of the following formal construction; cf. Ciarlet [1978].

5.8 Definition. A *finite element* is a triple (T, Π, Σ) with the following properties:

(i) T is a polyhedron in $\mathbb{R}^d$. (The parts of the surface ∂T lie on hyperplanes and
are called *faces*.)

(ii) Π is a subspace of $C(T)$ with finite dimension s. (Functions in Π are called
shape functions if they form a basis of Π.)

(iii) Σ is a set of s linearly independent functionals on Π. Every $p \in \Pi$ is uniquely
defined by the values of the s functionals in Σ. – Since usually the functionals
involve point evaluation of a function or its derivatives at points in T, we call
these *(generalized) interpolation conditions*.

In (ii) s is the *number of local degrees of freedom* or *local dimension*.

Although generally Π consists of polynomials, it is not enough to look only
at polynomial spaces, since otherwise we would exclude piecewise polynomial
elements such as the Hsieh–Clough–Tocher element. In fact, there are even finite
elements consisting of piecewise rational functions; see Wachspress [1971].

As a first example consider the finite element families $\mathcal{M}_0^k$. We have

$$\mathcal{M}_0^k = (T, \mathcal{P}^k, \Sigma^k),$$
$$\Sigma^k := \{p(z_i); \ i = 1, 2, \dots, \frac{(k+1)(k+2)}{2}\},$$

where the points z_i are defined in Remark 5.4 and depicted in Fig. 16 for $k \le 3$.

The condition for a smooth join between elements is dealt with in (iii), although
in fact, we actually need a still stronger formalization of this condition. However,

for the C^1 elements presented in Table 2, the meaning is clear. Thus, e.g., for the Argyris triangle, by Table 2 we have

$$\Pi := \mathcal{P}_5, \qquad \dim \Pi = 21,$$

$$\Sigma := \{p(a_i), \partial_x p(a_i), \partial_y p(a_i), \partial_{xx} p(a_i), \partial_{xy} p(a_i), \partial_{yy} p(a_i), \; i = 1, 2, 3,$$
$$\partial_n p(a_{12}), \partial_n p(a_{13}), \partial_n p(a_{23})\},$$

where a_1, a_2, a_3 are the vertices and $a_{ij} = \frac{1}{2}(a_i + a_j)$ are the midpoints of the sides.

Another example elucidates the role of the functionals in Σ. Fig. 19 shows three different finite elements with $\Pi = \mathcal{P}_1$. Only the first one belongs to $H^1(\Omega)$. Although the local degree of freedom is 3 in each case, the dimensions of the resulting finite element spaces are quite different; cf. Problem 5.13. Similarly, $\mathcal{M}_0^3$ and the cubic Hermite triangle are different elements with cubic polynomials shown in Fig. 20.

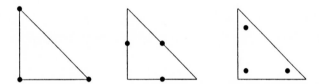

Fig. 19. The P_1 elements $\mathcal{M}_0^1$, $\mathcal{M}_*^1$, and $\mathcal{M}^1$. Here and in the other diagrams the points marked by a • refer to point evaluations from the set of functionals Σ associated to the finite element spaces in Definition 5.8(iii). The symbols for other functionals are found in Table 2 and Fig. 21.

5.9 The Cubic Hermite Triangle. The ten degrees of freedom of cubic polynomials can also be fixed in another way. In particular, we can choose the values of the polynomial and its first derivatives at the vertices $a_i, i = 1, 2, 3$, along with the value at the center $a_{123} = \frac{1}{3}(a_1 + a_2 + a_3)$. The cubic Hermite triangle is the triple $(T, \mathcal{P}_3, \Sigma_{HT})$ where

$$\Sigma_{HT} := \left\{ p(a_i), \frac{\partial}{\partial x} p(a_i), \frac{\partial}{\partial y} p(a_i), \; i = 1, 2, 3, \text{ and } p(a_{123}) \right\}. \qquad (5.9)$$

The functions in Σ_{HT} are linearly independent. To verify this, we consider an edge of the triangle between the vertices a_i and a_j ($i \neq j$). Let $q \in \mathcal{P}_3$ be the univariate polynomial which is the restriction of p to the edge. The values $q(a_i), q(a_j)$ and the derivatives at these points are given by p and the directional derivative.

Since the one-dimensional Hermite interpolation problem for two points and cubic polynomials has a unique solution, we can compute q. Hence, the values at the ten nodes shown in Fig. 16 for the Lagrange interpolation are uniquely determined.

Thus we have reduced the interpolation problem for the Hermite triangle to the usual Lagrange interpolation problem which is known to be solvable.

We emphasize that the derivatives are continuously joined only at the vertices [but not along the edges]. The cubic Hermite triangle is *not* a C^1 element. Nevertheless, we will see in Chapter VI that it provides appropriate nonconforming H^2 elements for the treatment of plates.

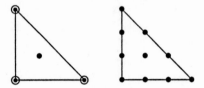

Fig. 20. Interpolation points for two elements with $\Pi = \mathcal{P}_3$, i.e. with piecewise cubic polynomials

5.10 The Bogner–Fox–Schmit rectangle. On the other hand there is a C^1-element with bicubic funtions. It is called the Bogner–Fox–Schmit element and depicted in Fig. 21.

$$\Pi_{\text{ref}} := \mathcal{Q}_3, \qquad \dim \Pi_{\text{ref}} = 16,$$
$$\Sigma := \{ p(a_i),\, \partial_x p(a_i),\, \partial_y p(a_i),\, \partial_{xy} p(a_i),\ i = 1, 2, 3, 4. \} \qquad (5.10)$$

Since the data in (5.10) refer to the tensor products of one-dimensional Hermite interpolation, the 16 functionals in Σ are linearly independent on $\mathcal{Q}_3$.

To verify C^1 continuity of the Bogner–Fox–Schmit element, consider the univarite polynomial on a vertical edge of the rectangle. Its restriction to the edge is a cubic polynomial in y which is determined by p and $\partial_y p$ at the two vertices. Similarly the normal derivative $\partial_x p$ is also a cubic polynomial and determined by $\partial_x p$ and $\partial_{xy} p$ at the vertices. Thus we have continuity of p and $\partial_x p$, i.e. C^1 continuity. $\qquad\square$

Fig. 21. C^1-element of Bogner–Fox–Schmit. The symbol $\nearrow$ refers to the mixed second derivative $\partial_{xy} p$.

5.11 Standard Elements. Strictly speaking the diagrams show the sets T and Σ from the triple (T, Π, Σ). Nevertheless, in many cases the associated family Π is considered clear and is sometimes just mentioned without a detailed specification. For example, diagrams as in Fig. 16 refer to $\Pi = \mathcal{M}_0^k$. If a point evaluation at the center of a triangle is added to $\mathcal{M}_0^1$ or $\mathcal{M}_0^2$, then the space is augmented by a bubble function as for the MINI element in section III.7 or the plate elements in section VI.6. Figures 17 and 18 show further standard elements with $\Pi = P_1$ and $\Pi = P_3$.

The functionals which are encountered with elements for scalar equations, are found in Table 2 and Fig. 21. The specification of vector valued elements often contains normal components or tangential components on the edges. Motivation is given in Problem 5.13, but applications are only contained in Chapter III and VI.

Definition 5.8 refers to a single element. The analysis of the finite element spaces can be obtained from results for a reference element, if all elements are constructed by affine transformations.

5.12 Definition. A family of finite element spaces S_h for partitions $\mathcal{T}_h$ of $\Omega \subset \mathbb{R}^d$ is called an *affine family* provided there exists a finite element $(T_{\text{ref}}, \Pi_{\text{ref}}, \Sigma)$ called the *reference element* with the following properties:

(iv) For every $T_j \in \mathcal{T}_h$, there exists an affine mapping $F_j : T_{\text{ref}} \longrightarrow T_j$ such that for every $v \in S_h$, its restriction to T_j has the form

$$v(x) = p(F_j^{-1}x) \quad \text{with } p \in \Pi_{\text{ref}}.$$

We have already encountered several examples of affine families. The families $\mathcal{M}_0^k$ and the rectangular elements considered so far are affine families. For example, $\mathcal{M}_0^k$ is defined by the triple $(\hat{T}, \mathcal{P}_k, \Sigma_k)$, where

$$\hat{T} := \{(\xi, \eta) \in \mathbb{R}^2; \ \xi \geq 0, \ \eta \geq 0, \ 1 - \xi - \eta \geq 0\} \tag{5.11}$$

is the unit triangle and $\Sigma_k := \{p(z_i); \ i = 1, 2, \ldots, s := k(k+1)/2\}$ is the set of nodal basis points z_i in Remark 5.6.

In our above discussion of the bilinear rectangular elements and the analogous biquadratic ones, it is clear how the transformation (iv) works (see e.g. (5.8)), but this is not the case for the complete polynomials on triangles. For rectangular elements, the unit square $[-1, +1]^2$ is the natural reference quadrilateral.

On the other hand, whenever conditions on the normal derivatives enter into the definition (e.g. in the definition of the Argyris triangle), then we do not have an affine family; see Fig. 22. This can be remedied by combining the normal derivatives with the tangential ones in the analysis. This has led to the theory of *almost-affine families*; see Ciarlet [1978].

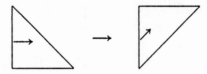

Fig. 22. Transformation of the unit triangle and one normal direction by the affine map $x \mapsto x$, $x + y \mapsto y$.

Choice of an Element

There are a large number of special elements which are useful for the treatment of systems of elliptic partial differential equations, see Chs. III and VI, or Ciarlet [1978], Bathe [1986]. It is useful to say something about how to choose an element even in the scalar case.

The choice of whether to use a triangular or rectangular partition depends primarily on the shape of the domain. Triangles are more flexible, but in solid mechanics, rectangular elements are generally preferred (cf. Ch. VI, §4).

For problems with a smooth behavior, we generally get better results using (bi-)quadratic elements than with (bi-)linear ones with the same number of free parameters. However, they do lead to linear systems with a larger bandwidth, and there is more work involved in setting up the stiffness matrices. This drawback is avoided when using standard finite element packages. Nevertheless, to save programming time and to get results as quickly as possible, linear elements are often used.

Problems

5.13 Consider the subset of all polynomials p in $\mathcal{P}_k$ for which
 a) the restriction of p to any edge lies in $\mathcal{P}_{k-1}$, or
 b) the restriction of the normal derivative $\partial_\nu p$ to any edge lies in $\mathcal{P}_{k-2}$.
Which of the two sets generates an affine family?

5.14 The completion of the space of vector-valued functions $C^\infty(\Omega)^n$ w.r.t. the norm
$$\|v\|^2 := \|v\|_{0,\Omega}^2 + \|\operatorname{div} v\|_{0,\Omega}^2$$
is denoted by $H(\operatorname{div}, \Omega)$. Obviously, $H^1(\Omega)^n \subset H(\operatorname{div}, \Omega) \subset L_2(\Omega)^n$. Show that a piecewise polynomial v is contained in $H(\operatorname{div}, \Omega)$ if and only if the components $v \cdot \nu$ in the direction of the normals are continuous on the inter-element boundaries. Hint: Apply Theorem 5.2 and use (2.22). — Similarly piecewise polynomials in the space $H(\operatorname{rot}, \Omega)$ are characterized by the continuity of the tangential components; see Problem VI.4.8.

5.15 Show that for a triangulation of a simply connected domain, *the number of triangles plus the number of nodes minus the number of edges* is always 1. Why doesn't this hold for multiply connected domains?

5.16 When considering the cubic Hermite triangle, there are three degrees of freedom per vertex and one per triangle. By the results of the previous problem, we know that the dimension must be smaller than for the standard Lagrange representation. Where are the missing degrees of freedom?

5.17 Let $f \in L_2(\Omega)$, and suppose u and u_h are the solutions of the Poisson equation $-\Delta u = f$ in $H_0^1(\Omega)$ and in a finite element space $S_h \subset C^0(\Omega)$, respectively. By construction ∇u and ∇u_h are at least L_2 functions. By the remark in Example 2.10, we know that the divergence of ∇u is an L_2 function. With the help of Problem 5.14, show that this no longer holds for the divergence of ∇u_h in general.

5.18 Show that the piecewise cubic continuous quadrilateral elements whose restrictions to the edges are quadratic polynomials, are exactly the serendipity class of eight node elements.
Hint: First consider a rectangle with sides parallel to the axes.

5.19 To construct triangular elements based on quadratic polynomials, consider the subspace of functions whose normal derivatives on the three edges are constant. Find the dimension of this space, distinguishing the cases when it is a right triangle or not.

5.20 The set of cubic polynomials whose restrictions to the edges of a triangle are quadratic form a 7-dimensional space. Give a basis for it on the unit triangle (5.9).
— We will later encounter the cubic bubble function B_3. The result of this problem can be identified with $\mathcal{P}_2 \oplus B_3$.

§ 6. Approximation Properties

In this section we give error bounds for finite element approximations. By Céa's lemma, in the energy norm it suffices to know how well the solution can be approximated by elements in the corresponding finite element space $S_h = S_h(\mathcal{T}_h)$. For general methods, a suitable framework is provided by the theory of *affine families*. We do not derive results for every individual element, but instead examine a *reference element*, and use transformation formulae to carry the results over to shape-regular grids.

We intend to provide error bounds in other norms besides the energy norm.

We will concentrate primarily on affine families of triangular elements. Clearly, the error for an interpolation method provides an upper bound for the error of the *best* approximation. It turns out that we actually get the correct order of approximation in this way. – We consider C^0 elements, which according to Theorem 5.2, are not contained in $H^m(\Omega)$ for $m > 1$, and so the higher Sobolev norms are not applicable. As substitutes, we use certain *mesh-dependent norms* which are tailored to the problem at hand. We do not use the symbols $\|\cdot\|_h$ and $\|\cdot\|_{m,h}$ for fixed norms, but allow the norm to change from case to case. Often mesh-dependent norms are *broken norms* as in (6.1) or norms with weight factors h^{-m} as in Problem III.1.9.

6.1 Notation. Given a partition $\mathcal{T}_h = \{T_1, T_2, \ldots, T_M\}$ of Ω and $m \geq 1$, let

$$\|v\|_{m,h} := \sqrt{\sum_{T_j \in \mathcal{T}_h} \|v\|_{m,T_j}^2} \, . \tag{6.1}$$

Clearly, $\|v\|_{m,h} = \|v\|_{m,\Omega}$ for $v \in H^m(\Omega)$.

Let $m \geq 2$. By the Sobolev imbedding theorem (see Remark 3.4) $H^m(\Omega) \subset C^0(\Omega)$, i.e. every $v \in H^m$ has a continuous representer. For every $v \in H^m$, there exists a uniquely defined interpolant in $S_h = S_h(\mathcal{T}_h)$ associated with the points in 5.6. We denote it by $I_h v$. The goal of this section is to estimate

$$\|v - I_h v\|_{m,h} \quad \text{by} \quad \|v\|_{t,\Omega} \quad \text{for } m \leq t.$$

The Bramble–Hilbert Lemma

First we obtain an error estimate for interpolation by polynomials. We begin by establishing the result for all domains which satisfy the hypotheses of the imbedding theorem. Later we shall apply it primarily to reference elements, i.e., on convex triangles and quadrilaterals.

6.2 Lemma. *Let $\Omega \subset \mathbb{R}^2$ be a domain with Lipschitz continuous boundary which satisfies a cone condition. In addition, let $t \geq 2$, and suppose $z_1, z_2, \ldots, z_s$ are $s := t(t + 1)/2$ prescribed points in $\bar{\Omega}$ such that the interpolation operator $I : H^t \to \mathcal{P}_{t-1}$ is well defined for polynomials of degree $\leq t - 1$. Then there exists a constant $c = c(\Omega, z_1, \ldots, z_s)$ such that*

$$\|u - Iu\|_t \leq c|u|_t \quad \text{for all } u \in H^t(\Omega). \tag{6.2}$$

Proof. We endow $H^t(\Omega)$ with the norm

$$\|\|v\|\| := |v|_t + \sum_{i=1}^{s} |v(z_i)|,$$

and show that the norms $\|\| \cdot \|\|$ and $\| \cdot \|_t$ are equivalent. Then (6.2) will follow from

$$\|u - Iu\|_t \leq c\|\|u - Iu\|\|$$
$$= c\Big[|u - Iu|_t + \sum_{i=1}^{s} |(u - Iu)(z_i)|\Big]$$
$$= c|u - Iu|_t = c|u|_t.$$

Here we have made use of the fact that Iu coincides with u at the interpolation points, since $D^\alpha Iu = 0$ for all $|\alpha| = t$.

One direction of the proof of equivalence of the norms is simple. By Remarks 3.4, the imbedding $H^t \hookrightarrow H^2 \hookrightarrow C^0$ is continuous. This implies

$$|v(z_i)| \leq c\|v\|_t \quad \text{for } i = 1, 2, \ldots, s,$$

and thus $\|\|v\|\| \leq (1 + cs)\|v\|_t$.

Suppose now that the converse

$$\|v\|_t \leq c\|\|v\|\| \quad \text{for all } v \in H^t(\Omega) \tag{6.3}$$

fails for every positive number c. Then there exists a sequence (v_k) in $H^t(\Omega)$ with

$$\|v_k\|_t = 1, \quad \|\|v_k\|\| \leq \frac{1}{k}, \quad k = 1, 2, \ldots .$$

By the Rellich selection theorem (Theorem 1.9), a subsequence of (v_k) converges in $H^{t-1}(\Omega)$. Without loss of generality, we can assume that the sequence itself converges. Then (v_k) is a Cauchy sequence in $H^{t-1}(\Omega)$. From $|v_k|_t \to 0$ and $\|v_k - v_\ell\|_t^2 \le \|v_k - v_\ell\|_{t-1}^2 + (|v_k|_t + |v_\ell|_t)^2$, we conclude that (v_k) is even a Cauchy sequence in $H^t(\Omega)$. Because of the completeness of the space, this establishes convergence in the sense of H^t to an element $v^* \in H^t(\Omega)$. By continuity considerations, we have

$$\|v^*\|_t = 1 \quad \text{and} \quad |||v^*||| = 0.$$

This is a contradiction, since $|v^*|_t = 0$ implies v^* is a polynomial in $\mathcal{P}_{t-1}$, and in view of $v^*(z_i) = 0$ for $i = 1, 2, \ldots, s$, v^* can only be the null polynomial. $\quad\square$

Using the lemma, we now immediately get the following result [Bramble and Hilbert 1970]. As usual, the kernel of a linear mapping L is denoted by $\ker L$, and $\|L\| := \sup\{\|Lv\|; \ \|v\| = 1\}$.

6.3 Bramble–Hilbert Lemma. *Let $\Omega \subset \mathbb{R}^2$ be a domain with Lipschitz continuous boundary. Suppose $t \ge 2$ and that L is a bounded linear mapping of $H^t(\Omega)$ into a normed linear space Y. If $\mathcal{P}_{t-1} \subset \ker L$, then there exists a constant $c = c(\Omega)\|L\| \ge 0$ such that*

$$\|Lv\| \le c|v|_t \quad \text{for all } v \in H^t(\Omega). \tag{6.4}$$

Proof. Let $I : H^t(\Omega) \to \mathcal{P}_{t-1}$ be an interpolation operator of the type appearing in the previous lemma. Using the lemma and the fact that $Iv \in \ker L$, we get

$$\|Lv\| = \|L(v - Iv)\| \le \|L\| \cdot \|v - Iv\|_t \le c\|L\| \cdot |v|_t,$$

where c is the constant in (6.2). $\quad\square$

For simplicity, we have restricted ourselves to bounded two-dimensional domains in order to be able to use the Lagrange interpolation polynomials. This restriction can be removed by utilizing other interpolation procedures; cf. 6.9 and Problem 6.16.

Triangular Elements with Complete Polynomials

We turn our attention once again to C^0 elements consisting of piecewise polynomials of degree $t - 1$ on triangles. Assume $t \ge 2$. Given a triangulation $\mathcal{T}_h$ and an associated family $S_h = \mathcal{M}_0^{t-1}(\mathcal{T}_h)$, by §5 there is a well-defined interpolation operator $I_h : H^t(\Omega) \to S_h$. Moreover, by Definition 5.1(3), $\mathcal{T}_h$ is associated with a shape parameter κ. The central result is the following approximation theorem.

6.4 Theorem. *Let $t \geq 2$, and suppose T_h is a shape-regular triangulation of Ω. Then there exists a constant $c = c(\Omega, \kappa, t)$ such that*

$$\|u - I_h u\|_{m,h} \leq c\, h^{t-m} |u|_{t,\Omega} \quad \text{for } u \in H^t(\Omega), \quad 0 \leq m \leq t, \qquad (6.5)$$

where I_h denotes interpolation by a piecewise polynomial of degree $t - 1$.

We present the proof of this approximation theorem in full generality later. For the moment we restrict ourselves to the case of a regular grid, i.e., to the case where all triangles are congruent as in Example 4.3. Each triangle can thus be considered to be a scaled version of a reference triangle T_1.

6.5 Remark. Let $t \geq 2$, and suppose

$$T_h := h T_1 = \{x = hy; \ y \in T_1\}$$

with $h \leq 1$. Then

$$\|u - Iu\|_{m,T_h} \leq c\, h^{t-m} |u|_{t,T_h}, \qquad (6.6)$$

for $0 \leq m \leq t$, where Iu is the polynomial in $\mathcal{P}_{t-1}$ which interpolates u (at the transformed points). Here c is the constant in Lemma 6.2.

Proof of the remark. Given a function $u \in H^t(T_h)$, we define $v \in H^t(T_1)$ by

$$v(y) = u(hy).$$

Then $\partial^\alpha v = h^{|\alpha|} \partial^\alpha u$ for $|\alpha| \leq t$. Since the transformation of the area in $\mathbb{R}^2$ yields an extra factor h^{-2}, we get

$$|v|_{\ell,T_1}^2 = \sum_{|\alpha|=\ell} \int_{T_1} (\partial^\alpha v)^2 dy = \sum_{|\alpha|=\ell} \int_{T_h} h^{2\ell} (\partial^\alpha u)^2 h^{-2} dx = h^{2\ell-2} |u|_{\ell,T_h}^2.$$

Assuming $h \leq 1$, after summation the smallest power dominates:

$$\|u\|_{m,T_h}^2 = \sum_{\ell \leq m} |u|_{\ell,T_h}^2 = \sum_{\ell \leq m} h^{-2\ell+2} |v|_{\ell,T_1}^2 \leq h^{-2m+2} \|v\|_{m,T_1}^2.$$

Now inserting $u - Iu$ in place of u in this formula, we get a result for the interpolation error. Combining the last two formulae with Lemma 6.2, we get

$$\begin{aligned}
\|u - Iu\|_{m,T_h} &\leq h^{-m+1} \|v - Iv\|_{m,T_1} \leq h^{-m+1} \|v - Iv\|_{t,T_1} \\
&\leq h^{-m+1} c |v|_{t,T_1} \\
&\leq h^{t-m} c |u|_{t,T_h},
\end{aligned}$$

for all $m \leq t$, and (6.6) is proved.   □

For a regular grid, the assertion of Theorem 6.4 is a direct consequence of Remark 6.5, since we get $\|u - I_h u\|_{m,\Omega}$ immediately by squaring the expressions in (6.6), and summing over all triangles.

We now examine triangular elements in more detail in preparation for the proof of the general case of Theorem 6.4, which follows the same lines as for the special case of a regular grid, although the technical difficulties are much greater.

6.6 Transformation Formula. *Let Ω and $\hat{\Omega}$ be affine equivalent, i.e., there exists a bijective affine mapping*

$$F : \hat{\Omega} \to \Omega,$$
$$F\hat{x} = x_0 + B\hat{x} \tag{6.7}$$

with a nonsingular matrix B. If $v \in H^m(\Omega)$, then $\hat{v} := v \circ F \in H^m(\hat{\Omega})$, and there exists a constant $c = c(\hat{\Omega}, m)$ such that

$$|\hat{v}|_{m,\hat{\Omega}} \le c \, \|B\|^m \, |\det B|^{-1/2} |v|_{m,\Omega}. \tag{6.8}$$

Proof. Consider the derivative of order m as a multilinear form, and write the chain rule in the form

$$D^m \hat{v}(\hat{x})(\hat{y}_1, \hat{y}_2, \ldots, \hat{y}_m) = D^m v(x)(B\hat{y}_1, B\hat{y}_2, \ldots, B\hat{y}_m).$$

Thus, $\|D^m \hat{v}\|_{\mathbb{R}^{nm}} \le \|B\|^m \|D^m v\|_{\mathbb{R}^{nm}}$. We apply this estimate to the partial derivatives $\partial_{i_1} \partial_{i_2} \ldots \partial_{i_m} v = D^m v(e_{i_1}, e_{i_2}, \ldots, e_{i_m})$. Taking the sum, we get

$$\sum_{|\alpha|=m} |\partial^\alpha \hat{v}|^2 \le n^m \max_{|\alpha|=m} |\partial^\alpha \hat{v}|^2 \le n^m \|D^m \hat{v}\|^2 \le n^m \|B\|^{2m} \|D^m v\|^2$$

$$\le n^{2m} \|B\|^{2m} \sum_{|\alpha|=m} |\partial^\alpha v|^2.$$

Finally we integrate, taking account of the *transformation formula* for multiple integrals:

$$\int_{\hat{\Omega}} \sum_{|\alpha|=m} |\partial^\alpha \hat{v}|^2 d\hat{x} \le n^{2m} \|B\|^{2m} \int_{\Omega} \sum_{|\alpha|=m} |\partial_\alpha v|^2 \cdot |\det B^{-1}| dx.$$

Taking the square root, we get (6.8). □

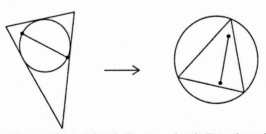

Fig. 23. An affine map from a triangle T_1 onto a triangle T_2 sends a pair of points on a circle inscribed in T_1 to points in a circle which contains T_2

The fact that transformations to and from shape-regular grids do not generate extra terms with powers of h can also be seen from simple geometric considerations. Let $F : T_1 \to T_2 : \hat{x} \mapsto B\hat{x} + x_0$ be a bijective affine mapping. We write ρ_i for the radius of the largest circle inscribed in T_i, and r_i for the radius of the smallest circle containing T_i. Given $x \in \mathbb{R}^2$ with $\|x\| \leq 2\rho_1$, we find two points $y_1, z_1 \in T_1$ with $x = y_1 - z_1$, see Fig. 23. Since $F(y_1)$, $F(z_1) \in T_2$, we have $\|Bx\| \leq 2r_2$. Thus,

$$\|B\| \leq \frac{r_2}{\rho_1}. \tag{6.9}$$

Now exchanging T_1 and T_2, we see that the inverse matrix satisfies $\|B^{-1}\| \leq r_1/\rho_2$, and thus

$$\|B\| \cdot \|B^{-1}\| \leq \frac{r_1 r_2}{\rho_1 \rho_2}. \tag{6.10}$$

Proof of Theorem 6.4. It suffices to establish the inequality

$$\|u - I_h u\|_{m,T_j} \leq ch^{t-m} |u|_{t,T_j} \quad \text{for all } u \in H^t(T_j)$$

for every triangle T_j of a shape-regular triangulation $\mathcal{T}_h$. To this end, choose a reference triangle (5.11) with $\hat{r} = 2^{-1/2}$ and $\hat{\rho} = (2 + \sqrt{2})^{-1} \geq 2/7$. Now let $F : T_{\text{ref}} \to T$ with $T = T_j \in \mathcal{T}_h$. Applying Lemma 6.2 on the reference triangle and using the transformation formula in both directions, we obtain

$$
\begin{aligned}
|u - I_h u|_{m,T} &\leq c\|B\|^{-m} |\det B|^{1/2} |\hat{u} - I_h \hat{u}|_{m,T_{\text{ref}}} \\
&\leq c\|B\|^{-m} |\det B|^{1/2} \cdot c|\hat{u}|_{t,T_{\text{ref}}} \\
&\leq c\|B\|^{-m} |\det B|^{1/2} \cdot c\|B\|^t \cdot |\det B|^{-1/2} |u|_{t,T} \\
&\leq c (\|B\| \cdot \|B^{-1}\|)^m \|B\|^{t-m} |u|_{t,T}.
\end{aligned}
$$

By the shape regularity, $r/\rho \leq \kappa$, and $\|B\| \cdot \|B^{-1}\| \leq (2 + \sqrt{2})\kappa$. Then (6.9) implies $\|B\| \leq h/\hat{\rho} \leq 4h$. Combining these facts, we have

$$|u - I_h u|_{\ell,T} \leq ch^{t-\ell} |u|_{t,T}.$$

Now squaring and summing over ℓ from 0 to m establishes the assertion. $\qquad\square$

Bilinear Quadrilateral Elements

For quadrilateral elements, we usually use tensor products instead of complete polynomials. Nevertheless, we can still make use of the techniques developed in the previous section to establish results on the order of approximation. The simple but important case of a bilinear element serves as a typical example.

Table 3. Error estimates for some finite elements

$\|u - I_h u\|_{m,h} \leq ch^{t-m}|u|_{t,\Omega}$	$0 \leq m \leq t$
C^0 elements	
linear triangle	$t = 2$
quadratic triangle	$2 \leq t \leq 3$
cubic triangle	$2 \leq t \leq 4$
bilinear quadrilateral	$t = 2$
serendipity element	$2 \leq t \leq 3$
9 node quadrilateral	$2 \leq t \leq 3$
C^1 elements	
Argyris element	$3 \leq t \leq 6$
Bell element	$3 \leq t \leq 5$
Hsieh–Clough–Tocher element	$3 \leq t \leq 4$ $(m \leq 2)$
reduc. Hsieh–Clough–Tocher element	$t = 3$ $(m \leq 2)$

6.7 Theorem. *Let $\mathcal{T}_h$ be a quasi-uniform decomposition of Ω into parallelograms. Then there exists a constant $c = c(\Omega, \kappa)$ such that*

$$\|u - I_h u\|_{m,\Omega} \leq ch^{2-m}|u|_{2,\Omega} \quad \text{for all } u \in H^2(\Omega),$$

where $I_h u$ interpolates u using bilinear elements.

Proof. For the same reasons as in the last proof, it suffices to show that for interpolation on the unit square $Q := [0, 1]^2$,

$$\|u - Iu\|_{2,Q} \leq c|u|_{2,Q} \quad \text{for all } u \in H^2(Q). \tag{6.11}$$

In view of the continuous imbedding $H^2(Q) \hookrightarrow C^0(Q)$, the function values of u at the 4 vertices are bounded by $c\|u\|_{2,Q}$. The interpolating polynomial Iu depends linearly on these 4 values, and thus $\|Iu\|_{2,Q} \leq c_1 \max_{x \in Q} |u(x)| \leq c_2\|u\|_{2,Q}$ and

$$\|u - Iu\|_2 \leq \|u\|_2 + \|Iu\|_2 \leq (c_2 + 1)\|u\|_2.$$

If u is a linear polynomial, then $Iu = u$, and $u - Iu = 0$. The Bramble–Hilbert lemma now guarantees (6.11). □

Analogously, for elements in the serendipity class, we have

$$\|u - I_h u\|_{m,\Omega} \leq ch^{t-m}|u|_{t,\Omega} \quad \text{for all } u \in H^t(\Omega), \ m = 0, 1 \text{ and } t = 2, 3.$$

The approximation properties for other triangular and quadrilateral elements are listed in Table 3.

Inverse Estimates

The above approximation theorems have the form

$$\|u - Iu\|_{m,h} \le ch^{t-m}\|u\|_t,$$

where m is *smaller* than t. For the moment we ignore the fact that on the right-hand side, the norm $\|\cdot\|_t$ may be replaced by the semi-norm $|\cdot|_t$. Thus, the approximation error is measured in a *coarser* norm than the given function. In a so-called *inverse estimate*, the reverse happens. The *finer* norm of the finite element functions will be estimated by a *coarser* one (obviously, this does not work for all functions in the Sobolev space).

6.8 Inverse Estimates. *Let* (S_h) *be an affine family of finite elements consisting of piecewise polynomials of degree k associated with uniform partitions. Then there exists a constant $c = c(\kappa, k, t)$ such that for all $0 \le m \le t$,*

$$\|v_h\|_{t,h} \le ch^{m-t}\|v_h\|_{m,h} \quad \text{for all } v_h \in S_h.$$

Sketch of the proof. We can reduce the proof to the discussion of a reference element by using the transformation formula 6.6. It suffices to show that

$$|v|_{t,T_{\text{ref}}} \le c|v|_{m,T_{\text{ref}}} \quad \text{for } v \in \Pi_{\text{ref}} \tag{6.12}$$

with $c = c(\Pi_{\text{ref}})$. The extension to elements of size h proceeds just as in the proof of Theorem 6.4. This leads to the factor ch^{m-t} in the estimate. Then summing the squares of the expressions over all triangles or quadrilaterals leads to the desired assertion.

To establish (6.12), we make use of the fact that the norms $\|\cdot\|_{t,T_{\text{ref}}}$ and $\|\cdot\|_{m,T_{\text{ref}}}$ are equivalent on the finite dimensional space $\Pi_{\text{ref}} \oplus \mathcal{P}_{m-1}$. Let $Iv \in \mathcal{P}_{m-1}$ be a polynomial that interpolates v at fixed points. Since $t > mi1$, we have $|Iv|_t = 0$. Combining these facts, we obtain from the Bramble–Hilbert lemma

$$\begin{aligned}
|v|_t = |v - Iv|_t &\le \|v - Iv\|_t \\
&\le c\|v - Iv\|_m \\
&= c'|v|_m,
\end{aligned}$$

and (6.12) is proved. $\qquad\qquad\Box$

In the Approximation Theorem 6.4 and the Inverse Estimate 6.8, the exponents in the term with h correspond to the difference between the orders of the Sobolev norms. This has been established by moving back and forth to and from the reference triangle. This technique is called a *scaling argument*.

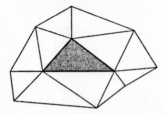

Fig. 24. The values of the Clément interpolant in the (shaded) triangle T depend on the values of the given function in its neighborhood $\tilde{\omega}_T$

Clément's Interpolation

The interpolation operator I_h in (6.5) can only be applied to H^2 functions. On the other hand, functions with less regularity can be approximated in some advanced theories. Clément [1975] has constructed an interpolation process which applies to H^1 functions. Typically this operator is used when features in H^1 and L^2 are to be combined. The crucial point is that the interpolation error depends only on the local mesh size. Thus, no power of h is lost, even if inverse estimates enter into the analysis.

The operator is defined *nearly locally*. Let $\mathcal{T}_h$ be a shape-regular triangulation of Ω. Given a node x_j, let

$$\omega_j := \omega_{x_j} := \bigcup \{T' \in \mathcal{T}_h; \, x_j \in T'\} \tag{6.13}$$

be the support of the shape function $v_j \in \mathcal{M}_0^1$. Here $v_j(x_k) = \delta_{jk}$. Furthermore, let

$$\tilde{\omega}_T := \bigcup \{\omega_j; \, x_j \in T\} \tag{6.14}$$

be a neighborhood of T. Since $\mathcal{T}_h$ is assumed to be shape regular, the area can be estimated by $\mu(\tilde{\omega}_T) \leq c(\kappa) \, h_T^2$. Moreover, the number of triangles that belong to $\tilde{\omega}_T$ is bounded.

6.9 Clément's Interpolation. *Let $\mathcal{T}_h$ be a shape-regular triangulation of Ω. Then there exists a linear mapping $I_h : H^1(\Omega) \rightarrow \mathcal{M}_0^1$ such that*

$$\begin{aligned}
\|v - I_h v\|_{m,T} &\leq c h_T^{1-m} \|v\|_{1,\tilde{\omega}_T} \quad \text{for } v \in H^1(\Omega), \; m = 0, 1, \; T \in \mathcal{T}_h \\
\|v - I_h v\|_{0,e} &\leq c h_T^{1/2} \|v\|_{1,\tilde{\omega}_T} \quad \text{for } v \in H^1(\Omega), \; e \in \partial T, \; T \in \mathcal{T}_h.
\end{aligned} \tag{6.15}$$

A simple construction is obtained by a combination of Clément's operator and the procedures of Scott and Zhang [1990] or Yserentant [1990]. The construction is performed in two steps. Given a nodal point x_j, let $Q_j : L_2(\omega_j) \to \mathcal{P}_0$ by the L_2-projection onto the constant functions. It follows from the Bramble–Hilbert lemma that

$$\|v - Q_j v\|_{0,\omega_j} \le ch_j |v|_{1,\omega_j}, \qquad (6.16)$$

where h_j is the diameter of ω_j. In order to cope with homogeneous Dirichlet boundary conditions on $\Gamma_D \subset \partial\Omega$ we modify the operator and set

$$\tilde{Q}_j v = \begin{cases} 0 & \text{if } x_j \in \Gamma_D, \\ Q_j v & \text{otherwise.} \end{cases} \qquad (6.17)$$

Here we get an analogous estimate to (6.16) by adapting the technique of the proof of Friedrich's inequality

$$\|v - \tilde{Q}_j v\|_{0,\omega_j} = \|v\|_{0,\omega_j} \le ch_j |v|_{1,\omega_j} \quad \text{if } x_j \in \Gamma_D. \qquad (6.18)$$

Next we define

$$I_h v := \sum_j (\tilde{Q}_j v) v_j \in \mathcal{M}_0^1. \qquad (6.19)$$

The shape functions v_j constitute a partition of unity. Specifically, for each x, $I_h v$ contains at most three nonzero terms. For each relevant term, $v - \tilde{Q}_j v$ can be estimated by (6.16) or (6.18). resp.

$$\|v - I_h v\|_{0,T} \le \sum_j \|(v - \tilde{Q}_j v) v_j\|_{0,T} \le \sum_j \|(v - \tilde{Q}_j v)\|_{0,\omega_j} \le 3ch_T \|v\|_{1,\tilde{\omega}_T}.$$

By summing the squares over all triangles, we obtain (6.15a) for $m = 0$. For the H^1-stability we refer to Corollary 7.8. $\qquad\qquad\qquad\qquad\qquad\qquad\qquad\square$

The construction is easily modified to get an analogous mapping from $H_0^1(\Omega)$ to $\mathcal{M}_0^1 \cap H_0^1(\Omega)$. If $x_j \in \partial\Omega$, then $P_j v$ may be set to zero and (6.16) follows from Friedrichs' inequality.

Appendix: On the Optimality of the Estimates

6.10 Remark. The inverse estimates show that the above approximation theorems are optimal (up to a constant). The assertions have the following structure:

Suppose the complete normed linear space X is compactly imbedded in Y. Then there exists a family (S_h) of subspaces of X satisfying the *approximation property*

$$\inf_{v_h \in S_h} \|u - v_h\|_Y \le ch^\alpha \|u\|_X \quad \text{for all } u \in X, \qquad (6.20)$$

and (with $\beta = \alpha$) the *inverse estimate*

$$\|v_h\|_X \le ch^{-\beta} \|v_h\|_Y \quad \text{for all } v_h \in S_h. \qquad (6.21)$$

An example of this is provided by $\| \cdot \|_X = \| \cdot \|_{2,h}$, $\| \cdot \|_Y = \| \cdot \|_{1,\Omega}$, $S_h = \mathcal{M}_0^1(\mathcal{T}_h)$, $\alpha = \beta = 1$.

A pair of inequalities of the form (6.20) and (6.21) involving an approximation property and an inverse estimate is called *optimal* provided that $\beta = \alpha$. We claim $\beta < \alpha$ is impossible. Indeed, otherwise there would exist a sequence of nested spaces $V_0 \subset V_1 \subset V_2 \subset \cdots$ with

$$\min_{v_n \in V_n} \|u - v_n\|_Y \le 2^{-\gamma n} \|u\|_X \quad \text{for all } u \in X \tag{6.22}$$

and

$$\|v_n\|_X \le 2^n \|v_n\|_Y.$$

Here $1 < \gamma < 2$. Choose $m \in \mathbb{N}$ with $2^{-(\gamma+1)m} < (1 - 2^{-\gamma-1})/5$. In view of the compact imbedding of X in Y, there exists an element $u \in X$ with $\|u\|_X = 1$ and $\|u\|_Y < \varepsilon := 2^{-\gamma m}$. Suppose (6.22) holds for $v_n \in V_n$. Set

$$w_m = v_m, \quad w_n = v_n - v_{n-1} \quad \text{for } n > m.$$

Then $\|w_m\|_Y \le \|u - w_m\|_Y + \|u\|_Y \le 2 \cdot 2^{-\gamma m}$, and

$$\|w_n\|_Y \le \|u - v_n\|_Y + \|u - v_{n-1}\|_Y \le 2^{-\gamma n} + 2^{-\gamma(n-1)} \le 5 \cdot 2^{-\gamma n}$$

for $n > m$. In view of the inverse inequality, it follows that $\|w_n\|_X \le 5 \cdot 2^{-(\gamma-1)n}$ for $n \ge m$, and

$$\|u\|_X = \| \sum_{n=m}^{\infty} w_n \|_X \le 5 \sum_{n=m}^{\infty} 2^{-(\gamma-1)n} < 1.$$

This is a contradiction. $\qquad\square$

6.11 Remark. The above proof also establishes that if

$$\inf_{v_h \in S_h} \|u - v_h\|_Y \le \text{const} \cdot h^\beta$$

for all $u \in Y$, and the inverse estimate (6.21) also holds, then u is contained in the subspace X. This result has far-reaching consequences for the practical use of finite elements. If it is known that the solution u of a boundary-value problem does not lie in a higher-order Sobolev space, then the finite element approximation has limited accuracy.

Here we should note that pairs of inequalities of the form (6.20) and (6.21) play a major role in classical approximation theory. The most widely known results along these lines deal with the approximation of 2π-periodic functions by trigonometric polynomials in $\mathcal{P}_{n,2\pi}$. Let $C^{k+\alpha}$ denote the space of functions whose k-th derivative is Hölder continuous with exponent α. Then by the theorems of Jackson,

$$\inf_{p \in \mathcal{P}_{n,2\pi}} \|f - p\|_{C^0} \le cn^{-k-\alpha}\|f\|_{C^{k+\alpha}},$$

while the Bernstein inequality

$$\|p\|_{C^{k+\alpha}} \le cn^{k+\alpha}\|p\|_{C^0} \quad \text{for all } p \in \mathcal{P}_{n,2\pi}$$

provides the corresponding inverse estimate.

Problems

6.12 Let $\mathcal{T}_h$ be a family of uniform partitions of Ω, and suppose S_h belong to an affine family of finite elements. Suppose the nodes of the basis are $z_1, z_2, \ldots, z_N$ with $N = N_h = \dim S_h$. Verify that for some constant c independent of h, the following inequality holds:

$$c^{-1}\|v\|_{0,\Omega}^2 \le h^2 \sum_{i=1}^{N} |v(z_i)|^2 \le c\|v\|_{0,\Omega}^2 \quad \text{for all } v \in S_h.$$

6.13 Under appropriate assumptions on the boundary of Ω, we showed that

$$\inf_{v \in S_h} \|u - v_h\|_{1,\Omega} \le c\, h\|u\|_{2,\Omega},$$

where for every $h > 0$, S_h is a finite-dimensional finite element space. Show that this implies the compactness of the imbedding $H^2(\Omega) \hookrightarrow H^1(\Omega)$. [Thus, the use of the compactness in the proof of the approximation theorem was not just a coincidence.]

6.14 Let $\mathcal{T}_h$ be a κ-regular partition of Ω into parallelograms, and let u_h be an associated bilinear element. Divide each parallelogram into two triangles, and let $\|\cdot\|_{m,h}$ be defined as in (6.1). Show that

$$\inf \|u_h - v_h\|_{m,\Omega} \le c(\kappa)h^{2-m}\|u_h\|_{2,\Omega}, \quad m = 0, 1,$$

where the infimum is taken over all piecewise linear functions on the triangles in $\mathcal{M}^1$.

6.15 For interpolation by piecewise linear functions, Theorem 6.4 asserts that

$$\|I_h v\|_{2,h} \le c\,\|v\|_{2,\Omega}.$$

Give a one-dimensional counterexample to show that

$$\|I_h v\|_{0,\Omega} \le c\,\|v\|_{0,\Omega}$$

is not possible with a constant c which is independent of h.

6.16 Prove the Bramble–Hilbert lemma for $t = 1$ by choosing Iv to be the constant function

$$Iv := \frac{\int_\Omega v\, dx}{\int_\Omega dx}.$$

6.17 Consider the situation as in the construction of Cléments' operator. We modifiy the definition of the operator $\bar{Q}_j : L_2(\omega_j) \rightarrow \mathcal{P}_0$ by the rule

$$\bar{Q}_j v := v(x_j) \quad \text{if } v|_{\omega_j} \in \mathcal{M}_0^1(\mathcal{T}_\ell), \tag{6.23}$$

i.e., if the restriction of v to ω_j is a finite element funtion with the present grid. Show that also in this case

$$\|v - \bar{Q}_j v\|_{0,\omega_j} \le h_j |v|_{1,\omega_j}$$

with c depending only on the shape parameter of the triangulation of ω_j.

Hint: The modification has an advantage. If the given function coincides with a piecewise linear function on a subdomain $\tilde{\Omega}$, then the projector reproduces v at the nodes in the interior of $\tilde{\Omega}$.

§ 7. Error Bounds for Elliptic Problems
of Second Order

Now we are ready to establish error estimates for finite element solutions. Usually error bounds are derived first with respect to the energy norm. The extension to the L_2-norm is performed by a duality technique that is often found in proofs of advanced results. We are looking for bounds of the form

$$\|u - u_h\| \le c\, h^p \tag{7.1}$$

for the difference between the true solution u and the approximate solution u_h in S_h. Here p is called the *order of approximation*. In general, it depends on the regularity of the solution, the degree of the polynomials in the finite elements, and the Sobolev norm in which the error is measured.

Remarks on Regularity

7.1 Definition. Let $m \ge 1$, $H_0^m(\Omega) \subset V \subset H^m(\Omega)$, and suppose $a(\cdot,\cdot)$ is a V-elliptic bilinear form. Then the variational problem

$$a(u, v) = (f, v)_0 \quad \text{for all } v \in V$$

is called H^s-*regular* provided that there exists a constant $c = c(\Omega, a, s)$ such that for every $f \in H^{s-2m}(\Omega)$, there is a solution $u \in H^s(\Omega)$ with

$$\|u\|_s \le c\, \|f\|_{s-2m}. \tag{7.2}$$

In this section we will make use of this definition only for $s \ge 2m$. We will drop this restriction later, in Ch. III, after norms with negative index are defined.

Regularity results for the Dirichlet problem of second order with zero boundary conditions can be found e.g. in Gilbarg and Trudinger [1983] and Kadlec [1964]. For simplicity, we do not present the most general results; see the remarks for Example 2.10 and Problem 7.12.

7.2 Regularity Theorem. *Let a be an H_0^1-elliptic bilinear form with sufficiently smooth coefficient functions.*
(1) If Ω is convex, then the Dirichlet problem is H^2-regular.
(2) If Ω has a C^s boundary with $s \ge 2$, then the Dirichlet problem is H^s-regular.

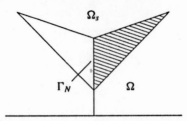

Fig. 25. Reflection of a convex domain Ω along the edge Γ_N on which a Neumann condition is given

We see from Example 2.1 with a domain with reentrant corner that the assumptions on the boundary cannot be dropped since the solution there is not in $H^2(\Omega)$.

We now give an example to show that the situation is more complicated if a Neumann condition is prescribed *on a part of the boundary*. Let Ω be the convex domain on the right-hand side of the y-axis shown in Fig. 25. Suppose the Neumann condition

$$\frac{\partial u}{\partial \nu} = 0$$

is prescribed on $\Gamma_N := \{(x, y) \in \partial\Omega; \ x = 0\}$, and that a Dirichlet boundary condition is prescribed on $\Gamma_D := \Gamma \backslash \Gamma_N$. The union of Ω with its reflection in Γ_N defines a symmetric domain Ω_s. Set

$$u(-x, y) = u(x, y) \quad \text{for } (x, y) \in \Omega_s \backslash \Omega.$$

Then its continuation is also a solution of a Dirichlet problem on Ω_s. But since Ω_s has a reentrant corner, the solution is not always in $H^2(\Omega_s)$, which means that $u \in H^2(\Omega)$ cannot hold for all problems on Ω with mixed boundary conditions.

Error Bounds in the Energy Norm

In the following, suppose that Ω is a polygonal domain. This means that it can be partitioned into triangles or quadrilaterals. In addition, in order to use Theorem 7.2, suppose Ω is convex.

7.3 Theorem. *Suppose $\mathcal{T}_h$ is a family of shape-regular triangulations of Ω. Then the finite element approximation $u_h \in S_h = \mathcal{M}_0^k \ (k \geq 1)$ satisfies*

$$\|u - u_h\|_1 \leq ch\|u\|_2 \tag{7.3}$$
$$\leq ch\|f\|_0.$$

Proof. By the convexity of Ω, the problem is H^2-regular, and $\|u\|_2 \leq c_1\|f\|_0$. By Theorem 6.4, there exists $v_h \in S_h$ with $\|u - v_h\|_{1,\Omega} = \|u - v_h\|_{1,h} \leq c_2 h\|u\|_{2,\Omega}$. Combining these facts with Céa's Lemma gives (7.3) with $c := (1 + c_1)c_2 c_3/\alpha$. $\qquad\square$

7.4 Remark. According to Theorem 6.4, we should get a higher-order error bound for quadratic triangular elements under the assumption of H^3-regularity. This observation is misleading, however, since – except in some special cases – smooth boundaries are required for H^3-regularity. But a domain Ω with smooth boundary cannot be decomposed into triangles, and the usual problems arise along the curved boundaries (cf. Ch. III, §1).

There is more regularity in the interior of the domain, and in most cases, the finite element approximations with quadratic or cubic triangles are so much better than with piecewise linear ones that it is worth the extra effort to use them.

The estimate (7.3) holds for any affine family of triangular elements which contains the P_1 elements as a subset. Moreover, by Theorem 6.7 analogous results hold if we use bilinear quadrilateral elements instead of linear triangles. Using the same arguments as in the proof of the previous theorem, we get

7.5 Theorem. *Suppose we are given a set of shape-regular partitions of Ω into parallelograms. Then the finite element approximation u_h by bilinear quadrilateral elements in S_h satisfies*

$$\|u - u_h\|_1 \le ch\|f\|_0. \tag{7.4}$$

L_2-Estimates

If the polynomial approximation error is measured in the L_2-norm (i.e., in the H^0-norm), then by Theorem 6.4 the order of approximation is better by one power of h. It is not at all obvious that this property carries over to finite element solutions. The proof uses the H^2-regularity a second time, and requires a *duality argument* which has been called *Nitsche's Trick*. We now present an abstract formulation of it; cf. Aubin [1967] and Nitsche [1968].

7.6 Aubin–Nitsche Lemma. *Let H be a Hilbert space with the norm $|\cdot|$ and the scalar product $(\cdot, \cdot)$. Let V be a subspace which is also a Hilbert space under another norm $\|\cdot\|$. In addition, let*

$$V \hookrightarrow H \quad \text{be continuous.}$$

Then the finite element solution in $S_h \subset V$ satisfies

$$|u - u_h| \le C\|u - u_h\| \sup_{g \in H} \left\{ \frac{1}{|g|} \inf_{v \in S_h} \|\varphi_g - v\| \right\} \tag{7.5}$$

where for every $g \in H$, $\varphi_g \in V$ denotes the corresponding unique (weak) solution of the equation

$$a(w, \varphi_g) = (g, w) \quad \text{for all } w \in V. \tag{7.6}$$

Proof. By duality, the norm of an element in a Hilbert space can be computed by

$$|w| = \sup_{g \in H} \frac{(g, w)}{|g|}. \tag{7.7}$$

Here and in (7.5), the supremum is taken only over those g with $g \neq 0$. We recall that u and u_h are given by

$$a(u, v) = \langle f, v \rangle \quad \text{for all } v \in V,$$
$$a(u_h, v) = \langle f, v \rangle \quad \text{for all } v \in S_h.$$

Hence, $a(u - u_h, v) = 0$ for all $v \in S_h$. Moreover, if we insert $w := u - u_h$ in (7.6), we get

$$(g, u - u_h) = a(u - u_h, \varphi_g)$$
$$= a(u - u_h, \varphi_g - v) \leq C\|u - u_h\| \cdot \|\varphi_g - v\|.$$

Here we have used the continuity of the bilinear form a, i.e., the fact that $a(u, v) \leq C\|u\| \cdot \|v\|$. It follows that

$$(g, u - u_h) \leq C\|u - u_h\| \inf_{v \in S_h} \|\varphi_g - v\|.$$

Now the duality argument (7.7) leads to

$$|u - u_h| = \sup_{g \in H} \frac{(g, u - u_h)}{|g|}$$
$$\leq C\|u - u_h\| \sup_{g \in H} \left\{ \inf_{v \in S_h} \frac{\|\varphi_g - v\|}{|g|} \right\}. \qquad \square$$

7.7 Corollary. *Under the hypotheses of either Theorem 7.3 or Theorem 7.5, if $u \in H^1(\Omega)$ is the solution of the associated variational problem, then*

$$\|u - u_h\|_0 \leq cCh\|u - u_h\|_1.$$

If in addition $f \in L_2(\Omega)$ so that $u \in H^2(\Omega)$, then

$$\|u - u_h\|_0 \leq cC^2h^2\|f\|_0.$$

Here c and C are the constants appearing in (7.3) and in (7.4)–(7.5), respectively.

Proof. Setting

$$H := H^0(\Omega), \quad |\cdot| := \|\cdot\|_0,$$
$$V := H_0^1(\Omega), \quad \|\cdot\| := \|\cdot\|_1,$$

we see that $V \subset H$, and the continuity of the imbedding is clear from $\|\cdot\|_0 \leq \|\cdot\|_1$. The Aubin–Nitsche Lemma is now applicable. In view of Theorem 7.3 or 7.5, the quantity in the curly brackets in (7.5) is at most ch, and the lemma immediately implies the desired result. $\qquad \square$

A Simple L_∞-Estimate

The above estimates do not exclude the possibility that the error is large at certain points. To prevent this, we need to work with the L_∞-norm $\|v\|_{\infty,\Omega} := \sup_{x\in\Omega} |v(x)|$. For problems in two-dimensional domains with H^2-regularity, we have

$$\|u - u_h\|_\infty \leq c\,h^2 |\log h|^{3/2} \|D^2 u\|_\infty.$$

A proof of this fact based on weighted norms can be found e.g. in Ciarlet [1978]; see also Frehse and Rannacher [1978]. Here we restrict ourselves to proving the much weaker assertion

$$\|u - u_h\|_\infty \leq c\,h\,|u|_2. \tag{7.8}$$

Given a function $v \in H^2(T_{\text{ref}})$, let Iv be its interpolant in the polynomial space Π_{ref}. Since $H^2 \subset C^0$, by the Bramble–Hilbert Lemma we have

$$\|v - Iv\|_{\infty,T_{\text{ref}}} \leq c\,|v|_{2,T_{\text{ref}}}. \tag{7.9}$$

Let u be the solution of the variational problem, and let $I_h u$ be its interpolant in S_h. Pick an element T from the triangulation (which we assume to be uniform). Let $\hat{u}$ be the affine transformation of $u|_T$ onto the reference triangle. By (7.9) and the transformation formula (6.8), we get

$$\begin{aligned}
\|u - I_h u\|_{\infty,T} = \|\hat{u} - I\hat{u}\|_{\infty,T_{\text{ref}}} &\leq c|\hat{u}|_{2,T_{\text{ref}}} \\
&\leq ch|u|_{2,T} \leq ch|u|_{2,\Omega}.
\end{aligned} \tag{7.10}$$

Taking the maximum over all triangles, we have

$$\|u - I_h u\|_{\infty,\Omega} \leq c\,h|u|_{2,\Omega}.$$

Similarly, by an affine argument, we get the inverse estimate

$$\|v_h\|_{\infty,\Omega} \leq c\,h^{-1}\|v_h\|_{0,\Omega} \quad \text{for all } v_h \in S_h.$$

Now by Theorems 6.4 and 7.7, it follows that $\|u_h - I_h u\|_{0,\Omega} \leq c\,h^2|u|_{2,\Omega}$ for $u_h - I_h u = (u - I_h u) - (u - u_h) \in S_h$. Using the inverse estimate, we now get

$$\begin{aligned}
\|u - u_h\|_{\infty,\Omega} &\leq \|u - I_h u\|_{\infty,\Omega} + \|u_h - I_h u\|_{\infty,\Omega} \\
&\leq \|u - I_h u\|_{\infty,\Omega} + c\,h^{-1}\|u_h - I_h u\|_{0,\Omega},
\end{aligned}$$

and the result follows from (6.5) and (7.10). $\qquad\square$

The L_2-Projector

The norm of the L_2-projector onto a finite element space is not always bounded by an h-independent number when it is considered in H^1. The boundedness is easily derived in the case in which we obtain an L_2-estimate by a duality argument. We will encounter the same technique of proof in Ch. III, §6 and Ch. VI, §6.

7.8 Corollary. *Assume that the hypothesis of Theorem 7.3 (or Theorem 7.5) are satisfed, and that $\{T_h\}$ is a family of uniform triangulations of Ω. Let Q_h be the L_2-projector onto $S_h \subset H^1(\Omega)$. Then*

$$\|Q_h v\|_1 \leq c \|v\|_1 \quad \text{for all } v \in H^1(\Omega) \tag{7.11}$$

holds with a constant c which is independent of h.

Proof. Given $v \in H^1(\Omega)$ let $v_h \in S_h$ be the solution of the variational problem

$$(\nabla v_h, \nabla w)_0 + (v_h, w)_0 = \langle \ell, w \rangle \quad \text{for all } w \in S_h$$

with $\langle \ell, w \rangle := (\nabla v, \nabla w)_0 + (v, w)_0$. Obviously, $\|v_h\|_1 \leq \|v\|_1$. An essential ingredient is the L_2-error estimate from Corollary 7.7

$$\|v - v_h\|_0 \leq c_1 h \|v - v_h\|_1 \leq 2c_1 h \|v\|_1. \tag{7.12}$$

Combining this with an inverse estimate, we get

$$\begin{aligned}
\|Q_h v\|_1 &\leq \|Q_h v - v_h\|_1 + \|v_h\|_1 \\
&\leq c_2 h^{-1} \|Q_h(v - v_h)\|_0 + \|v_h\|_1 \leq c_2 h^{-1} \|v - v_h\|_0 + \|v\|_1 \\
&\leq c_2 h^{-1} 2c_1 h \|v\|_1 + \|v\|_1.
\end{aligned}$$

This proves the assertion with $c = 1 + 2c_1 c_2$. □

The assumptions of the corollary are very restrictive and in some cases the stability of the L_2-projector is wanted in locally refined meshes; see e.g. Ch. V, §5 and Yserentant [1990].

7.9 Lemma. *Let T_h be a shape-regular triangulation of Ω and Q_h be the L_2-projector onto $\mathcal{M}_0^1$. Then*

$$\|Q_h v\|_1 \leq c \|v\|_1 \quad \text{for all } v \in H_0^1(\Omega) \tag{7.13}$$

holds with a constant c which is independent of h.

Proof. We start with Clément's interpolation operator and obtain

$$\|v - I_h v\|_0^2 \leq c \sum_T h_T^2 \|v\|_{1,T}^2. \tag{7.14}$$

Since the triangulation is assumed to be shape-regular, we could estimate the diameters of all triangles in ω_T by ch_T, i.e. the diameter of T. The estimates are still local. The minimal property of the L_2-projector Q_h implies

$$\|v - Q_h v\|_0^2 \le c \sum_T h_T^2 \|v\|_{1,T}^2 . \tag{7.15}$$

Next from the Bramble–Hilbert lemma and a standard scaling argument we know that there is a piecewise constant function $w_h \in \mathcal{M}^0(\mathcal{T}_h)$ such that

$$\|v - w_h\|_{0,T} \le ch_T |v|_{1,T} .$$

Now we apply an inverse estimate on each triangle:

$$\begin{aligned}
|Q_h v|_1^1 &= \sum_T |Q_h v|_{1,T}^2 = \sum_T |Q_h v - w_h|_{1,T}^2 \\
&\le c \sum_T h_T^{-2} \|Q_h v - w_h\|_{0,T}^2 \\
&\le c \sum_T 2h_T^{-2}\left(\|Q_h v - v\|_{0,T}^2 + \|v - w_h\|_{0,T}^2\right) \\
&\le c \sum_T \|v\|_{1,T}^2 = c\|v\|_1^2 .
\end{aligned}$$

Since $\|Q_h v\|_0 \le \|v\|_0 \le \|v\|_1$, the proof is complete. □

Note that Problem 6.15 illustrates that one has to be careful when dealing with a projector for one norm and considering stability for another one.

Problems

7.10 Consider solving the boundary-value problem

$$-\Delta u = 0 \quad \text{in } \Omega := (-1, +1)^2 \subset \mathbb{R}^2,$$
$$u(x, y) = xy \quad \text{on } \partial\Omega$$

using linear triangular elements on a regular triangular grid with $2/h \in \mathbb{N}$ as in the model problem 4.3. When the reduction to homogeneous boundary conditions as in (2.21) is performed with

$$u_0(x, y) := \begin{cases} 1 + x - y & \text{for } x \ge y, \\ 1 + y - x & \text{for } x \le y, \end{cases}$$

the finite element approximation at the grid points is

$$u_h(x_i, y_i) = u(x_i, y_i) = x_i y_i . \tag{7.16}$$

Verify that the minimal value for the variational functional on S_h is

$$J(u_h) = \frac{8}{3} + \frac{4}{3}h^2,$$

and hence, $J(u_h)$ is only an approximation.

7.11 Let $\Omega = (0, 2\pi)^2$ be a square, and suppose $u \in H_0^1(\Omega)$ is a weak solution of $-\Delta u = f$ with $f \in L_2(\Omega)$. Using Problem 1.16, show that $\Delta u \in L_2(\Omega)$, and then use the Cauchy–Schwarz inequality to show that all second derivatives lie in L_2, and thus u is an H^2 function.

7.12 *(A superconvergence property)* The boundary value problem with the ordinary differential equation

$$-u''(x) = f(x), \quad x \in (0, 1),$$
$$u(0) = u(1) = 0$$

characterizes the solution of a variational problem with the bilinear form

$$a(u, v) := \int_0^1 u'v'dx.$$

Let u_h be the solution in the set of piecewise linear functions on a partition of $(0, 1)$, and let v_h be the interpolant of u in the same set. Show that $u_h = v_h$ by verifying $a(u_h - v_h, w_h) = 0$ for all piecewise linear w_h.

§ 8. Computational Considerations

The computation of finite element approximations can be divided into two parts:
1. construction of a grid by partitioning Ω, and setting up the stiffness matrix.
2. solution of the system of equations.

The central topic of this section is the computation of the stiffness matrix. The solution of the system of equations will be treated in Chapters IV and V.

Assembling the Stiffness Matrix

For finite elements with a nodal basis, such as the linear and quadratic triangular elements, the stiffness matrix can be assembled elementwise. This can be seen from the associated quadratic form. For simplicity, we consider only the principal part:

$$a(u, v) = \int_\Omega \sum_{k,l} a_{kl} \partial_k u \, \partial_l v \, dx.$$

Then

$$
\begin{aligned}
A_{ij} = a(\psi_i, \psi_j) &= \int_\Omega \sum_{k,l} a_{kl} \partial_k \psi_i \, \partial_l \psi_j \, dx \\
&= \sum_{T \in \mathcal{T}_h} \int_T \sum_{k,l} a_{kl} \partial_k \psi_i \, \partial_l \psi_j \, dx.
\end{aligned}
\tag{8.1}
$$

In forming the sum, we need only take account of those triangles which overlap the support of both ψ_i and ψ_j.

Normally, we do not compute this matrix by locating the triangles involved for a given set of node indices i, j. Although we used this type of *node-oriented* approach for the model problem in §4, in practice it wastes too much time in repeated calculations.

Table 4. Shape functions (nodal basis functions) for linear (left) and quadratic (right) elements

$N_1 = 1 - \xi - \eta$
$N_2 = \xi$
$N_3 = \eta$

$N_1 = (1 - \xi - \eta)(1 - 2\xi - 2\eta)$
$N_2 = \xi(2\xi - 1)$
$N_3 = \eta(2\eta - 1)$
$N_4 = 4\xi(1 - \xi - \eta)$
$N_5 = 4\xi\eta$
$N_6 = 4\eta(1 - \xi - \eta)$

It turns out that it is much better to use an *element-oriented* approach. For every element $T \in \mathcal{T}_h$, we find the additive contribution from (8.1) to the stiffness matrix. If every element contains exactly s nodes, this requires finding an $s \times s$ submatrix. We transform the triangle T under consideration to the reference triangle T_{ref}. Let $F : T_{\text{ref}} \rightarrow T$, $\xi \mapsto B\xi + x_0$ be the corresponding linear mapping. Then the contribution of T is given by the integral

$$\frac{\mu(T)}{\mu(T_{\text{ref}})} \int_{T_{\text{ref}}} \sum_{\substack{k,l \\ k',l'}} a_{kl}(B^{-1})_{k'k}(B^{-1})_{l'l} \, \partial_{k'} N_i \, \partial_{l'} N_j \, d\xi. \tag{8.2}$$

Here $\mu(T)$ is the area of T. After transformation, every function in the nodal basis coincides with one of the normed *shape functions* $N_1, N_2, \ldots, N_s$ on the reference triangle. These are listed in Table 4 for linear and quadratic elements.[6] For the model problem 4.3, using a right triangle T (with right angle at point number 1), we get

$$a(u, u)|_T = \frac{1}{2}(u_1 - u_2)^2 + \frac{1}{2}(u_1 - u_3)^2,$$

where u_i is the coefficient of u in the N_i expansion. This gives

$$a(\psi_i, \psi_j)|_T = \frac{1}{2} \begin{pmatrix} 2 & -1 & -1 \\ -1 & 1 & \\ -1 & & 1 \end{pmatrix}$$

for the stiffness matrix on the element level. For linear elements, it is also easy to find the so-called *mass matrix* whose elements are $(\psi_i, \psi_j)_{0,T}$. For an arbitrary triangle,

$$(\psi_i, \psi_j)_{0,T} = \frac{\mu(T)}{12} \begin{pmatrix} 2 & 1 & 1 \\ 1 & 2 & 1 \\ 1 & 1 & 2 \end{pmatrix}. \tag{8.3}$$

For differential equations with variable coefficients, the evaluation of the integrals (8.2) is usually accomplished using a Gaussian quadrature formula for multiple

---

[6] To avoid indices, in Table 4 we have written ξ and η instead of ξ_1 and ξ_2.

In addition, we note that for the quadratic triangular elements, the basis functions N_1, N_2 and N_3 can be replaced by the corresponding nodal basis functions of linear elements. The coefficients in the expansion $\sum_{i=1}^{6} z_i N_i$ then have a different meaning: z_1, z_2 and z_3 are still the values at the vertices, but z_4, z_5 and z_6 become the deviations at the midpoints of the sides from the linear function which interpolates at the vertices.

This basis is not a purely nodal basis, although the correspondence is very simple. However, it has two advantages: we get simpler integrands in (8.2), and the condition number of the system matrix is generally lower (cf. hierarchical bases).

integrals; cf. Table 5. For equations with constant coefficients, we are usually integrating polynomials which can be computed in closed form by making use of the following formula for the unit triangle (5.9):

$$I_{pqr} := \int\int_{\substack{\xi,\eta\geq 0 \\ 1-\xi-\eta\geq 0}} \xi^p \eta^q (1-\xi-\eta)^r d\xi d\eta = \frac{p!\,q!\,r!}{(p+q+r+2)!}. \qquad (8.4)$$

The formula (8.4) can be applied to triangles in arbitrary position by replacing ξ, η, and $1-\xi-\eta$ by the barycentric coordinates. – Note that for linear elements, the integrands in (8.2) are actually constants.

Table 5. Sample points (ξ_i, η_i) and weights w_i for Gaussian quadrature formulae for polynomials up to degree 5 over the unit triangle

i	ξ_i	η_i	w_i
1	1/3	1/3	9/80
2	$(6+\sqrt{15})/21$	$(6+\sqrt{15})/21$	
3	$(9-2\sqrt{15})/21$	$(6+\sqrt{15})/21$	$(155+\sqrt{15})/2400$
4	$(6+\sqrt{15})/21$	$(9-2\sqrt{15})/21$	
5	$(6-\sqrt{15})/21$	$(6-\sqrt{15})/21$	
6	$(9+2\sqrt{15})/21$	$(6-\sqrt{15})/21$	$(155-\sqrt{15})/2400$
7	$(6-\sqrt{15})/21$	$(9+2\sqrt{15})/21$	

Static Condensation

Although the stiffness matrix can be assembled additively from $s \times s$ submatrices, the bandwidth is much larger than s (cf. Example 4.3). On the other hand, the variables corresponding to interior nodes of elements are easily treated. Both the nine-point rectangular element and the cubic ten-point triangular element have one interior node, for example.

The elimination of a variable corresponding to an interior node changes only those matrix elements for the nodes of the same element. In particular, *no* zeros are filled in. The work required for the elimination is equivalent to the work needed by the Cholesky method for the elimination of variables in an $s \times s$ matrix, i.e., in a small matrix.

The process of elimination of the variables for all these nodes is called *static condensation*.

Complexity of Setting up the Matrix

In setting up the system matrix, we need to perform Ms^2 matrix element calculations, where M is the number of elements, and s is the number of local degrees of freedom. Thus, clearly one tries to avoid calculating with finite elements that have a large number of local degrees, if possible. Only recently computations with polynomials of high degree have impact on the design of finite element programs. They are so designed that their good approximation properties more than compensate for the increase of the computational effort; see Schwab [1998].

It is for this reason that in practice C^1 elements are not used for systems of partial differential equations. For planar C^1 elements, it is well known that we need at least 12 degrees of freedom per function. Thus, for elliptic systems with three variables, we would have to set up a

$$36 \times 36 \text{ matrix}$$

for each element.

Effect on the Choice of a Grid

Once we have selected an element type, the work required to set up the stiffness matrix is approximately proportional to the number of unknowns. However, the work required for the solution of the corresponding system of linear equations using classical methods increases faster than linearly. For large systems, this can quickly lead to memory problems.

These considerations suggest individually tailoring the grid to the problem in order to reduce the number of variables as much as possible.

With the development of newer methods for solving systems of equations, such as the ones in Chapters IV and V, this problem has become less critical, and once again assembling the matrix constitutes the main part of the work. Thus, it makes more sense to save computation time there if possible. One way to do this is to build the grid so that the elements are all translations of a few basic ones. If the coefficients of the differential equation are piecewise constant functions, the computational effort can be reduced. Dividing each triangle into four congruent parts means here that the matrix elements for the subtriangles can be obtained from those of the original triangles with just a few calculations.

Local Mesh Refinement

A triangle can easily be decomposed into four congruent subtriangles. Thus, using bisection we can easily perform a global grid refinement to halve the mesh size. This process leaves the regularity parameter κ (the maximum ratio of circumcircle radius to the radius of an inscribed circle) unchanged.

Sometimes, as in the following situation, it may be preferable to perform a refinement on only part of a domain Ω:

1. In some subdomain the derivatives (which determine the order of approximation) are much greater than in the rest of the domain. This may be clear from
the nature of the problem, or from the computation of error estimators which
will be dealt with in Ch. III, §7. In this case, refining this part of the grid can
lead to a reduction of the error in the entire domain.
2. We would like to start with a very coarse grid, and let the final grid be determined by automatic refinements. Often it is appropriate for the given problem
that the amount of refinement is different in different parts of the domain.
3. We want to compute the solution to higher accuracy in some subdomain.

The fact that in the ideal case it is even possible to carry out a refinement in
the direction of an edge or of a vertex using only *similar triangles* is illustrated in
Figs. 12 and 13. However, these are exceptional cases. Some care is required in
order to generate finer grids from coarser ones automatically. In particular, if more
than one level of refinement is used, we have to be careful to avoid thin triangles.

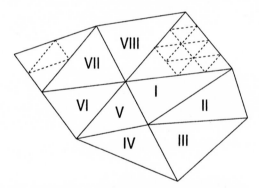

Fig. 26. Coarse grid (solid lines) and a refinement (dotted lines)

The following *refinement rule*, which can be found e.g. in the multigrid algorithm of Bank [1990], guarantees that each of the angles in the original triangulation
is bisected at most once. We may think of starting with a triangulation as in Fig. 26.
This triangulation contains several *hanging nodes* (cf. Fig. 11) which must be converted to non-hanging nodes.

8.1 Refinement Rules.

(1) If an edge of a triangle T contains two or more vertices of other triangles (not
counting its own vertices), then the triangle T is divided into four congruent
subtriangles. This process is repeated until such triangles no longer exist.
(2) Every triangle which contains a vertex of another triangle at the midpoint of
one of its edges is divided into two parts. We call the new edge a *green edge*.

(3) If a further refinement is desired, we first eliminate the green edges before proceeding.

For the triangulation in Fig. 26, we first apply rule (1) to the triangles I and VIII. This requires using the rule twice on triangle VII. Next, we construct green edges in the triangles II, V, VI, and in three subtriangles.

Despite its recursive nature, we claim that this procedure stops after a finite number of iterations. Let m denote the maximal number of levels in the desired refinement, where the maximum is to be taken over all elements (in the example, $m = 2$). Then every element will be divided at most m times. This gives an upper bound on the number of steps in (1).

Refinements of Partitions of 3-Dimensional Domain

A triangle is easily divided into four congruent subtriangles, but the situation is more involved in $\mathbb{R}^3$. A tetrahedron cannot be partitioned into eight congruent subtetrahedra; cf. Problem 8.7.

There is a partitioning that was described by Freudenthal [1942] although it is usually called *Kuhn's triangulation*. For its definition, first a cube is decomposed into $3! = 6$ tetrahedra. On the other hand the cube consists of eight subcubes which again can be partitioned into tetrahedra. The latter ones provide a decomposition of the original tetrahedra. This process also shows that six types of tetrahedra are sufficient even if the refinement procedure is repeated several times.

There is another technique due to Rivara [1984] that is more convenient than the implementation of Kuhn's triangulation. In the two-dimensional case, it works with splitting triangles by *halving* their longest sides.

Problems

8.2 Set up the system matrix A_Q for solving the Poisson equation using bilinear quadrilateral elements on the unit square. Note that with a cyclic numbering, by invariance at the element level, we get the form

$$\begin{pmatrix} \alpha & \beta & \gamma & \beta \\ \beta & \alpha & \beta & \gamma \\ \gamma & \beta & \alpha & \beta \\ \beta & \gamma & \beta & \alpha \end{pmatrix}$$

with $\alpha + 2\beta + \gamma = 0$.

Clearly, the entire matrix is determined from the stiffness matrices of all elements. In particular, for a regular grid, we get the difference star

$$\frac{8}{3} \begin{bmatrix} -1 & -1 & -1 \\ -1 & 8 & -1 \\ -1 & -1 & -1 \end{bmatrix}_* .$$

Conversely, can the above matrix be computed from the difference star?

Because of the cyclic structure, the vectors

$$(1, i^k, (-1)^k, (-i)^k), \quad k = 0, 1, 2, 3,$$

are eigenvectors. Is it possible to find a corresponding set of real eigenvectors?

Fig. 27. a) Criss, b) Cross, and c) Criss-Cross Grids

8.3 Suppose for the model problem 4.3 that we combine two triangles in a square into a *macro-element*. Clearly, we get the same stiffness matrix as for the refinement shown in Figs. 25a and 25b. Now if we symmetrize the problem and take the function which is the average of the initial two, we get the so-called *criss-cross grid*; see Fig. 27c. Find the corresponding system matrix.

8.4 Consider the model problem, and compare the stiffness matrix A_Q in Problem 8.2 with those obtained for two standard triangles A_{2T}, and for the criss-cross grid A_{cc}. How large are the condition numbers of the matrices $A_Q^{-1} A_{2T}$, $A_Q^{-1} A_{cc}$, and $A_{cc}^{-1} A_{2T}$? In particular, show that A_{2T} is stiffer than A_Q, i.e., $A_{2T} - A_Q$ is positive semidefinite.

8.5 The above figure shows a line with a refinement, as could be found along a vertical grid line in Fig. 12. Extend this to a triangulation consisting of right isosceles triangles which connect to a coarse grid.

8.6 Suppose we want to solve the elliptic differential equation

$$\mathrm{div}[a(x) \, \mathrm{grad} \, u] = f \text{ in } \Omega$$

with suitable boundary conditions using linear triangular elements from $\mathcal{M}_0^1$. Show that we get the same solution if $a(x)$ is replaced by a function which is constant on each triangle. How can we find the right constants?

8.7 In $\mathbb{R}^2$ we can obviously decompose every triangle into four congruent subtriangles. With the help of a sketch, verify that the analogous assertion for a tetrahedron in $\mathbb{R}^3$ does not hold.

8.8 Let $\lambda_1, \lambda_2, \lambda_3$ be the barycentric coordinates of a triangle T with vertices z_1, z_2, z_3. Show that

$$p(z_i) = \frac{3}{\mu(T)} \int_T (3\lambda_i - \lambda_j - \lambda_k) p \, dx \quad \text{for } p \in \mathcal{P}_1,$$

if i, j, k is a permutation of $1, 2, 3$.

Chapter III
Nonconforming and Other Methods

In the theory of conforming finite elements it is assumed that the finite element spaces lie in the function space in which the variational problem is posed. Moreover, we also require that the given bilinear form $a(\cdot, \cdot)$ can be computed exactly on the finite element spaces. However, these conditions are too restrictive for many real-life problems.

1. In general, homogeneous boundary conditions cannot be satisfied exactly for curved boundaries.
2. When we have variable coefficients or curved boundaries, we can only compute approximations to the integrals needed to assemble the stiffness matrix.
3. For plate problems and in general for fourth order elliptic differential equations, conforming methods require C^1 elements, and this leads to very large systems of equations.
4. We may want to enforce constraints only in the weak sense. A typical example is the Stokes problem, where the variational problem is posed in the space of divergence-free flows,

$$\{v \in H_0^1(\Omega)^n; \; (\text{div } v, \lambda)_{0,\Omega} = 0 \quad \text{for all } \lambda \in L_2(\Omega)\}.$$

The constraint leads to saddle point problems, and we can only take into account finitely many of the infinitely many constraints.

In this chapter we show that these types of deviations from the theory of conforming elements are admissible and do not spoil convergence. In admitting them, we are committing what are called *variational crimes*.

In §1 we establish generalizations of Céa's lemma, and examine its use by looking at two applications. Then we give a short description of isoparametric elements. §§3 and 4 contain deep functional analytic methods which are of particular importance for the mixed methods of mechanics. We illustrate them in §§6 and 7 for the Stokes problem.

§8 will be concerned with a posteriori error estimates for finite element solutions. Here arguments from the theory of nonconforming elements enter even if we deal with conforming elements.

We should mention that the theory described in §3 has also recently been used to establish the convergence of difference methods and finite volume methods.

§ 1. Abstract Lemmas
and a Simple Boundary Approximation

If the finite element spaces being used to solve an H^m-elliptic problem do not lie in the Sobolev space $H^m(\Omega)$, we refer to them as *nonconforming elements*. In this case, convergence is by no means obvious. Moreover, in addition to the approximation error, there is now an error called the *consistency error*. To analyze the situation, we need certain generalizations of Céa's lemma. We shall apply these to a simple nonconforming element. We also show how they can be used when the conformity fails in a completely different way and the boundary conditions are relaxed.

Generalizations of Céa's Lemma

As usual, let $H_0^m(\Omega) \subset V \subset H^m(\Omega)$. We replace the given variational problem

$$a(u, v) = \langle \ell, v \rangle \quad \text{for all } v \in V \tag{1.1}$$

by a sequence of finite-dimensional problems: *Find $u_h \in S_h$ with*

$$a_h(u_h, v) = \langle \ell_h, v \rangle \quad \text{for all } v \in S_h. \tag{1.2}$$

Here the bilinear forms a_h are assumed to be uniformly elliptic, i.e., there exists a constant $\alpha > 0$ independent of h such that

$$a_h(v, v) \geq \alpha \|v\|_{m,\Omega}^2 \quad \text{for all } v \in S_h. \tag{1.3}$$

Our error estimates for nonconforming methods are based on the following generalizations of Céa's lemma. For the first generalization, we do not require that a_h be defined for all functions in V. In particular, we permit the evaluation of a_h using quadrature formulae involving point evaluation functionals which are not defined for H^1 functions. However, we still require that $S_h \subset V$.

1.1 First Lemma of Strang. *Under the above hypotheses, there exists a constant c independent of h such that*

$$\|u - u_h\| \leq c\Big(\inf_{v_h \in S_h} \{\|u - v_h\| + \sup_{w_h \in S_h} \frac{|a(v_h, w_h) - a_h(v_h, w_h)|}{\|w_h\|}\} $$
$$+ \sup_{w_h \in S_h} \{\frac{\langle \ell, w_h \rangle - \langle \ell_h, w_h \rangle}{\|w_h\|}\}\Big).$$

Proof. Let $v_h \in S_h$. For convenience, set $u_h - v_h = w_h$. Then by the uniform continuity and (1.2)–(1.3), we have

$$\alpha \|u_h - v_h\|^2 \le a_h(u_h - v_h, u_h - v_h) = a_h(u_h - v_h, w_h)$$
$$= a(u - v_h, w_h) + [a(v_h, w_h) - a_h(v_h, w_h)] + [a_h(u_h, w_h) - a(u, w_h)]$$
$$= a(u - v_h, w_h) + [a(v_h, w_h) - a_h(v_h, w_h)]$$
$$- [\langle \ell, w_h \rangle - \langle \ell_h, w_h \rangle].$$

Dividing through by $\|u_h - v_h\| = \|w_h\|$ and using the continuity of a, we get

$$\|u_h - v_h\| \le C \Big(\|u - v_h\| + \frac{|a(v_h, w_h) - a_h(v_h, w_h)|}{\|w_h\|} + \frac{|\langle \ell_h, w_h \rangle - \langle \ell, w_h \rangle|}{\|w_h\|} \Big).$$

Since v_h is an arbitrary element in S_h, the assertion follows from the triangle inequality

$$\|u - u_h\| \le \|u - v_h\| + \|u_h - v_h\|.$$

□

Dropping the conformity condition $S_h \subset V$ has several consequences. In particular, the H^m-norm might not be defined for all elements in S_h, and we have to use mesh-dependent norms $\| \cdot \|_h$ as discussed e.g. in II.6.1.

We assume that the bilinear forms a_h are defined for functions in V and in S_h, and that we have ellipticity and continuity:

$$a_h(v, v) \ge \alpha \|v\|_h^2 \qquad \text{for all } v \in S_h,$$
$$|a_h(u, v)| \le C \|u\|_h \|v\|_h \quad \text{for all } u \in V + S_h, \ v \in S_h, \tag{1.4}$$

with some positive constants α and C independent of h.

The following lemma is often denoted as second lemma of Strang.

1.2 Lemma of Berger, Scott, and Strang. *Under the above hypotheses there exists a constant c independent of h such that*

$$\|u - u_h\|_h \le c \Big(\inf_{v_h \in S_h} \|u - v_h\|_h + \sup_{w_h \in S_h} \frac{|a_h(u, w_h) - \langle \ell_h, w_h \rangle|}{\|w_h\|_h} \Big).$$

Remark. The first term is called the *approximation error*, and the second one is called the *consistency error*.

Proof. Let $v_h \in S_h$. From (1.4) we see that

$$\alpha \|u_h - v_h\|_h^2 \le a_h(u_h - v_h, u_h - v_h)$$
$$= a_h(u - v_h, u_h - v_h) + [\langle \ell_h, u_h - v_h \rangle - a_h(u, u_h - v_h)].$$

Dividing by $\|u_h - v_h\|_h$ and replacing $u_h - v_h$ by w_h, we have

$$\|u_h - v_h\|_h \le \alpha^{-1} \Big(C \|u - v_h\|_h + \frac{|a_h(u, w_h) - \langle \ell_h, w_h \rangle|}{\|w_h\|_h} \Big).$$

The assertion now follows from the triangle inequality as in the proof of the first lemma. □

1.3 Remark. Using a variant of the second lemma of Strang, we no longer need the requirement that the bilinear form a_h be defined on V. The formal extension of S_h to $S_h + V$ contains pitfalls, but according to the proof, it suffices to estimate the linear form

$$a_h(v_h, w_h) - \langle \ell_h, w_h \rangle \quad \text{for all } w_h \in S_h \tag{1.5}$$

for elements $v_h \in S_h$ whose distance from u is small. Indeed, in view of (1.2), this form coincides with $a_h(v_h - u_h, w_h)$. To evaluate (1.5), we can insert a term which can be interpreted as $a_h(u, w_h)$. The advantage is that this can be done with an individually chosen function.

Duality Methods

In using duality methods in the context of nonconforming elements, we get two additional terms as compared with the Aubin–Nitsche lemma.

1.4 Lemma. *Suppose that the Hilbert spaces V and H satisfy the hypotheses of the Aubin–Nitsche lemma. In addition, suppose $S_h \subset H$ and that the bilinear form a_h is defined on $V \cup S_h$ so that it coincides with a on V. Then the finite element solution u_h of (1.2) satisfies*

$$
\begin{aligned}
|u - u_h| \leq \sup_{g \in H} \frac{1}{|g|} \Big\{ & c\|u - u_h\|_h \|\varphi_g - \varphi_{g,h}\|_h \\
& + |a_h(u - u_h, \varphi_g) - (u - u_h, g)| \\
& + |a_h(u, \varphi_g - \varphi_{g,h}) - \langle \ell, \varphi_g - \varphi_{g,h} \rangle| \Big\}.
\end{aligned} \tag{1.6}
$$

Here $\varphi_g \in V$ and $\varphi_{g,h} \in S_h$ are the weak solutions of $a_h(w, \varphi) = (w, g)$ for given $g \in H$.

Proof. By the definition of u_h, φ_g and $\varphi_{g,h}$, for every $g \in H$ we have

$$
\begin{aligned}
(u - u_h, g) &= a_h(u, \varphi_g) - a_h(u_h, \varphi_{g,h}) \\
&= a_h(u - u_h, \varphi_g - \varphi_{g,h}) \\
&\quad + a_h(u_h, \varphi_g - \varphi_{g,h}) + a_h(u - u_h, \varphi_{g,h}) \\
&= a_h(u - u_h, \varphi_g - \varphi_{g,h}) \\
&\quad - [a_h(u - u_h, \varphi_g) - (u - u_h, g)] \\
&\quad - [a_h(u, \varphi_g - \varphi_{g,h}) - \langle \ell, \varphi_g - \varphi_{g,h} \rangle].
\end{aligned}
$$

The last equality is most easily verified by replacing the linear functionals in the square brackets by terms involving the bilinear form a_h, and then comparing terms. The assertion now follows from (II.7.7) and the continuity of a_h.   □

The extra terms in (1.6) are basically of the same form as those in the second lemma of Strang. We shall see that in applications, the main effort is to verify that the hypotheses of the lemma hold.

The Crouzeix–Raviart Element

The Crouzeix–Raviart element is the simplest nonconforming element for the discretization of second order elliptic boundary-value problems. It is also called the *nonconforming P_1 element*.

Fig. 28. The Crouzeix–Raviart element or nonconforming P_1-element

$$\mathcal{M}^1_* := \{v \in L_2(\Omega); \ v|_T \text{ is linear for every } T \in \mathcal{T}_h,$$
$$v \text{ is continuous at the midpoints of the triangle edges}\}, \qquad (1.7)$$
$$\mathcal{M}^1_{*,0} := \{v \in \mathcal{M}^1_*; \ v = 0 \text{ at the midpoints of the edges on } \partial\Omega\}.$$

To solve the Poisson equation, let

$$a_h(u, v) := \sum_{T \in \mathcal{T}_h} \int_T \nabla u \cdot \nabla v \, dx \quad \text{for all } u, v \in H^1(\Omega) + \mathcal{M}^1_{*,0},$$

$$\|v\|_h := \sqrt{a_h(v, v)} \qquad \text{for all } v \in H^1(\Omega) + \mathcal{M}^1_{*,0}.$$

By definition $\|v\|_h^2 := \sum_{T \in \mathcal{T}_h} |v|_{1,T}^2$, and it is called a *broken H^1 semi-norm*.

For simplicity, suppose Ω is a convex polyhedron. Then the problem is H^2-regular, and $u \in H^2(\Omega)$.

Given $v \in H^2(\Omega)$, let $Iv \in \mathcal{M}^1_{*,0} \cap C^0(\Omega)$ be the continuous piecewise linear function which interpolates v at the vertices of the triangles. We denote edges of the triangles by the letter e.

To apply Lemma 1.2, we compute

$$L_u(w_h) : = a_h(u, w_h) - \langle \ell, w_h \rangle$$

$$= \sum_{T \in \mathcal{T}_h} \int \nabla u \nabla w_h \, dx - \int_\Omega f w_h \, dx$$

$$= \sum_{T \in \mathcal{T}_h} \left(\int_{\partial T} \partial_\nu u \, w_h ds - \int_T \Delta u \, w_h dx \right) - \int_\Omega f w_h dx$$

$$= \sum_{T \in \mathcal{T}_h} \int_{\partial T} \partial_\nu u \, w_h ds,$$

for $w_h \in \mathcal{M}^1_{*,0}$. Here we have used the fact that $-\Delta u = f$ holds in the weak sense; cf. Example II.2.10. In addition, note that each interior edge appears twice in the sum. Thus, the values of the integrals do not change if we subtract the integral mean value $\overline{w_h(e)}$ on each edge e:

$$L_u(w_h) = \sum_T \sum_{e \in \partial T} \int_e \partial_\nu u (w_h - \overline{w_h(e)}) ds.$$

It follows from the definition of $\overline{w_h(e)}$ that $\int_e (w_h - \overline{w_h(e)}) ds = 0$. The values of the integrals also do not change if we subtract an arbitrary constant function from $\partial_\nu u$ on each edge e. This can be $\partial_\nu I u$ in particular, and we get

$$L_u(w_h) = \sum_T \sum_{e \in \partial T} \int_e \partial_\nu (u - Iu)(w_h - \overline{w_h(e)}) ds.$$

It follows from the Cauchy–Schwarz inequality that

$$|L_u(w_h)| \leq \sum_T \sum_{e \in \partial T} \left[\int_e |\nabla(u - Iu)|^2 ds \int_e |w_h - \overline{w_h(e)}|^2 ds \right]^{1/2}. \tag{1.8}$$

We now derive bounds for the integrals in (1.8). By the trace theorem and the Bramble–Hilbert lemma,

$$\int_{\partial T_{\text{ref}}} |\nabla(v - Iv)|^2 ds \leq c\|\nabla(v - Iv)\|_{1,T_{\text{ref}}}^2 \leq c\|v - Iv\|_{2,T_{\text{ref}}}^2 \leq c'|v|_{2,T_{\text{ref}}}^2,$$

for $v \in H^2(T_{\text{ref}})$. Using the transformation formulas from Ch. 2, §6, we see that

$$\int_{\partial T} |\nabla(v - Iv)|^2 ds \leq ch|v|_{2,T}^2 \tag{1.9}$$

for $T \in \mathcal{T}_h$. Similarly, for each edge e of ∂T_{ref},

$$\int_e |w_h - \overline{w_h(e)}|^2 ds \leq c\|w_h\|_{1,T_{\text{ref}}}^2 \leq c'|w_h|_{1,T_{\text{ref}}}^2 \quad \text{for all } w_h \in \mathcal{P}_1.$$

Here the Bramble–Hilbert lemma applies because the left-hand side vanishes for constant functions. For $e \in T \in \mathcal{T}_h$, the transformation theorems yield

$$\int_e |w_h - \overline{w_h(e)}|^2 ds \leq ch|w_h|_{1,T}^2 \quad \text{for all } w_h \in \mathcal{M}_{*,0}^1. \tag{1.10}$$

We now insert the estimates (1.9) and (1.10) into (1.8), and use the Cauchy–Schwarz inequality for Euclidean scalar products:

$$\begin{aligned}
|L_u(w_h)| &\leq \sum_T 3\, ch |u|_{2,T} |w_h|_{1,T} \\
&\leq c'h \left[\sum_T |u|_{2,T}^2 \sum_T |w_h|_{1,T}^2 \right]^{1/2} \\
&= c'h|u|_{2,\Omega} \|w_h\|_h .
\end{aligned} \tag{1.11}$$

Finally, we observe that the conforming P_1 elements are contained in $\mathcal{M}^1_{*,0}$. This means that we do not need to establish a new approximation theorem for $\mathcal{M}^1_{*,0}$, and it follows that

$$\|u - u_h\|_h \leq ch|u|_{2,\Omega} \leq ch\|f\|_0. \tag{1.12}$$

We now apply Lemma 1.4 to the Crouzeix–Raviart element. Let $V = H^1(\Omega)$ and $H = L_2(\Omega)$. In particular, to estimate the first term, we regard $\varphi_g - \varphi_{g,h}$ as the discretization error for the problem $a(w, \varphi) = (w, g)_{0,\Omega}$. We make use of (1.12) and the regularity of the problem:

$$\|\varphi_g - \varphi_{g,h}\|_h \leq ch|\varphi_g|_2 \leq c''h\|g\|_0.$$

An essential observation is that the formula (1.11) holds for all $w \in \mathcal{M}^1_{*,0} + H^1_0$, as can be seen immediately by examining the derivation of the formula. It follows that the extra terms in (1.6) satisfy

$$
\begin{aligned}
|a_h(u - u_h, \varphi_g) - (u - u_h, g)| &= |L_{\varphi_g}(u - u_h)| \\
&\leq c'h\,|\varphi_g|_2\,\|u - u_h\|_h \\
&\leq c'h\,\|g\|_0\|u - u_h\|_h,
\end{aligned}
$$
$$
\begin{aligned}
|a_h(u, \varphi_g - \varphi_{g,h}) - (f, \varphi_g - \varphi_{g,h})| &= |L_u(\varphi_g - \varphi_{g,h})| \\
&\leq c'h\,|u|_2\,\|\varphi_g - \varphi_{g,h}\|_h.
\end{aligned}
$$

Combining the last three estimates, we have

$$
\begin{aligned}
|(u - u_h, g)| &\leq c\,h(\|u - u_h\|_h + h|u|_2)\|g\|_0 \\
&\leq c\,h^2|u|_2\,\|g\|_0.
\end{aligned}
$$

Combining these duality calculations with (1.12), we obtain

1.5 Theorem. *Suppose Ω is convex or that it has a smooth boundary. -Then using the Crouzeix–Raviart elements to discretize the Poisson equation, we have*

$$\|u - u_h\|_0 + h\|u - u_h\|_h \leq c\,h^2|u|_2.$$

Remark. The result closely resembles the result in Ch. II, §7, but there is a difference. While for conforming methods the H^2-regularity was used only quantitatively, here it also enters qualitatively in the convergence proof. This corresponds with the practical observation that nonconforming elements are much more sensitive to "near singularities" i.e., to the appearance of large H^2 norms.

A Simple Approximation to Curved Boundaries

We consider a second order differential equation on a domain Ω with smooth boundary. This means that for every point on $\Gamma = \partial\Omega$, there exist orthogonal coordinates (ξ, η) so that in a neighborhood, the boundary can be described as the graph of a C^2 function g. Suppose the domain Ω is decomposed into elements so that every element T has three vertices, at least one of which is an interior point of Ω. If two vertices of T lie on Γ, then the boundary piece of Γ with endpoints at these vertices is an edge of the element. Suppose all other edges of the elements are straight lines. We refer to these elements as *curved triangles*; see Fig. 29.

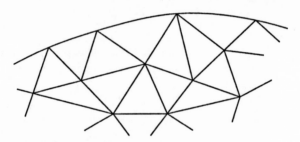

Fig. 29. Part of a triangulation of a domain with curved boundary

If we replace the boundary curves between two neighboring vertices by a line segment, we get a polygonal approximation Ω_h of Ω. The partition $\mathcal{T}_h$ of Ω induces a triangulation of Ω_h. We suppose that it is admissible. We call $\mathcal{T}_h$ *uniform* or *shape regular*, provided that the induced triangulation of Ω_h possesses the respective property.

We choose the finite elements to be the linear triangular elements, where the zero boundary conditions are enforced only at the nodes on Γ:

$$S_h := \{v \in C^0(\Omega);\ v|_T \text{ is linear for every } T \in \mathcal{T}_h,$$
$$v(z) = 0 \text{ for every node } z \in \Gamma\}.$$

Thus, $S_h \not\subset H_0^1(\Omega)$. Nevertheless, since $S_h \subset H^1(\Omega)$, it is not necessary to work with a new (mesh-dependent) norm, and we can set $a_h(u, v) = a(u, v)$.

1.6 Lemma. *Let Ω be a domain with C^2 boundary, and let $\mathcal{T}_h$ be a sequence of shape-regular triangulations. Then*

$$\|v\|_{0,\Gamma} \le c h^{3/2} |v|_{1,\Omega} \quad \text{for all } v \in S_h. \tag{1.13}$$

Proof. Let T be an element with a curved edge $\Gamma_1 := T \cap \Gamma$. We shall show that

$$\int_{\Gamma_1} v^2 ds \le c h_T^3 \int_T (|\partial_1 v|^2 + |\partial_2 v|^2) dx. \tag{1.14}$$

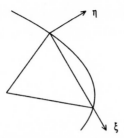

Fig. 30. Local coordinates for a curved element

Then the assertion follows after summing over all triangles of $\mathcal{T}_h$.

Suppose we choose the coordinate system so that the ξ-axis passes through the two vertices of T lying on Γ; see Fig. 30. Let $(\xi_1, 0)$, $(\xi_2, 0)$ be the coordinates of the vertices, and suppose the boundary is given by $\eta = g(\xi)$. From $g(\xi_1) = g(\xi_2) = 0$, $|\xi_1 - \xi_2| \leq h_T$, and $|g''(\xi)| \leq c$, it follows that

$$|g(\xi)| \leq ch_T^2 \quad \text{for all } \xi_1 \leq \xi \leq \xi_2. \tag{1.15}$$

Since $v \in S_h$ is linear in T and vanishes at two points on the ξ-axis, v has the form

$$v(\xi, \eta) = b\eta$$

on T. The gradient is constant in T, $|\nabla u| = b$, and the area of T can be bounded from below by that of the largest inscribed circle. Its radius is at least h_T/κ, and thus

$$\int_T |\nabla v|^2 dx \geq \pi (h_T/\kappa)^2 b^2.$$

On the other hand,

$$\int_\Gamma v^2 ds = \int_{\xi_1}^{\xi_2} [bg(\xi)]^2 \sqrt{1 + [g'(\xi)]^2} \, d\xi$$
$$\leq [bch_T^2]^2 \int_{\xi_1}^{\xi_2} 2 \, d\xi$$
$$= 2c^2 b^2 h_T^5.$$

The assertion now follows by comparing the last two estimates.   □

We remark that in view of (1.15), if we replace a piece of the curved boundary by a straight line, we cut off a domain $T'' := T \cap (\Omega \setminus \Omega_h)$ with an area

$$\mu(T'') \leq ch\mu(T). \tag{1.16}$$

Now let u_h be the weak solution in S_h, i.e.

$$a(u_h, v) = (f, v)_{0,\Omega} \quad \text{for all } v \in S_h.$$

In addition, suppose $u \in H^2(\Omega) \cap H_0^1(\Omega)$ is the solution of the Dirichlet problem (II.2.7). Then $Lu = f$ in the sense of $L_2(\Omega)$, and integrating by parts, we have

$$(f, v)_{0,\Omega} = (Lu, v)_{0,\Omega}$$
$$= a(u, v) - \int_\Gamma \sum_{k,\ell} a_{k\ell} \, \partial_k u \, v \cdot v_\ell ds$$

for $v \in S_h \subset H^1(\Omega)$. Applying the Cauchy–Schwarz inequality, the trace theorem, and the previous lemma, we get

$$|a(u, v) - (f, v)_{0,\Omega}| \leq c \|\nabla u\|_{0,\Gamma} \|v\|_{0,\Gamma}$$
$$\leq c \|u\|_{2,\Omega} h^{3/2} \|v\|_{1,\Omega}.$$

The second lemma of Strang gives a term of order $h^{3/2}$, which is small in comparison with the usual term of order h.

1.7 Theorem. *Let Ω be a domain with C^2 boundary, and suppose we use linear triangular elements on shape-regular triangulations. Then the finite element approximation satisfies*

$$\|u - u_h\|_{1,\Omega} \leq c \, h \, \|u\|_{2,\Omega} \tag{1.17}$$
$$\leq c \, h \, \|f\|_{0,\Omega}.$$

The estimate remains correct if we replace a by

$$a_h(u, v) := \int_{\Omega_h} \sum_{k,\ell} a_{k\ell} \, \partial_k u \, \partial_\ell v \, dx.$$

In particular, $|a_h(u, v) - a(u, v)| \leq c \|u\|_{1,\Omega} \|v\|_{1,\Omega \setminus \Omega_h}$, and since ∇v is constant on every element T, (1.16) implies

$$\|v\|_{1,\Omega \setminus \Omega_h} \leq ch \|v\|_{1,\Omega} \quad \text{for all } v \in S_h.$$

Modifications of the Duality Argument

The general Lemma 1.4 is not applicable here because the estimate (1.13) does not hold for all $v \in S_h + H_0^1$. For simplicity, we now apply the duality method along with the tools which we have already developed. Then even for L_2 estimates, we get an extra term of order $h^{3/2}$ which is no longer small compared with the main term. Using a more refined argument, this extra term could be avoided [Blum 1991].

In order to avoid having to work with a double sum in the boundary integrals, we restrict our attention to the Poisson equation, and note that the supremum in (II.7.7) is attained for $g = w$.

With $w := u - u_h$, let φ be the solution of the equation (II.7.6); i.e., let

$$-\Delta\varphi = w \text{ in } \Omega,$$

$$\varphi = 0 \text{ on } \Gamma.$$

Since Ω is smooth, the problem is H^2-regular. Hence, $\varphi \in H^2(\Omega) \cap H^1_0(\Omega)$ and

$$\|\varphi\|_{2,\Omega} \leq c\|w\|_{0,\Omega}.$$

Since $w \notin H^1_0(\Omega)$, in contrast to the calculations with conforming elements, we get boundary terms when applying Green's formula:

$$\|w\|^2_{0,\Omega} = (w, -\Delta\varphi)_{0,\Omega}$$
$$= a(w, \varphi) - (w, \partial_\nu\varphi)_{0,\Gamma}. \tag{1.18}$$

Let v_h be an arbitrary element in S_h. Then $a(u - u_h, -v_h) = (\partial_\nu u, -v_h)_{0,\Gamma}$, and since $\varphi \in H^1_0(\Omega)$, the last term can be replaced by $(\partial_\nu u, \varphi - v_h)_{0,\Gamma}$. By (1.18),

$$\|w\|^2_{0,\Omega} = a(w, \varphi - v_h) - (\partial_\nu u, \varphi - v_h)_{0,\Gamma} - (w, \partial_\nu\varphi)_{0,\Gamma}. \tag{1.19}$$

Now we select v_h to be the interpolant of φ in S_h.

We estimate the first term in the same way as for conforming elements:

$$a(w, \varphi - v_h) \leq C\|w\|_{1,\Omega}\|\varphi - v_h\|_{1,\Omega}$$
$$\leq C\|w\|_{1,\Omega}\,ch\|\varphi\|_{2,\Omega}$$
$$\leq ch\|w\|_{1,\Omega}\|w\|_{0,\Omega}.$$

To deal with the second term in (1.19), we need the estimate $\|\varphi - I\varphi\|_{0,\Gamma} \leq ch^{3/2}\|\varphi\|_{2,\Omega}$, whose proof (which we do not present here) is based on a scaling argument. Now we apply the trace theorem to ∇u:

$$|(\partial_\nu u, \varphi - v_h)_{0,\Gamma}| \leq \|\nabla u\|_{0,\Gamma}\|\varphi - v_h\|_{0,\Gamma}$$
$$\leq c\|u\|_{2,\Omega}\,ch^{3/2}\|\varphi\|_{2,\Omega}$$
$$\leq ch^{3/2}\|u\|_{2,\Omega}\|w\|_{0,\Omega}.$$

Next we apply Lemma 1.6 and the trace theorem to the last term to get

$$|(w, \partial_\nu\varphi)_{0,\Gamma}| \leq \|w\|_{0,\Gamma}\|\nabla\varphi\|_{0,\Gamma}$$
$$\leq \|u_h\|_{0,\Gamma}\|\nabla\varphi\|_{0,\Gamma}$$
$$\leq ch^{3/2}\|u_h\|_{1,\Omega}\|\varphi\|_{2,\Omega}$$
$$\leq ch^{3/2}(\|u\|_{1,\Omega} + \|u - u_h\|_{1,\Omega})\|w\|_{0,\Omega}$$
$$\leq ch^{3/2}\|u\|_{2,\Omega}\|w\|_{0,\Omega}.$$

Combining the above, we have

$$\|w\|^2_{0,\Omega} \leq c\|w\|_{0,\Omega}\{h\|w\|_{1,\Omega} + h^{3/2}\|u\|_{2,\Omega}\}.$$

Recalling that $w = u - u_h$, we have

1.8 Theorem. *Under the hypotheses of Theorem 1.7,*

$$\|u - u_h\|_{0,\Omega} \le ch^{3/2}\|u\|_{2,\Omega}.$$

The error term $\mathcal{O}(h^{3/2})$ arises from the pointwise estimate of the finite-element functions $|u_h(x)| \le ch^2|\nabla u_h(x)|$ for all $x \in \Gamma$; cf. (1.15). If we approximate the boundary with quadratic (instead of linear) functions, giving a one higher power of h, the final result is improved by the same factor. This can be achieved using isoparametric elements, for example.

Problems

1.9 Let S_h be an affine family of C^0 elements. Show that in both the approximation and inverse estimates, $\|\cdot\|_{2,h}$ can be replaced by the mesh-dependent norm

$$\|\|v\|\|_h^2 := \sum_{T_j} \|v\|_{2,T_j}^2 + h^{-1} \sum_{\{e_m\}} \int_{e_m} \lceil \frac{\partial v}{\partial \nu} \rceil^2 ds.$$

Here $\{e_m\}$ is the set of inter-element boundaries, and $\lceil \cdot \rceil$ denotes the jump of a function.
Hint: In $H^2(T_{\text{ref}})$, $\|v\|_{2,T_{\text{ref}}}$ and $\left(\|v\|_{2,T_{\text{ref}}}^2 + \int_{\partial T_{\text{ref}}} |\nabla v|^2 ds\right)^{1/2}$ are equivalent norms.

1.10 The linear functional L_u appearing in the analysis of the Crouzeix–Raviart element vanishes on the subset $H_0^1(\Omega)$ by the definition of weak solutions. What is wrong with the claim that L_u vanishes for all $w \in L_2(\Omega)$ because of the density of $H_0^1(\Omega)$ in $L_2(\Omega)$?

1.11 If the stiffness matrices are computed by using numerical quadrature, then only approximations a_h of the bilinear form are obtained. This holds also for conforming elements. Estimate the influence on the error of the finite element solution, given the estimate

$$|a(v, v) - a_h(v, v)| \le \varepsilon(h) \|v\|_1^2 \quad \text{for all } v \in S_h.$$

1.12 The Crouzeix–Raviart element has locally the same degrees of freedom as the conforming P_1 element $\mathcal{M}_0^1$, i. e. the Courant triangle. Show that the (global) dimension of the finite element spaces differ by a factor that is close to 3 if a rectangular domain as in Fig. 9 is partitioned.

§ 2. Isoparametric Elements

For the treatment of second order elliptic problems on domains with curved boundaries, we need to use elements with curved sides if we want to get higher accuracy.
For many problems of fourth order, we even have to do a good job of approximating the boundary in the C^1-norm just to get convergence. For this reason, certain
so-called *isoparametric families* of finite elements were developed. They are a
generalization of the *affine* families.

For triangulations, isoparametric elements actually play a role only near the
boundary. On the other hand, (simple) isoparametric quadrilaterals are often used
in the interior since this allows us to generate arbitrary quadrilaterals, rather than
just parallelograms.

We restrict our attention to planar domains, and consider families of elements
where every $T \in \mathcal{T}_h$ is generated by a bijective mapping F:

$$
\begin{aligned}
T_{\text{ref}} &\longrightarrow T \\
(\xi, \eta) &\longmapsto (x, y) = F(\xi, \eta) = (p(\xi, \eta), q(\xi, \eta)).
\end{aligned}
\tag{2.1}
$$

This framework includes the affine families when p and q are required to be
linear functions. When p and q are polynomials of higher degree, we get the more
general situation of isoparametric elements. More precisely, the polynomials in
the parametrization are chosen from the same family Π as the shape functions of
the element (T, Π, Σ).

Isoparametric Triangular Elements

The important case where p and q are quadratic polynomials is shown in Fig. 31.
By Remark II.5.4, we know that six points P_i, $1 \le i \le 6$, can be prescribed. Then
p and q as polynomials of degree 2 are uniquely defined by the coordinates of the
points $P_1, \ldots, P_6$. In particular, if P_4, P_5, and P_6 are nodes at the midpoints of
the edges of the triangle whose vertices are P_1, P_2, and P_3, then obviously we get
a linear mapping.

The introduction of isoparametric elements raises the following questions:
1. Can isoparametric elements be combined with affine ones without losing the
   desired additional degrees of freedom?
2. How are the concepts "uniform" and "shape regular" to be understood so that
   the results for affine families can be carried over to isoparametric ones?

In order to keep the computational costs down, we should use elements with
straight edges in the interior of Ω. This is why elements with only one curved side

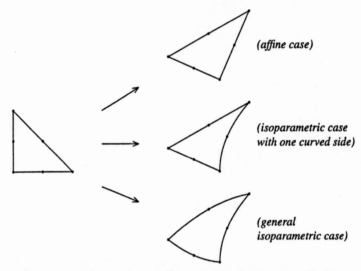

(affine case)

(isoparametric case with one curved side)

(general isoparametric case)

Fig. 31. *Isoparametric elements with linear and quadratic parametrization*

are of special interest. Suppose the edges of T_{ref} with $\xi = 0$ and with $\eta = 0$ are mapped to the straight edges of T. It is useful to choose the centers of these edges as the images of the corresponding center points on the edges of the reference triangle. Then in the quadratic case we have

$$
\begin{aligned}
p(\xi, \eta) &= a_1 + a_2\xi + a_3\eta + a_4\xi\eta, \\
q(\xi, \eta) &= b_1 + b_2\xi + b_3\eta + b_4\xi\eta.
\end{aligned}
\tag{2.2}
$$

The restrictions of p and q to the edges $\xi = 0$ and $\eta = 0$ are linear functions, which results in a continuous match with neighboring affine elements without taking any special measures.

For affine families, the condition of shape regularity can be formulated in various ways, and a number of equivalent definitions can be found in the literature. The corresponding conditions for isoparametric elements are not completely independent, and cannot be replaced by *one* simple condition.

2.1 Definition. A family of isoparametric partitions $\mathcal{T}_h$ is called *shape regular* provided there exists a constant κ such that:

(i) For every parametrization $F : T_{\text{ref}} \longrightarrow T \in \mathcal{T}_h$,

$$
\frac{\sup\{\|DF(\zeta) \cdot z\|; \ \zeta \in T_{\text{ref}}, \ \|z\| = 1\}}{\inf\{\|DF(\zeta) \cdot z\|; \ \zeta \in T_{\text{ref}}, \ \|z\| = 1\}} \le \kappa.
$$

(ii) For every $T \in \mathcal{T}_h$, there exists an inscribed circle with radius ρ_T such that

$$
diameter(T) \le \kappa\rho_T.
$$

If in addition

$$diameter(T) \leq 2h \quad \text{and} \quad \rho_T \geq h/\kappa,$$

then $\mathcal{T}_h$ is called *uniform*.

Isoparametric Quadrilateral Elements

Isoparametric quadrilaterals are also of use in the interior since only parallelograms can be obtained from a square with affine mappings (see Ch. II, §5).

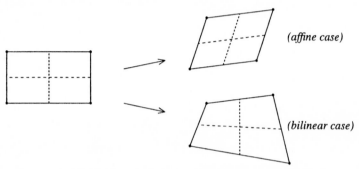

Fig. 32. *Isoparametric quadrilaterals with bilinear parametrization*

Let $T_{\mathrm{ref}} = [0, 1]^2$ be the unit square. Then

$$F : \begin{cases} p(\xi, \eta) = a_1 + a_2\xi + a_3\eta + a_4\xi\eta \\ q(\xi, \eta) = b_1 + b_2\xi + b_3\eta + b_4\xi\eta \end{cases} \qquad (2.3)$$

maps T_{ref} to a general quadrilateral. From the theory of bilinear quadrilateral elements, we know that the two sets of four parameters are uniquely determined by the eight coordinates of the four corners of the image of T_{ref}.

In addition, it is clear that when ξ and η are both constant, the parametrization F is a linear function of the arc length. It follows that the image is a quadrilateral with straight edges. The vertices are numbered so that the orientation is preserved. Because of the linearity of the parametrization on the edges, connecting the element to its neighbors is no problem.

2.2 Remark. A family of partitions $\mathcal{T}_h$ involving general quadrilaterals with bilinear parametrizations is *shape regular* provided there exists a constant $\kappa > 1$ such that the following conditions are satisfied:

  (i) For every quadrilateral T, the ratio of maximal to minimal edge lengths is bounded by κ.

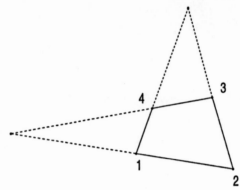

Fig. 33. General quadrilateral

(ii) Every T contains an inscribed circle with radius $\rho_T \geq h_T / \kappa$, where h_T is the diameter of T.

(iii) All angles are smaller than $\pi - \varphi_0$ with some $\varphi_0 > 0$. [We note that the second condition implies that all angles are greater than φ_0 with some $\varphi_0 > 0$.]

Moreover, we note that DF and also $\det(DF)$ depend linearly on the parameters. In particular, the determinants attain their maximum and minimum values at vertices of the quadrilateral. For the quadrilateral shown in Fig. 33, P_2 and P_4 are the extremal points since the intersections of the sides ly on their extension through P_4.

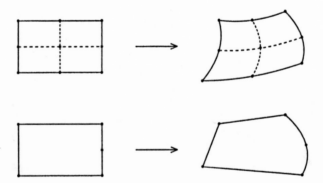

Fig. 34. Isoparametric quadrilateral elements with 9 and 5 parameters

Fig. 34 shows curved quadrilaterals which arise from biquadratic parametrizations. The parametrization for 9 prescribed points corresponds exactly with the 9 node element in Ch. II, §5. The 5 point case is of particular practical importance since it can model one curved side. Here the correct parametrization is

$$x = a_1 + a_2\xi + a_3\eta + a_4\xi\eta + a_5\xi\eta(1 - \eta),$$
$$y = b_1 + b_2\xi + b_3\eta + b_4\xi\eta + b_5\xi\eta(1 - \eta).$$

At first glance, we are tempted to replace the shape functions corresponding to the coefficients a_5 and b_5 by the simpler (quadratic) expressions $a_5\eta(1 - \eta)$ and $b_5\eta(1 - \eta)$. This would be possible for interpolation at the 5 points, but would not lead to a linear expression on the edge $\xi = 0$.

Problems

2.3 Suppose we have a program for generating quadrilateral elements, and now want to use it to build triangular elements. We map quadrilaterals (bilinearly) to triangles by identifying pairs of vertices in the image. What triangular elements do we get using the bilinear, 8 point, and 9 point elements, respectively?

2.4 Suppose that in setting up the system matrix we use a quadrature formula with positive weights. Show that in spite of the error in the numerical integration, the matrix is at least positive semidefinite. Describe a case in which the matrix is singular.

§ 3. Further Tools from Functional Analysis

In Chapter II the existence of solutions to the variational problem was established using the Lax–Milgram theorem. There the symmetry and the ellipticity of the bilinear form $a(\cdot, \cdot)$ were essential hypotheses. In order to treat saddle point problems, we need a more general approach which does not require that the quadratic form be positive definite.

In this connection, a detailed discussion of linear functionals on Sobolev spaces is useful. This leads us to the so-called *negative norms*.

Negative Norms

Let V be a Hilbert space. By the Riesz representation theorem, every continuous linear functional $\ell \in V'$ can be identified with an element from V itself. Thus, often it is not necessary to distinguish between V and V'. However, in the variational calculus, this can obscure certain important aspects of the problem. Before discussing methods of functional analysis, we first orient ourselves with a simple example.

Consider the *Helmholtz equation*

$$\begin{aligned}
-\Delta u + u &= f \quad \text{in } \Omega, \\
u &= 0 \quad \text{on } \partial\Omega,
\end{aligned} \tag{3.1}$$

with $f \in L_2(\Omega)$. The weak solution $u \in H_0^1(\Omega)$ is characterized by

$$(u, v)_1 = (f, v)_0 \quad \text{for all } v \in H_0^1(\Omega), \tag{3.2}$$

where $(\cdot, \cdot)_1$ is the scalar product on the Hilbert space $H^1(\Omega)$. The problem (3.1) can thus be formulated as follows: *Given f, find the Riesz representation for the functional*

$$\ell : H_0^1(\Omega) \longrightarrow \mathbb{R}, \quad \langle \ell, v \rangle := (f, v)_0. \tag{3.3}$$

If we identify $H_0^1(\Omega)$ with its dual, then there is nothing to do to solve the variational problem. (Note that the analogous representation for the space $H^0(\Omega) = L_2(\Omega)$ is indeed trivial.)

There is another formulation for the dual space which fits better with the form (3.3) of the given functional. The equation (3.1) is then defined for all functions f in the associated completion of $L_2(\Omega)$.

3.1 Definition. Let $m \geq 1$. Given $u \in L_2(\Omega)$, define the norm

$$\|u\|_{-m,\Omega} := \sup_{v \in H_0^m(\Omega)} \frac{(u, v)_{0,\Omega}}{\|v\|_{m,\Omega}}.$$

We define $H^{-m}(\Omega)$ to be the completion of $L_2(\Omega)$ w.r.t. $\|\cdot\|_{-m,\Omega}$.

For the Sobolev spaces built on $L^2(\Omega)$, we identify the dual space of $H_0^m(\Omega)$ with $H^{-m}(\Omega)$. Moreover, by the definition of H^{-m}, there is a dual pairing $\langle u, v \rangle$ for all $u \in H^{-m}$, $v \in H_0^m$, i.e. $\langle u, v \rangle$ is a bilinear form, and

$$\langle u, v \rangle = (u, v)_{0,\Omega}, \quad \text{whenever } u \in L_2(\Omega), \; v \in H_0^m(\Omega).$$

Clearly,

$$\ldots \supset H^{-2}(\Omega) \supset H^{-1}(\Omega) \supset L_2(\Omega) \supset H_0^1(\Omega) \supset H_0^2(\Omega) \supset \ldots$$

$$\cdots \leq \|u\|_{-2,\Omega} \leq \|u\|_{-1,\Omega} \leq \|u\|_{0,\Omega} \leq \|u\|_{1,\Omega} \leq \|u\|_{2,\Omega} \leq \cdots$$

H^{-m} was defined to be the dual space of H_0^m and not of H^m. Thus we obtain an improvement of II.2.9 only for Dirichlet problems.

3.2 Remark. Let a be an H_0^m-elliptic bilinear form. Then with the notation of the proof of the Existence Theorem II.2.9, we have

$$\|u\|_m \leq \alpha^{-1}\|f\|_{-m}. \tag{3.4}$$

Proof. By Definition 3.1, $(u, v)_0 \leq \|u\|_{-m}\|v\|_m$. Substituting $v = u$ in the weak equation gives

$$\alpha\|u\|_m^2 \leq a(u, u) = (f, u)_0 \leq \|f\|_{-m}\|u\|_m,$$

and the assertion follows after dividing by $\|u\|_m$.   □

This asserts that the Dirichlet problem is H^m-regular in the sense of Definition II.7.1.

3.3 Remark. Let $V \subset U$ be Hilbert spaces, and suppose the imbedding $V \hookrightarrow U$ is continuous and dense. In addition, suppose we identify U' with U via the Riesz representation. Then

$$V \subset U \subset V'$$

is called a *Gelfand triple*. We have already encountered the following Gelfand triples:

$$H^m(\Omega) \subset L_2(\Omega) \subset H^m(\Omega)',$$

and

$$H_0^m(\Omega) \subset L_2(\Omega) \subset H^{-m}(\Omega).$$

Adjoint Operators

Let X and Y be Banach spaces whose dual spaces are X' and Y', respectively. The dual pairings will usually be written as $\langle \cdot, \cdot \rangle$ without reference to the spaces. Let $L : X \longrightarrow Y$ be a bounded linear operator. Given $y^* \in Y'$,

$$x \longmapsto \ell_{y^*}(x) := \langle y^*, Lx \rangle$$

defines a continuous linear functional on X. The associated linear mapping

$$L' : Y' \longrightarrow X',$$
$$y^* \longmapsto \ell_{y^*}, \quad \text{i.e. } \langle L'y^*, x \rangle := \langle y^*, Lx \rangle,$$

is called the *adjoint* of L.

Often the adjoint operator can be used to determine the image of L. More generally, let V be a closed subspace of X. Then

$$V^0 := \{\ell \in X'; \ \langle \ell, v \rangle = 0 \quad \text{for all } v \in V\}$$

is called the *polar set* of V. Since in the Hilbert space case we cannot always identify the dual space X' with X, we must distinguish between the polar set V^0 and the *orthogonal complement*

$$V^\perp = \{x \in X; \ (x, v) = 0 \quad \text{for all } v \in V\}.$$

In the sequel we shall make multiple use of the following *closed range theorem* (see e.g. Yosida [1971]). We give a proof at the end of this section.

3.4 Theorem. *Let X and Y be Banach spaces, and let $L : X \longrightarrow Y$ be a bounded linear mapping. Then the following assertions are equivalent:*
  (i) The image $L(X)$ is closed in Y,
 (ii) $L(X) = (\ker L')^0$.

An Abstract Existence Theorem

Let U and V be Hilbert spaces, and suppose $a : U \times V \longrightarrow \mathbb{R}$ is a bilinear form. We define an associated linear operator $L : U \longrightarrow V'$ by

$$\langle Lu, v \rangle := a(u, v) \quad \text{for all } v \in V.$$

The variational problems discussed above had the following structure: Given $f \in V'$, find $u \in U$ so that

$$a(u, v) = \langle f, v \rangle \quad \text{for all } v \in V. \tag{3.5}$$

We can formally write $u = L^{-1} f$.

3.5 Definition. Let U and V be normed linear spaces. A linear mapping L is an *isomorphism* provided it is bijective and L and L^{-1} are continuous.

The importance of the following theorem for the finite element theory was pointed out by Babuška [1971]; see also Babuška and Aziz [1972]. It can be traced back to Nečas and Nirenberg; cf. Babuška [1971].

3.6 Theorem. *Let U and V be Hilbert spaces. Then a linear mapping $L : U \longrightarrow V'$ is an isomorphism if and only if the associated form $a : U \times V \longrightarrow \mathbb{R}$ satisfies the following conditions:*
(ii) (Continuity). There exists $C \geq 0$ such that

$$|a(u, v)| \leq C \|u\|_U \|v\|_V . \tag{3.6}$$

(ii) (Inf-sup condition). There exists $\alpha > 0$ such that

$$\sup_{v \in V} \frac{a(u, v)}{\|v\|_V} \geq \alpha \|u\|_U \quad \text{for all } u \in U. \tag{3.7}$$

(iii) For every $v \in V$, there exists $u \in U$ with

$$a(u, v) \neq 0. \tag{3.8}$$

Supplement. If we assume only (i) and (ii), then

$$L : U \longrightarrow \{v \in V; \ a(u, v) = 0 \ \text{for all } u \in U\}^0 \subset V' \tag{3.9}$$

is an isomorphism. Moreover, (3.7) is equivalent to

$$\|Lu\|_{V'} \geq \alpha \|u\|_U \quad \text{for all } u \in U. \tag{3.10}$$

The name for condition (3.7) comes from the equivalent formulation

$$\inf_{u \in U} \sup_{v \in V} \frac{a(u, v)}{\|u\|_U \|v\|_V} \geq \alpha > 0. \tag{3.7'}$$

Proof of Theorem 3.6. The equivalence of the continuity of $L : U \longrightarrow V'$ with (3.6) follows from a simple calculation.

From (3.7) we immediately deduce that L is injective. Suppose $Lu_1 = Lu_2$. Then by definition, $a(u_1, v) = a(u_2, v)$ for all $v \in V$. Thus, $\sup_v a(u_1 - u_2, v) = 0$, and (3.7) implies $u_1 - u_2 = 0$.

Given $f \in L(U)$, by the injectivity there exists a unique inverse $u = L^{-1} f$. We now apply (3.7) a second time:

$$\alpha \|u\|_U \leq \sup_{v \in V} \frac{a(u, v)}{\|v\|_V} = \sup_{v \in V} \frac{\langle f, v \rangle}{\|v\|_V} = \|f\|. \tag{3.11}$$

This is (3.10), and L^{-1} is continuous on the image of L.

The continuity of L and L^{-1} implies that $L(U)$ is closed. Now (3.9) follows from Theorem 3.4. This establishes the supplement to Theorem 3.6.

Finally, condition (iii) ensures the surjectivity of L. Indeed, by (3.9) $L(U)$ is the polar set of the null element, and so coincides with V'. Hence, the conditions (i), (ii), and (iii) are sufficient to ensure that $L : U \longrightarrow V'$ is an isomorphism.

In view of (3.11), the necessity of the conditions is immediate. $\square$

The Lax–Milgram theorem (for linear spaces) follows as a special case. Indeed, if a is a continuous V-elliptic bilinear form, then the inf-sup condition follows from

$$\sup_v \frac{a(u, v)}{\|v\|} \geq \frac{a(u, u)}{\|u\|} \geq \alpha \|u\|.$$

In particular, the differential operator in (II.2.5) can be regarded as a bijective mapping $L : H_0^1(\Omega) \longrightarrow H^{-1}(\Omega)$. The converse follows from Problem 3.8. [However, the assertion that for H^2-regular problems, $L : H^2(\Omega) \cap H_0^1(\Omega) \longrightarrow H^0(\Omega)$ is also an isomorphism cannot be obtained in this framework.]

In the proof of Theorem 3.6 we used the closedness of the image of L. At first glance this appears to be just a technicality which allows the application of Theorem 3.4. However, the counterexample II.2.7 and Problem II.2.14 (see also Remark 6.5) show how important this point is.

An Abstract Convergence Theorem

To solve equation (3.5) numerically we are led naturally to a Galerkin method. Let $U_h \subset U$ and $V_h \subset V$ be finite-dimensional spaces. Then given $f \in V'$, we seek $u_h \in U_h$ such that

$$a(u_h, v) = \langle f, v \rangle \quad \text{for all } v \in V_h. \tag{3.5$_h$}$$

In order to carry over Céa's lemma, we now require that the spaces U_h and V_h fit together.

3.7 Lemma. *Suppose the bilinear form $a : U \times V \longrightarrow \mathbb{R}$ satisfies the hypotheses of Theorem 3.6. Suppose the subspaces $U_h \subset U$ and $V_h \subset V$ are chosen so that (3.7′) and (3.8) also hold when U and V are replaced by U_h and V_h, respectively. Then*

$$\|u - u_h\| \leq (1 + \frac{C}{\alpha}) \inf_{w_h \in U_h} \|u - w_h\|.$$

Remark. We say that the subspaces U_h and V_h satisfy the *Babuška condition* or an *inf-sup condition* provided (3.7′) holds for U_h and V_h as stated in Lemma 3.7.

Proof of Lemma 3.7. By (3.5) and $(3.5)_h$,

$$a(u - u_h, v) = 0 \quad \text{for all } v \in V_h.$$

Let w_h be an arbitrary element in U_h. Then

$$a(u_h - w_h, v) = a(u - w_h, v) \quad \text{for all } v \in V_h.$$

For $\langle \ell, v \rangle := a(u - w_h, v)$, we have $\|\ell\| \leq C\|u - w_h\|$. By assumption, the mapping $L_h : U_h \longrightarrow V_h'$ generated by $a(u_h - w_h, \cdot)$ satisfies $\|(L_h)^{-1}\| \leq 1/\alpha$. Thus,

$$\|u_h - w_h\| \leq \alpha^{-1}\|\ell\| \leq \alpha^{-1}C\|u - w_h\|.$$

The assertion follows from the triangle inequality $\|u - u_h\| \leq \|u - w_h\| + \|u_h - w_h\|$. □

We mention that the theory described in this section has also recently been used to establish the convergence of difference methods and finite volume methods.

Proof of Theorem 3.4

For completeness we now prove Theorem 3.4.

It suffices to establish the identity

$$\overline{L(X)} = (\ker L')^0. \tag{3.12}$$

By the definition of the polar set and of the adjoint operator, we have

$$\begin{aligned}
(\ker L')^0 &= \{y \in Y; \ \langle y^*, y \rangle = 0 \quad \text{for } y^* \in \ker L'\} \\
&= \{y \in Y; \ \langle y^*, y \rangle = 0 \quad \text{for } y^* \in Y' \text{ with } \langle L'y^*, x \rangle = 0 \quad \text{for } x \in X\} \\
&= \{y \in Y; \ \langle y^*, y \rangle = 0 \quad \text{for } y^* \in Y' \text{ with } \langle y^*, Lx \rangle = 0 \quad \text{for } x \in X\}.
\end{aligned} \tag{3.13}$$

Hence, $L(X) \subset (\ker L')^0$. Since the polar set is the intersection of closed sets, it is itself closed, and consequently so is $\overline{L(X)} \subset (\ker L')^0$.

Suppose that there exists $y_0 \in (\ker L')^0$ with $y_0 \notin \overline{L(X)}$. Then the distance of y_0 from $L(X)$ is positive, and there exist a small open sphere centered at the point y_0 which is disjoint from the convex set $L(X)$. By the separation theorem for convex sets, there exist a functional $y^* \in Y'$ and a number a with

$$\langle y^*, y_0 \rangle > a,$$
$$\langle y^*, Lx \rangle \leq a \quad \text{for all } x \in X.$$

Since L is linear, this is possible only if $\langle y^*, Lx \rangle = 0$ for all $x \in X$. Thus, $a > 0$ and $\langle y^*, y_0 \rangle \neq 0$. This would contradict (3.13), and so (3.12) must hold. □

Problems

3.8 Let $a : V \times V \to \mathbb{R}$ be a positive symmetric bilinear form satisfying the hypotheses of Theorem 3.6. Show that a is elliptic, i.e., $a(v, v) \geq \alpha_1 \|v\|_V^2$ for some $\alpha_1 > 0$.

3.9 [Nitsche, private communication] Show the following converse of Lemma 3.7: Suppose that for every $f \in V'$, the solution of (3.5) satisfies

$$\lim_{h \to 0} u_h = u := L^{-1} f.$$

Then

$$\inf_h \inf_{u_h \in U_h} \sup_{v_h \in V_h} \frac{a(u_h, v_h)}{\|u_h\|_U \|v_h\|_V} > 0.$$

Hint: Use (3.10) and apply the principle of uniform boundedness.

3.10 Show that

$$\|v\|_0^2 \leq \|v\|_m \|v\|_{-m} \quad \text{for all } v \in H_0^m(\Omega),$$
$$\|v\|_1^2 \leq \|v\|_0 \|v\|_2 \quad \text{for all } v \in H^2(\Omega) \cap H_0^1(\Omega).$$

Hint: To prove the second relation, use the Helmholtz equation $-\Delta u + u = f$.

3.11 Let L be an H^1-elliptic differential operator. In which Sobolev spaces $H^s(\Omega)$ is the set

$$\{u \in H^1(\Omega); \; Lu = f \in L_2(\Omega), \; \|f\|_0 \leq 1\}$$

compact?

3.12 (Fredholm Alternative) Let H be a Hilbert space. Assume that the linear mapping $A : H \to H'$ can be decomposed in the form $A = A_0 + K$, where A_0 is H-elliptic, and K is compact. Show that either A satisfies the inf-sup condition, or there exists an element $x \in H$, $x \neq 0$, with $Ax = 0$.

§ 4. Saddle Point Problems

We now turn to variational problems with constraints. Let X and M be two Hilbert spaces, and suppose

$$a : X \times X \longrightarrow \mathbb{R}, \quad b : X \times M \longrightarrow \mathbb{R}$$

are continuous bilinear forms. Let $f \in X'$ and $g \in M'$. We denote both the dual pairing of X with X' and that of M with M' by $\langle \cdot, \cdot \rangle$. We consider the following minimum problem.

Problem (M). Find the minimum over X of

$$J(u) = \frac{1}{2}a(u, u) - \langle f, u \rangle \tag{4.1}$$

subject to the constraint

$$b(u, \mu) = \langle g, \mu \rangle \quad \text{for all } \mu \in M. \tag{4.2}$$

Saddle Points and Minima

Our starting point is the same as in the classical theory of Lagrange extremal problems. If $\lambda \in M$, then J and the *Lagrange function*

$$\mathcal{L}(u, \lambda) := J(u) + [b(u, \lambda) - \langle g, \lambda \rangle] \tag{4.3}$$

have the same values on the set of all points which satisfy the constraints. Instead of finding the minimum of J, we can seek a minimum of $\mathcal{L}(\cdot, \lambda)$ with fixed λ. This raises the question of whether $\lambda \in M$ can be selected so that the minimum of $\mathcal{L}(\cdot, \lambda)$ over the space X is assumed by an element which satisfies the given constraints. Since $\mathcal{L}(u, \lambda)$ contains only bilinear and quadratric expressions in u and λ, we are led to the following *saddle point problem*:

Problem (S). Find $(u, \lambda) \in X \times M$ with

$$\begin{aligned} a(u, v) + b(v, \lambda) &= \langle f, v \rangle \quad \text{for all } v \in X, \\ b(u, \mu) &= \langle g, \mu \rangle \quad \text{for all } \mu \in M. \end{aligned} \tag{4.4}$$

It is easy to see that every solution (u, λ) of Problem (S) must satisfy the *saddle point property*

$$\mathcal{L}(u, \mu) \leq \mathcal{L}(u, \lambda) \leq \mathcal{L}(v, \lambda) \quad \text{for all } (v, \mu) \in X \times M.$$

Here only the nonnegativity of $a(v, v) \geq 0$ is needed (cf. the Characterization Theorem II.2.2). The first component of a saddle point (u, λ) provides a solution of Problem (M).

The converse of this assertion is by no means obvious. Even if the minimum problem has a solution, we can ensure the existence of Lagrange multipliers only under additional hypotheses. We can see this already for a simple finite-dimensional example.

4.1 Example. Consider the following minimum problem in $\mathbb{R}^2$:

$$x^2 + y^2 \longrightarrow \text{min!}$$
$$x + y = 2.$$

Clearly, $x = y = 1$, $\lambda = -2$ provides a saddle point for the Lagrange function $\mathcal{L}(x, y, \lambda) = x^2 + y^2 + \lambda(x + y - 2)$, and $x = y = 1$ is a solution of Problem (M).

A formal doubling of the constraints clearly leads to a minimum problem with the same minimum:

$$x^2 + y^2 \longrightarrow \text{min!}$$
$$x + y = 2,$$
$$3x + 3y = 6.$$

However, the Lagrange multipliers for

$$\mathcal{L}(x, y, \lambda, \mu) = x^2 + y^2 + \lambda(x + y - 2) + \mu(3x + 3y - 6)$$

are no longer uniquely defined. Every combination with $\lambda + 3\mu = -2$ leads to a saddle point. Moreover, arbitrarily small perturbations of the data on the right-hand side can lead to a problem with no solution.

The inf-sup Condition

As we saw in Ch. II, §2, in infinite-dimensional spaces we have to correctly formulate the definiteness condition for the form a. The same holds for the constraints; it does not suffice to require their linear independence. An inf-sup condition provides the correct framework, similar to its appearance in Theorem 3.6. Equation (4.4) defines a linear mapping

$$L : X \times M \longrightarrow X' \times M'$$
$$(u, \lambda) \longmapsto (f, g). \tag{4.5}$$

To show that L is an isomorphism we need the inf-sup condition (3.7). Following Brezzi [1974], we express this in terms of properties of the forms a and b. We introduce special notation for the affine space of admissible elements and for the corresponding linear spaces:

$$
\begin{aligned}
V(g) &:= \{v \in X; \ b(v, \mu) = \langle g, \mu \rangle \quad \text{for all } \mu \in M\}, \\
V &:= \{v \in X; \ b(v, \mu) = 0 \qquad \text{for all } \mu \in M\}.
\end{aligned}
\tag{4.6}
$$

Since b is continuous, V is a closed subspace of X.

It is often easier to handle the saddle point equation (4.4) if we reformulate it as an operator equation. To this end, we associate the mapping

$$
A : X \longrightarrow X',
$$
$$
\langle Au, v \rangle = a(u, v) \quad \text{for all } v \in X,
$$

with the bilinear form a. Thus, the mapping A is defined by the action of the functional $Au \in X'$ on each $v \in X$. Similarly, we associate a mapping B and its adjoint mapping B' with the form b:

$$
B : X \longrightarrow M', \qquad\qquad\qquad B' : M \longrightarrow X',
$$
$$
\langle Bu, \mu \rangle = b(u, \mu) \quad \text{for all } \mu \in M, \qquad \langle B'\lambda, v \rangle = b(v, \lambda) \quad \text{for all } v \in X.
$$

Then (4.4) is equivalent to

$$
\begin{aligned}
Au + B'\lambda &= f, \\
Bu &= g.
\end{aligned}
\tag{4.7}
$$

4.2 Lemma. *The following assertions are equivalent:*
(i) There exists a constant $\beta > 0$ with

$$
\inf_{\mu \in M} \ \sup_{v \in X} \ \frac{b(v, \mu)}{\|v\| \|\mu\|} \geq \beta.
\tag{4.8}
$$

(ii) The operator $B : V^{\perp} \longrightarrow M'$ is an isomorphism, and

$$
\|Bv\| \geq \beta \|v\| \quad \text{for all } v \in V^{\perp}.
\tag{4.9}
$$

(iii) The operator $B' : M \longrightarrow V^0 \subset X'$ is an isomorphism, and

$$
\|B'\mu\| \geq \beta \|\mu\| \quad \text{for all } \mu \in M.
\tag{4.10}
$$

Proof. The equivalence of (i) and (iii) is just the assertion of the supplement to Theorem 3.6.

Suppose condition (iii) is satisfied. Then for given $v \in V^{\perp}$, we define a functional $g \in V^0$ by $w \longmapsto (v, w)$. Since B' is an isomorphism, there exists $\lambda \in M$ with

$$b(w, \lambda) = (v, w) \quad \text{for all } w. \tag{4.11}$$

By the definition of the functional g, we have $\|g\| = \|v\|$, and (4.10) implies $\|v\| = \|g\| = \|B'\lambda\| \geq \beta\|\lambda\|$. Now substituting $w = v$ in (4.11), we get

$$\sup_{\mu \in M} \frac{b(v, \mu)}{\|\mu\|} \geq \frac{b(v, \lambda)}{\|\lambda\|} = \frac{(v, v)}{\|\lambda\|} \geq \beta\|v\|.$$

Thus $B : V^{\perp} \longrightarrow M'$ satisfies the three conditions of Theorem 3.6, and the mapping is an isomorphism.

Suppose condition (ii) is satisfied, i.e., $B : V^{\perp} \longrightarrow M'$ is an isomorphism. For given $\mu \in M$, we determine the norm via duality:

$$\|\mu\| = \sup_{g \in M'} \frac{\langle g, \mu \rangle}{\|g\|} = \sup_{v \in V^{\perp}} \frac{\langle Bv, \mu \rangle}{\|Bv\|}$$
$$= \sup_{v \in V^{\perp}} \frac{b(v, \mu)}{\|Bv\|} \leq \sup_{v \in V^{\perp}} \frac{b(v, \mu)}{\beta\|v\|}.$$

But then condition (i) is satisfied, and everything is proved.   □

Another condition which is equivalent to the inf-sup condition can be found in Problem 4.18, where we also interpret the condition 4.2(ii) as a decomposition property.

After these preparations, we are ready for the main theorem for saddle point problems [Brezzi 1974]. The condition (ii) in the following theorem is often referred to as the *Brezzi condition*. – Recall that as in (4.6), the kernel of B is denoted by V.

4.3 Theorem. *(Brezzi's splitting theorem) For the saddle point problem (4.4), the mapping (4.5) defines an isomorphism $L : X \times M \longrightarrow X' \times M'$ if and only if the following two conditions are satisfied:*

(i) The bilinear form a is V-elliptic, i.e.,

$$a(v, v) \geq \alpha\|v\|^2 \quad \text{for all } v \in V,$$

where $\alpha > 0$, and V is as in (4.6).

(ii) The bilinear form b satisfies the inf-sup condition (4.8).

Proof. Suppose the conditions on a and b are satisfied. We first show that for every pair of functionals $(f, g) \in X' \times M'$, there is exactly one solution (u, λ) of the saddle point problem satisfying

$$
\begin{aligned}
\|u\| &\leq \alpha^{-1}\|f\| &&+ \beta^{-1}(1 + \frac{C}{\alpha})\|g\|, \\
\|\lambda\| &\leq \beta^{-1}(1 + \frac{C}{\alpha})\|f\| + \beta^{-1}(1 + \frac{C}{\alpha})\frac{C}{\beta}\|g\|.
\end{aligned}
\tag{4.12}
$$

$V(g)$ is not empty for $g \in M'$. Indeed, by Lemma 4.2(ii), there exists $u_0 \in V^{\perp}$ with

$$
Bu_0 = g.
$$

Moreover, $\|u_0\| \leq \beta^{-1}\|g\|$.

With $w := u - u_0$, (4.4) is equivalent to

$$
\begin{aligned}
a(w, v) + b(v, \lambda) &= \langle f, v \rangle - a(u_0, v) &&\text{for all } v \in X, \\
b(w, \mu) &= 0 &&\text{for all } \mu \in M.
\end{aligned}
\tag{4.13}
$$

By the V-ellipticity of a, the function

$$
\frac{1}{2} a(v, v) - \langle f, v \rangle + a(u_0, v)
$$

attains its minimum for some $w \in V$ with

$$
\|w\| \leq \alpha^{-1}(\|f\| + C\|u_0\|).
$$

In particular, the characterization theorem II.2.2 implies

$$
a(w, v) = \langle f, v \rangle - a(u_0, v) \quad \text{for all } v \in V.
\tag{4.14}
$$

The equations (4.13) will be satisfied if we can find $\lambda \in M$ such that

$$
b(v, \lambda) = \langle f, v \rangle - a(u_0 + w, v) \quad \text{for all } v \in X.
$$

The right-hand side defines a functional in X', which in view of (4.14) lies in V^0. By Lemma 4.2(iii), this functional can be represented as $B'\lambda$ with $\lambda \in M$, and

$$
\|\lambda\| \leq \beta^{-1}(\|f\| + C\|u\|).
$$

This establishes the solvability. Since $\|w\|$ can be estimated in the usual way via (4.14), we get the inequalities (4.12) by inserting the bounds on $\|u_0\|$ and $\|\lambda\|$ and using $\|u\| \leq \|u_0\| + \|w\|$.

The solution is unique, as can be seen from the homogeneous equation. If we insert $f = 0$, $g = 0$, $v = u$, $\mu = -\lambda$ in (4.4) and add, we get $a(u, u) = 0$. Since $u \in V$, the V-ellipticity implies $u = 0$. Moreover,

$$\sup_v |b(v, \lambda)| = 0,$$

and $\lambda = 0$ follows from (4.8). Thus, L is injective and surjective, and (4.12) asserts that L^{-1} is continuous.

Conversely, suppose that L is an isomorphism. In particular, suppose $\|L^{-1}\| \leq C$. By the Hahn–Banach theorem, every functional $f \in V'$ has an extension $\tilde{f} : X \longrightarrow \mathbb{R}$ with $\|\tilde{f}\| = \|f\|$. Set $(u, \lambda) = L^{-1}(\tilde{f}, 0)$. Then u is a minimum of $\frac{1}{2} a(v, v) - \langle f, v \rangle$ in V. The mapping $f \longmapsto u \in V$ is bounded, and thus a is V-elliptic.

Finally, for every $g \in M'$, we associate $u \in X$ with $\|u\| \leq c\|g\|$ via $(u, \lambda) = L^{-1}(0, g)$. Given $u \in X$, let $u^{\perp} \in V^{\perp}$ be the projection. Since $\|u^{\perp}\| \leq \|u\|$, the mapping $g \longmapsto u \longmapsto u^{\perp}$ is bounded, and $Bu^{\perp} = g$. Hence, $B : V^{\perp} \longrightarrow M'$ is an isomorphism, and by Lemma 4.2(ii), b satisfies the inf-sup condition. $\square$

We note that coercivity of a was assumed only on the kernel of B and not on the entire space X. We will need this weak assumption in most applications. An exception will be the Stokes problem. Here coercivity is not restricted to the divergence-free functions. — Note that the norm of the operator B does not enter into the a priori estimate (4.12).

Mixed Finite Element Methods

We now discuss a natural approach to the numerical solution of saddle point problems: Choose finite-dimensional subspaces $X_h \subset X$ and $M_h \subset M$, and solve

Problem (S_h). Find $(u_h, \lambda_h) \in X_h \times M_h$ such that

$$\begin{aligned} a(u_h, v) + b(v, \lambda_h) &= \langle f, v \rangle \quad \text{for all } v \in X_h, \\ b(u_h, \mu) \qquad\quad &= \langle g, \mu \rangle \quad \text{for all } \mu \in M_h. \end{aligned} \tag{4.15}$$

This approach is called a mixed method. In view of Lemma 3.7, we need to choose finite element spaces which satisfy requirements similar to those on X and M in Theorem 4.3, see Brezzi [1974] and Fortin [1977]. This is not always easy to do in practice. For fluid mechanics, the coercivity is trivial, and only the inf-sup condition is critical. For problems in elasticity theory, however, making finite element spaces satisfy both conditions can often be difficult, and requires that the finite element spaces X_h and M_h fit together. Practical experience shows that enforcing these conditions is of the utmost importance for the stability of the finite element computation.

It is useful to introduce the following notation which is analogous to (4.6):

$$V_h := \{v \in X_h; \ b(v, \mu) = 0 \quad \text{for all } \mu \in M_h\}.$$

4.4 Definition. A family of finite element spaces X_h, M_h is said to satisfy the *Babuška–Brezzi condition* provided there exist constants $\alpha > 0$ and $\beta > 0$ independent of h such that

(i) The bilinear form a is V_h-elliptic with ellipticity constant $\alpha > 0$.

(ii)

$$\sup_{v \in X_h} \frac{b(v, \lambda_h)}{\|v\|} \geq \beta \|\lambda_h\| \quad \text{for all } \lambda_h \in M_h. \tag{4.16}$$

The terminology in the literature varies. Often the condition (ii) alone is called the *Brezzi condition*, the *Ladyshenskaja–Babuška–Brezzi condition*, or for short the *LBB condition*. This condition is the more important of the two, and we will usually call it the *inf-sup condition*.

It is clear that – possibly after a reduction in α and β – we can take the same constants in 4.3 and 4.4.

The following result is an immediate consequence of Lemma 3.7 and Theorem 4.3.

4.5 Theorem. *Suppose the hypotheses of Theorem 4.3 hold, and suppose X_h, M_h satisfy the Babuška–Brezzi conditions. Then*

$$\|u - u_h\| + \|\lambda - \lambda_h\| \leq c\Big\{ \inf_{v_h \in X_h} \|u - v_h\| + \inf_{\mu_h \in M_h} \|\lambda - \mu_h\| \Big\}. \tag{4.17}$$

In general, $V_h \not\subset V$. We get a better result in the special case of conforming approximation where $V_h \subset V$. We note that in this case also the finite element approximation of $V(g)$ may be nonconforming for $g \neq 0$.

4.6 Definition. The spaces $X_h \subset X$ and $M_h \subset M$ satisfy condition (C) provided $V_h \subset V$, i.e. if for every $v_h \in X_h$, $b(v_h, \mu_h) = 0$ for all $\mu_h \in M_h$ implies $b(v_h, \mu) = 0$ for all $\mu \in M$.

4.7 Theorem. *Suppose the hypotheses of Theorem 4.5 are satisfied along with the condition (C). Then the solution of Problem (S_h) satisfies*

$$\|u - u_h\| \leq c \inf_{v_h \in X_h} \|u - v_h\|.$$

Proof. Let $v_h \in V_h(g)$. Then in the usual way, we have

$$a(u_h - v_h, v) = a(u_h, v) - a(u, v) + a(u - v_h, v)$$
$$= b(v, \lambda - \lambda_h) + a(u - v_h, v)$$
$$\leq C\|u - v_h\| \cdot \|v\|$$

for all $v \in V_h$, since $b(v, \lambda - \lambda_h)$ vanishes because of condition (C). With $v :=$ $u_h - v_h$, we have $\|u_h - v_h\|^2 \leq \alpha^{-1} C \|u_h - v_h\| \cdot \|u - v_h\|$, and the assertion follows after dividing by $\|u_h - v_h\|$. $\qquad\qquad\qquad\qquad\qquad\qquad$ $\square$

For completeness we mention that the assumption $X_h \subset X$ may be abandoned. Also mesh-dependent norms may be used. In these cases the theory above has to be combined with arguments that we encountered with Strang's lemmas.

Fortin Interpolation

We continue with our treatment of abstract saddle point problems with a tool due to Fortin [1977] which is useful for verifying that the inf-sup condition holds.

4.8 Fortin's Criterion. *Suppose that the bilinear form $b : X \times M \longrightarrow \mathbb{R}$ satisfies the inf-sup condition. In addition, suppose that for the subspaces X_h, M_h, there exists a bounded linear projector $\Pi_h : X \longrightarrow X_h$ such that*

$$b(v - \Pi_h v, \mu_h) = 0 \quad \text{for } \mu_h \in M_h. \tag{4.18}$$

If $\|\Pi_h\| \leq c$ for some constant c independent of h, then the finite element spaces X_h, M_h satisfy the inf-sup condition.

Proof. By assumption, for $\mu_h \in M_h$

$$\beta \|\mu_h\| \leq \sup_{v \in X} \frac{b(v, \mu_h)}{\|v\|} = \sup_{v \in X} \frac{b(\Pi_h v, \mu_h)}{\|v\|} \leq c \sup_{v \in X} \frac{b(\Pi_h v, \mu_h)}{\|\Pi_h v\|}$$
$$= c \sup_{v_h \in X_h} \frac{b(v_h, \mu_h)}{\|v_h\|},$$

since $\Pi_h v \in X_h$. $\qquad\qquad\qquad\qquad\qquad\qquad\qquad\qquad\qquad\qquad\qquad$ $\square$

Note that the condition in Fortin's criterion can be checked without referring explicitly to the norm of the Lagrangian multipliers. This is an advantage when the space of the Lagrangian multipliers is equipped with an exotic norm, and is for example thus used when the Lagrangian multipliers belong to trace spaces.

4.9 Remark. There is a converse statement to Fortin's criterion. *If the finite element spaces X_h, M_h satisfy the inf-sup condition, then there exists a bounded linear projector $\Pi_h : X \to X_h$ such that (4.18) holds.*

Indeed, given $v \in X$, define $u_h \in X_h$ as the solution of the equations

$$\begin{aligned} (u_h, w) + b(w, \lambda_h) &= (v, w) \quad \text{for all } w \in X_h, \\ b(u_h, \mu) &= b(v, \mu) \quad \text{for all } \mu \in M_h. \end{aligned} \tag{4.19}$$

Since the inner product in X is coercive by definition, the problem is stable, and from Theorem 4.3 it follows that

$$\|u_h\| \leq c\|v\|.$$

Moreover, a linear mapping is defined by $v \longmapsto \Pi v := u_h$, and the proof is complete.   □

The linear process defined above is called *Fortin interpolation*.

As a corollary we obtain a relationship between the approximation with the constraint induced by the bilinear form b and the approximation in the larger finite element space X_h.

4.10 Remark. If the spaces X_h and M_h satisfy the inf-sup condition, then there exists a constant c independent of h such that for every $u \in V(g)$,

$$\inf_{v_h \in V_h(g)} \|u - v_h\| \leq c \inf_{w_h \in X_h} \|u - w_h\|.$$

Proof. We make use of Fortin interpolation. Obviously, $\Pi_h w_h = w_h$ for each $w_h \in X_h$. Given $u \in V(g)$ we have $\Pi_h u \in V_h(g)$ and

$$\|u - \Pi_h u\| = \|u - w_h - \Pi_h(u - w_h)\| \leq (1 + c)\|u - w_h\|.$$

Since this holds for all $w_h \in X_h$, the proof is complete.   □

Sometimes error estimates are wanted for some norms for which not all hypotheses of Theorem 4.3 hold. In this context we note that the norm of the bilinear form b does not enter into the a priori estimate (4.12). This fact is used for the estimate of $\|\lambda - \lambda_h\|$ when an estimate of $\|u - u_h\|$ has been established by applying other tools.

Saddle Point Problems with Penalty Term

To conclude this section, we consider a variant of Problem (S) which plays a role in elasticity theory. We want to treat so-called *problems with a small parameter t* in such a way that we get convergence as $h \to 0$ which is uniform in the parameter t. This can often be achieved by formulating a saddle point problem with penalty term. – Readers who are primarily interested in the Stokes problem may want to skip the rest of this section.

Suppose that in addition to the bilinear forms a and b,

$$c : M_c \times M_c \longrightarrow \mathbb{R}, \quad c(\mu, \mu) \geq 0 \quad \text{for all } \mu \in M_c \qquad (4.20)$$

is a bilinear form on a dense set $M_c \subset M$. Moreover, let t be a small real-valued parameter. Now we modify (4.4) by adding a *penalty term*:

Problem (S_t). Find $(u, \lambda) \in X \times M_c$ with

$$
\begin{aligned}
a(u, v) + b(v, \lambda) &= \langle f, v \rangle \quad \text{for all } v \in X, \\
b(u, \mu) - t^2 c(\lambda, \mu) &= \langle g, \mu \rangle \quad \text{for all } \mu \in M_c.
\end{aligned}
\tag{4.21}
$$

The associated bilinear form on the product space is

$$
A(u, \lambda; v, \mu) := a(u, v) + b(v, \lambda) + b(u, \mu) - t^2 c(\lambda, \mu).
$$

First we consider the case where c is bounded [Braess and Blömer 1990]. Then c can be extended continuously to the entire space $M \times M$, and we can suppose $M_c = M$.

4.11 Theorem. *Suppose that the hypotheses of Theorem 4.3 are satisfied. In addition, let $c : M \times M \longrightarrow \mathbb{R}$ be a continuous bilinear form with $c(\mu, \mu) \geq 0$ for all $\mu \in M$. Then (4.21) defines an isomorphism $L : X \times M \longrightarrow X' \times M'$, and L^{-1} is uniformly bounded for $0 \leq t \leq 1$.*

In Theorem 4.11 it is essential that the solution of the saddle point problem with penalty term is uniformly bounded in t for all $0 \leq t \leq 1$. We can think of the penalty term as a perturbation. It is often supposed to have a stabilizing effect. Surprisingly, this is not always true, and the norm of the form c enters into the constant in the inf-sup condition. The following example shows that this is not just an artifact of the proof, which we present at the end of this section.

4.12 Example. Let $X = M := L_2(\Omega)$, $a(u, v) := 0$, $b(v, \mu) := (v, \mu)_{0,\Omega}$, and $c(\lambda, \mu) := K \cdot (\lambda, \mu)_{0,\Omega}$. Clearly, the solution of

$$
\begin{aligned}
b(v, \lambda) &= (f, v)_{0,\Omega}, \\
b(u, \mu) - t^2 c(\lambda, \mu) &= (g, \mu)_{0,\Omega}
\end{aligned}
$$

is $\lambda = f$ and $u = g + t^2 K f$. Thus, the solution grows as $K \to \infty$ and we cannot expect a bounded solution for an unbounded bilinear form c.

In plate theory we frequently encounter saddle point problems with penalty terms which represent *singular perturbations*, i.e., which stem from a differential operator of higher order. Then we introduce a semi-norm on M_c, and define the corresponding norm

$$
\begin{aligned}
|\mu|_c &:= \sqrt{c(\mu, \mu)}, \\
\||(v, \mu)\|| &:= \|v\|_X + \|\mu\|_M + t|\mu|_c,
\end{aligned}
\tag{4.22}
$$

on $X \times M_c$; see Huang [1990]. On the other hand, this now requires the ellipticity of a on the entire space X, rather than just on the kernel V as in Theorem 4.3. It is clear from the previous example that we indeed need some additional assumption of this kind.

4.13 Theorem. *Suppose the hypotheses of Theorem 4.3 are satisfied and that a is elliptic on X. Then the mapping L defined by the saddle point problem with penalty term satisfies the inf-sup condition*

$$\inf_{(u,\lambda)\in X\times M_c} \sup_{(v,\mu)\in X\times M_c} \frac{A(u,\lambda;v,\mu)}{|||(u,\lambda)|||\cdot|||(v,\mu)|||} \geq \gamma > 0, \qquad (4.23)$$

for all $0 \leq t \leq 1$, where γ is independent of t.

These two theorems are consequences of the following lemma [Kirmse 1990] whose hypotheses appear to be very technical at first glance. However, by Problem 4.23, the condition (4.25) below is equivalent to the Babuška condition for the X-components,

$$\sup_{(v,\mu)} \frac{A(u,0;v,\mu)}{|||(v,\mu)|||} \geq \alpha'\|u\|_X, \qquad (4.24)$$

with suitable α'. In particular, it is therefore also necessary for stability.

4.14 Lemma. *Suppose that the hypotheses of Theorem 4.3 are satisfied, and suppose that*

$$\frac{a(u,u)}{\|u\|_X} + \sup_{\mu\in M_c} \frac{b(u,\mu)}{\|\mu\|_M + t|\mu|_c} \geq \alpha\|u\|_X \qquad (4.25)$$

for all $u \in X$ and some $\alpha > 0$. Then the inf-sup condition (4.23) holds.

Proof. We consider three cases.

Case 1. Let $\|u\|_X + \|\lambda\|_M \leq \delta^{-1}t|\lambda|_c$, where $\delta > 0$ will be chosen later. Then

$$A(u,\lambda;u,-\lambda) = a(u,u) + t^2 c(\lambda,\lambda)$$
$$\geq \frac{1}{2}t^2|\lambda|_c^2 + \frac{1}{2}t^2|\lambda|_c^2$$
$$\geq \frac{1}{2}\delta^2\{(\|u\|_X + \|\lambda\|_M)^2 + t^2|\lambda|_c^2\} \geq \frac{1}{4}\delta^2|||(u,\lambda)|||^2.$$

Dividing through by $|||(u,\lambda)|||$, we have

$$\frac{1}{4}\delta^2|||(u,\lambda)||| \leq \frac{A(u,\lambda;u,-\lambda)}{|||(u,\lambda)|||} \leq \sup_{(v,\mu)} \frac{A(u,\lambda;v,\mu)}{|||(v,\mu)|||}.$$

Case 2. Let $\|u\|_X + \|\lambda\|_M > \delta^{-1}t|\lambda|_c$ and $\|u\|_X \leq \frac{\beta}{2\|a\|}\|\lambda\|_M$. By the inf-sup condition (4.8),

$$\beta\|\lambda\|_M \leq \sup_v \frac{b(v,\lambda)}{\|v\|_X} = \sup_v \frac{A(u,\lambda;v,0) - a(u,v)}{\|v\|_X}$$
$$\leq \sup_{(v,\mu)} \frac{A(u,\lambda;v,\mu)}{|||(v,\mu)|||} + \|a\|\,\|u\|_X$$
$$\leq \sup_{(v,\mu)} \frac{A(u,\lambda;v,\mu)}{|||(v,\mu)|||} + \frac{1}{2}\beta\|\lambda\|_M.$$

Now we can estimate $\|\lambda\|_M$, and in view of the case distinction $\|u\|_X$ and $t|\lambda|_c$ as well, by the first term on the right-hand side.

Case 3. Let $\|u\|_X + \|\lambda\|_M > \delta^{-1}t|\lambda|_c$ and $\|u\|_X \geq \frac{\beta}{2\|a\|}\|\lambda\|_M$. Then $\delta\|\|(u,\lambda)\|\| \leq \|u\|_X$, where δ depends only on α, β and δ. By hypothesis (4.25),

$$\alpha\delta\|\|(u,\lambda)\|\| \leq \alpha\|u\|_X$$

$$\leq \frac{a(u,u)}{\|u\|_X} + \sup_\mu \frac{A(u,\lambda;0,\mu) + t^2c(\lambda,\mu)}{\|\mu\|_M + t|\mu|_c}$$

$$\leq \frac{A(u,\lambda;u,-\lambda)}{\|u\|_X} + \sup_\mu \frac{A(u,\lambda;0,\mu)}{\|\|(0,\mu)\|\|} + t|\lambda|_c$$

$$\leq (\frac{1}{\delta}+1)\sup_{(v,\mu)} \frac{A(u,\lambda;v,\mu)}{\|\|(v,\mu)\|\|} + t|\lambda|_c.$$

With $\delta \leq \frac{\alpha\beta}{4\|a\|+2\beta}$ we have $t|\lambda|_c \leq \frac{1}{2}\alpha\|u\|_X$, and the second term in the sum can be absorbed by a factor of 2.

This establishes the assertion in all cases. $\square$

The previous two theorems now follow immediately. The ellipticity on the entire space X in Theorem 4.13 implies $a(u,u) \geq \alpha\|u\|_X^2$, and (4.25) is clear. – On the other hand, Theorem 4.3 ensures that the Babuška condition holds for the pair $(u,0)$, and combining it with the Cauchy–Schwarz inequality gives

$$\gamma\|u\|_X \leq \sup_{(v,\mu)} \frac{a(u,v) + b(u,\mu)}{\|v\|_X + \|\mu\|_M}$$

$$\leq \sup_v \frac{a(u,v)}{\|v\|_X} + \sup_\mu \frac{b(u,\mu)}{\|\mu\|_M}$$

$$\leq [\|a\|\,a(u,u)]^{1/2} + (1 + \|c\|)\sup_\mu \frac{b(u,\mu)}{\|\mu\|_M + |\mu|_c} \qquad (4.26)$$

$$\leq \frac{\|a\|\,a(u,u)}{\|u\|_X} + 2(1 + \|c\|)\sup_\mu \frac{b(u,\mu)}{\|\mu\|_M + |\mu|_c}.$$

Here we have used the fact that the form c in Theorem 4.11 was assumed to be bounded, and have applied the same kind of argument as used in Problem 4.22. $\square$

The uniform boundedness of the solution implies that the solution is a continuous function of the parameter.

4.15 Corollary. *Let the conditions of Theorem 4.11 prevail. Then, given $f \in X'$ and $g \in M'$, the solution (u,λ) of Problem (S_t) depends continuously on t.*

Proof. Let (u_t,λ_t) and (u_s,λ_s) be the solutions for the parameters t and s respectively. Then we have

$$a(u_t - u_s, v) + b(v, \lambda_t - \lambda_s) = 0 \qquad\qquad \text{for all } v \in X,$$

$$b(u_t - u_s, \mu) - t^2c(\lambda_t - \lambda_s, \mu) = -(t^2 - s^2)c(\lambda_s, \mu) \text{ for all } \mu \in M.$$

The stability now implies $\|u_t - u_s\|_X + \|\lambda_t - \lambda_s\|_M \le$ const $|t^2 - s^2|$, and we have continuity in the parameter. $\qquad\qquad\qquad\qquad\qquad\qquad\qquad\qquad\qquad\qquad$ □

Problems

4.16 Show that the inf-sup condition (4.8) is equivalent to the following decomposition property: For every $u \in X$ there exists a decomposition

$$u = v + w$$

with $v \in V$ and $w \in V^\perp$ such that

$$\|w\|_X \le \beta^{-1}\|Bu\|_{M'},$$

where $\beta > 0$ is a constant independent of u.

4.17 Let X, M, and the maps a, b, f, g be as in the saddle point problem (S). Given $\rho \in M$, let $\rho^\perp := \{\mu \in M;\ (\rho, \mu) = 0\}$. We now minimize the expression (4.1) subject to the restricted set of constraints

$$b(u, \mu) = \langle g, \mu \rangle \quad \text{for all } \mu \in \rho^\perp.$$

Show that the solution is characterized by

$$
\begin{aligned}
a(u, v) + b(v, \lambda) \qquad\quad &= \langle f, v \rangle && \text{for all } v \in X, \\
b(u, \mu) \qquad\quad + (\sigma, \mu) &= \langle g, \mu \rangle && \text{for all } \mu \in M, \\
(\tau, \lambda) \qquad\quad &= 0 && \text{for all } \tau \in \text{span}[\rho]
\end{aligned}
$$

with $u \in X$, $\lambda \in M$, $\sigma \in \text{span}[\rho]$.

4.18 Suppose that the subspaces X_h, M_h satisfy the Babuška–Brezzi condition, and suppose we

$$\text{increase or decrease } X_h \text{ or } M_h.$$

Which of the conditions in Definition 4.4 have to be rechecked?

4.19 When $M = L_2$, we can identify M with its dual space, and simply write $b(v, \mu) = (Bv, \mu)_0$. The solution of the saddle point problem does not change if $a(u, v)$ is replaced by

$$a_t(u, v) := a(u, v) + t^{-2}(Bu, Bv)_0, \quad t > 0.$$

This is called the method of the *augmented Lagrange function*; see Fortin and Glowinski [1983].

(a) Show that a_t is elliptic on the entire space X under the hypotheses of Theorem 4.3.

(b) Suppose we ignore the explicit constraints, and introduce $\lambda = t^2 Bu$ as a new variable. Show that this leads to a saddle point problem with penalty term.

4.20 As e.g. in (4.19) and (4.21), a saddle point problem is often stable for two pairings X_1, M_1 and X_2, M_2. Now suppose $X_1 \subset X_2$ and

$$\|v\|_{X_1} \geq \|v\|_{X_2} \quad \text{on } X_1.$$

Show that, conversely,

$$\|\lambda\|_{M_1} \leq c \, \|\lambda\|_{M_2} \quad \text{on } M_1 \cap M_2,$$

where $c \geq 0$. – If M_1 is also dense in M_2, then $M_1 \supset M_2$.

4.21 The pure Neumann Problem (II.3.8)

$$-\Delta u = f \quad \text{in } \Omega,$$
$$\frac{\partial u}{\partial v} = g \quad \text{on } \partial\Omega$$

is only solvable if $\int_\Omega f \, dx + \int_\Gamma g \, ds = 0$. This compatibility condition follows by applying Gauss' integral theorem to the vector field ∇u. Since u+const is a solution whenever u is, we can enforce the constraint

$$\int_\Omega u \, dx = 0.$$

Formulate the associated saddle point problem, and use the trace theorem and the second Poincaré inequality to show that the hypotheses of Theorem 4.3 are satisfied.

4.22 Let a, b, and c be positive numbers. Show that $a \leq b + c$ implies $a \leq b^2/a + 2c$.

4.23 Show the equivalence of the conditions (4.24) and (4.25). For the nontrivial direction, use the same argument as in the derivation of (4.26); cf. Braess [1996].

4.24 Let u be a (classical) solution of the biharmonic equation

$$\Delta^2 u = f \quad \text{in } \Omega,$$
$$u = \frac{\partial u}{\partial v} = 0 \quad \text{on } \partial\Omega.$$

Show that $u \in H_0^1$ together with $w \in H^1$ is a solution of the saddle point problem

$$(w, \eta)_{0,\Omega} + (\nabla\eta, \nabla u)_{0,\Omega} = 0 \qquad \text{for all } \eta \in H^1,$$
$$(\nabla w, \nabla v)_{0,\Omega} \qquad\quad = (f, v)_{0,\Omega} \quad \text{for all } v \in H_0^1.$$

Suitable elements and analytic methods can be found e.g. in Ciarlet [1978] and in Babuška, Osborn, and Pitkäranta [1980].

§ 5. Mixed Methods for the Poisson Equation

The treatment of the Poisson equation by mixed methods already elucidates some characteristic features. For example, there are two different pairs of spaces for which the saddle point problem is stable in the sense of Babuška and Brezzi. It is interesting that different boundary conditions turn out to be natural conditions in the two cases.

The method often called the *dual mixed method*, is well established for a long time. On the other hand, the *primal mixed method* has recently attracted a lot of interest since it shows that mixed methods are often related to a softening of the energy functional and how elasticity problems with a small parameter can be treated in a robust way.

Moreover there are special results if X or M coincides with an L_2-space.

The Poisson Equation as a Mixed Problem

The Laplace equation or the Poisson equation $\Delta u = \operatorname{div} \operatorname{grad} u = -f$ can be written formally as the system

$$
\begin{aligned}
\operatorname{grad} u &= \sigma, \\
\operatorname{div} \sigma &= -f.
\end{aligned}
\tag{5.1}
$$

Let $\Omega \subset \mathbb{R}^d$. Then (5.1) leads directly to the following saddle point problem: Find $(\sigma, u) \in L_2(\Omega)^d \times H_0^1(\Omega)$ such that

$$
\begin{aligned}
(\sigma, \tau)_{0,\Omega} - (\tau, \nabla u)_{0,\Omega} &= 0 && \text{for all } \tau \in L_2(\Omega)^d, \\
-(\sigma, \nabla v)_{0,\Omega} &= -(f, v)_{0,\Omega} && \text{for all } v \in H_0^1(\Omega).
\end{aligned}
\tag{5.2}
$$

These equations can be treated in the general framework of saddle point problems with

$$
\begin{aligned}
X &:= L_2(\Omega)^d, \quad M := H_0^1(\Omega), \\
a(\sigma, \tau) &:= (\sigma, \tau)_{0,\Omega}, \quad b(\tau, v) := -(\tau, \nabla v)_{0,\Omega}.
\end{aligned}
\tag{5.3}
$$

The linear forms are continuous, and a is obviously L_2-elliptic. To check the inf-sup condition, we use Friedrichs' inequality in a similar way as for the original minimum problem in Ch. II, §2. Given $v \in H_0^1(\Omega)$, consider the quotient appearing in the condition for $\tau := -\nabla v \in L_2(\Omega)^d$:

$$
\frac{b(\tau, v)}{\|\tau\|_0} = \frac{-(\tau, \nabla v)_{0,\Omega}}{\|\tau\|_0} = \frac{(\nabla v, \nabla v)_{0,\Omega}}{\|\nabla v\|_0} = |v|_1 \geq \frac{1}{c}\|v\|_1.
$$

Since c comes from Friedrichs' inequality and depends only on Ω, the saddle point problem (5.2) is stable.

It is easy to construct suitable finite elements for a triangulation $\mathcal{T}_h$. Choose $k \geq 1$, and set

$$X_h := (\mathcal{M}^{k-1})^d = \{\sigma_h \in L_2(\Omega)^d; \ \sigma_h|_T \in \mathcal{P}_{k-1} \text{ for } T \in \mathcal{T}_h\},$$
$$M_h := \mathcal{M}_{0,0}^k \ \ = \{v_h \in H_0^1(\Omega); \ v_h|_T \in \mathcal{P}_k \ \ \text{ for } T \in \mathcal{T}_h\}.$$

Note that only the functions in M_h are continuous. Since $\nabla M_h \subset X_h$, we can verify the inf-sup condition in the same way as for the continuous problem.

The saddle point problem with a different pairing is more important for practical computations. It refers to the space encountered in Problem II.5.14:

$$H(\text{div}, \Omega) := \{\tau \in L_2(\Omega)^d; \ \text{div} \, \tau \in L_2(\Omega)\}$$

with the graph norm of the divergence operator,

$$\|\tau\|_{H(\text{div},\Omega)} := (\|\tau\|_0^2 + \|\,\text{div}\,\tau\|_0^2)^{1/2}. \tag{5.4}$$

We seek $(\sigma, u) \in H(\text{div}, \Omega) \times L_2(\Omega)$ such that

$$
\begin{aligned}
(\sigma, \tau)_{0,\Omega} + (\text{div} \, \tau, u)_{0,\Omega} &= 0 && \text{for all } \tau \in H(\text{div}, \Omega), \\
(\text{div} \, \sigma, v)_{0,\Omega} &= -(f, v)_{0,\Omega} && \text{for all } v \in L_2(\Omega).
\end{aligned}
\tag{5.5}
$$

To apply the general theory, we set

$$X := H(\text{div}, \Omega), \qquad M := L_2(\Omega),$$
$$a(\sigma, \tau) := (\sigma, \tau)_{0,\Omega}, \quad b(\tau, v) := (\text{div} \, \tau, v)_{0,\Omega}.$$

Clearly, the linear forms are continuous. Then since $\text{div} \, \tau = 0$ for τ in the kernel V, we have

$$a(\tau, \tau) = \|\tau\|_0^2 = \|\tau\|_0^2 + \|\,\text{div}\,\tau\|_0^2 = \|\tau\|_{H(\text{div},\Omega)}^2.$$

This establishes the ellipticity of a on the kernel. Moreover, for given $v \in L_2$ there exists $w \in C_0^\infty(\Omega)$ with $\|v - w\|_{0,\Omega} \leq \frac{1}{2}\|v\|_{0,\Omega}$. Set $\xi := \inf\{x_1; \ x \in \Omega\}$ and

$$\tau_1(x) = \int_\xi^{x_1} w(t, x_2, \ldots, x_n) dt,$$
$$\tau_i(x) = 0 \quad \text{for } i > 1.$$

Then obviously $\text{div} \, \tau = \partial \tau_1 / \partial x_1 = w$, and the same argument as in the proof of Friedrichs' inequality gives $\|\tau\|_0 \leq c\|w\|_0$. Hence,

$$\frac{b(\tau, v)}{\|\tau\|_{H(\text{div},\Omega)}} \geq \frac{(w, v)_{0,\Omega}}{(1+c)\|w\|_{0,\Omega}} \geq \frac{1}{2(1+c)}\|v\|,$$

and so the inf-sup condition is satisfied.

By Theorem 4.3, (5.5) defines a stable saddle point problem. At first glance, it appears that a solution exists only in $u \in L_2$. However, $u \in H_0^1(\Omega)$, and since $C_0^\infty(\Omega)^d \subset H(\text{div}, \Omega)$, the first equation of (5.5) says in particular that

$$\int_\Omega u \frac{\partial \tau_i}{\partial x_i} \, dx = -\int_\Omega \sigma_i \tau_i dx \quad \text{for } \tau_i \in C_0^\infty(\Omega).$$

Thus, in view of Definition II.1.1, u possesses a weak derivative $\frac{\partial u}{\partial x_i} = \sigma_i$, and hence $u \in H^1(\Omega)$. Now (5.5) together with Green's formula (II.2.9) and $\nabla u = \sigma$ implies

$$\int_{\partial\Omega} u \cdot \tau v \, ds = \int_\Omega \nabla u \cdot \tau \, dx + \int_\Omega \text{div } \tau u \, dx$$
$$= \int_\Omega \sigma \cdot \tau \, dx + \int_\Omega \text{div } \tau u \, dx = 0. \tag{5.7}$$

Since this holds for all $\tau \in C^\infty(\Omega)^d$, we have $u = 0$ on the boundary in the generalized sense, i.e., in fact $u \in H_0^1(\Omega)$.

In the standard case the natural boundary condition is $\frac{\partial u}{\partial v} = 0$, but here the natural boundary condition is $u = 0$.

We note that the equations (5.2) characterize the solution of the variational problem

$$\frac{1}{2}(\sigma, \sigma)_0 - (f, u) \to \text{min!}$$
$$\nabla u - \sigma = 0$$

Here the Lagrange multiplier coincides with σ, and can be eliminated from the equations. — On the other hand, (5.5) arises from the variational problem

$$\frac{1}{2}(\sigma, \sigma)_0 \to \text{min!}$$
$$\text{div } \sigma + f = 0.$$

Here the Lagrange multiplier coincides with u from (5.1).

Sometimes (5.2) with $X := L_2$ and $M := H_0^1$ is called a *primal* mixed method while (5.5) with $X := H(\text{div})$ and $M := L_2$ is called a *dual* mixed method.

The dual mixed variational problem is related to an error estimate.

5.1 Theorem of Prager and Synge. *Let $\sigma \in H(\text{div})$, $v \in H_0^1(\Omega)$, and assume that $\text{div } \sigma + f = 0$. Furtheremore, let u be the solution of the Poisson equation $\Delta u = -f$ with zero boundary conditions. Then,*

$$|u - v|_1^2 + \| \text{grad } u - \sigma \|_0^2 = \| \text{grad } v - \sigma \|_0^2.$$

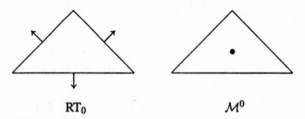

$$\text{RT}_0 \hspace{5cm} \mathcal{M}^0$$

Fig. 35. Raviart–Thomas element for $k = 0$: One normal component is prescribed on each edge.

Proof. By applying Green's formula we obtain

$$\int_\Omega \text{grad}(u - v)(\text{grad}\, u - \sigma)dx = -\int_\Omega (u - v)(\Delta u - \text{div}\,\sigma)dx = 0.$$

From this orthogonality relation we conclude that

$$\| \text{grad}\, v - \sigma \|_0^2 = \| \text{grad}(v - u) \|_0^2 + \| \text{grad}\, u - \sigma \|_0^2$$

which by the definition of the $| \cdot |_1$-semi-norm yields the desired equation. $\qquad \square$

The Raviart–Thomas Element

The elements of Raviart and Thomas [1977] are suitable for the saddle point problem (5.5). Let $k \geq 0$, $\Omega \subset \mathbb{R}^2$, and suppose $\mathcal{T}_h$ is a shape-regular triangulation and that

$$X_h := \text{RT}_k$$
$$:= \{\tau \in L_2(\Omega)^2; \ \tau|_T = \begin{pmatrix} a_T \\ b_T \end{pmatrix} + c_T \begin{pmatrix} x \\ y \end{pmatrix}, \ a_T, b_T, c_T \in \mathcal{P}_k \text{ for } T \in \mathcal{T}_h,$$
$$\tau \cdot \nu \ \text{ is continuous on the inter-element boundaries}\},$$
$$M_h := \mathcal{M}^k(\mathcal{T}_h) = \{v \in L_2(\Omega); \ v|_T \in \mathcal{P}_k \text{ for } T \in \mathcal{T}_h\}.$$
$$(5.8)$$

The continuity of the normal components on the boundaries ensures the conformity $X_h \subset H(\text{div}, \Omega)$; cf. Problem II.5.14.

For convenience, we consider the Raviart–Thomas element only for $k = 0$. Its construction heavily depends on the following assertion. The functions in $(\mathcal{P}_1)^2$ which have the form

$$p = \begin{pmatrix} a \\ b \end{pmatrix} + c \begin{pmatrix} x \\ y \end{pmatrix}$$

are characterized by the fact that $n \cdot p$ is constant on each line $\alpha x + \beta y = $ const whenever n is orthogonal to the line. Therefore, given a triangle T, the normal component is constant and can be prescribed on each edge of T (see Fig. 35). Formally, the Raviart–Thomas element is the triple

$$(T, (\mathcal{P}_0)^2 + x \cdot \mathcal{P}_0, n_i p(z_i), i = 1, 2, 3 \text{ with } z_i \text{ being the midpoint of edge } i).$$

The solvability of the interpolation problem is easily verified. Given a vertex a_i of T, we can find a vector $v_i \in \mathbb{R}^2$ such that its projections onto the normals of the adjacent edges have the prescribed values. Now determine $p \in (\mathcal{P}_1)^2$ such that

$$p(a_i) = v_i, \quad i = 1, 2, 3.$$

It is immediate from $p \in (\mathcal{P}_1)^2$ that the normal components are linear on each edge of the triangle. They are even constant, since by construction they attain the same value at both vertices of the edge. Thus the function constructed indeed belongs to the specified subset of $(\mathcal{P}_1)^2$.

A proof of the inf-sup condition will be given below.

The Raviart–Thomas element and the similar BDM elements due to Brezzi, Douglas, and Marini [1985] are frequently used for the discretization of problems in $H(\text{div}, \Omega)$. Analogous elements for 3-dimensional problems have been described by Brezzi, Douglas, Durán, and Fortin [1987].

Interpolation by Raviart–Thomas elements

Due to Theorem 4.5 the error of the finite element solution for the discretization with the Raviart–Thomas element can be expressed in terms of approximation by the finite element functions. As usual the latter is estimated via interpolation. To this end an interpolation operator is defined which is based on the degrees of freedom specified in the definition of the element.

5.2 An Interpolation Operator. Let $k \geq 0$ and T be a triangle. Define

$$\rho_T : H^1(T) \to RT_k(T)$$

by

$$\int_e (q - \rho_T q) \cdot n p_k \, ds = 0 \qquad \forall p_k \in \mathcal{P}_k \text{ and each edge } e \in \partial T, \quad (5.9a)$$

$$\int_T (q - \rho_T q) \cdot p_{k-1} \, dx = 0 \qquad \forall p_{k-1} \in \mathcal{P}_{k-1}^2 \quad (\text{if } k \geq 1). \qquad (5.9b)$$

Given a triangulation $\mathcal{T}$ on Ω, define $\rho_\Omega : H^1(\Omega) \to RT_k$ locally by

$$(\rho_\Omega q)_{|T} = \rho_T(q_{|T}) \qquad \forall T \in \mathcal{T}.$$

We restrict ourselves to the case $k = 0$. We recall that the normal component of $v \in RT_0$ is constant on each edge. Equation (5.9a) states that it coincides with the mean value of the normal component of the given function. This holds for the solution of the interpolation problem.

From Gauss' integral theorem we conclude now that

$$\int_T \mathrm{div}(q - \rho_T q)dx = \sum_{e \in \partial T} \int_e (q - \rho_T q) \cdot nds = 0. \tag{5.10}$$

On the other hand, the Raviart–Thomas element is piecewise linear and $\alpha :=$ $\mathrm{div}\,\rho_T q$ is constant on T. By (5.10) α is the mean value of $\mathrm{div}\,v$ on T. Therefore α is the constant with the least L_2 deviation from $\mathrm{div}\,q$. So we have established the following property for $k = 0$. A proof for $k > 0$ can be found in Brezzi and Fortin [1990].

5.3 Minimal Property. *Given a triangulation $\mathcal{T}$ on Ω. Let Π_k be the L^2-projection onto $\mathcal{M}^k$. Then we have for all $q \in H^1(\Omega)$*

$$\mathrm{div}(\rho_\Omega q) = \Pi_k \,\mathrm{div}\,q. \tag{5.11}$$

This equation is often called the *commuting diagram property*

$$
\begin{array}{ccc}
H_1(\Omega) & \xrightarrow{\mathrm{div}} & L_2(\Omega) \\
\rho_\Omega \downarrow & & \downarrow \Pi_k \\
RT_k & \xrightarrow{\mathrm{div}} & \mathcal{M}^k.
\end{array}
$$

The proof of the inf-sup condition is related to the following.

5.4 Lemma. *The mapping*

$$\mathrm{div} : RT_0 \to \mathcal{M}^0$$

is surjective.

Proof. After enlarging Ω by finitely many triangles if necessary, we may assume that Ω is convex. Given $f \in \mathcal{M}^0$, there is a $u \in H^2(\Omega) \cap H_0^1(\Omega)$ such that $\Delta u = f$. Set $q := \mathrm{grad}\,u$. By Gauss' integral formula we have

$$\int_{\partial T} q \cdot nds = \int_T \mathrm{div}\,qdx = \int_T fdx.$$

From (5.9a) we conclude that $\int_T \mathrm{div}\,\rho_\Omega qdx = \int_{\partial T}(\rho_\Omega q) \cdot nds = \int_T fdx$. Since $\mathrm{div}\,\rho_\Omega q$ and f are constant in T, it follows that $\mathrm{div}\,\rho_\Omega q = f$. $\qquad\square$

Finally we note that the mapping $\mathcal{M}^0 \to RT_0$ in the construction above is bounded. Therefore, recalling Fortin's criterion we see that the inf-sup condition has been established simultaneously.

The error of the finite element solution will be derived from the approximation error.

5.5 Lemma. *Let $\mathcal{T}_h$ be a shape regular triangulation of Ω. Then*

$$\|q - \rho_\Omega q\|_{H(\mathrm{div},\Omega)} \le ch \, |q|_1 + \inf_{v_h \in \mathcal{M}^0} \| \mathrm{div} \, q - v_h \|_0. \tag{5.12}$$

Proof. We first consider the interpolation on a triangle. By the trace theorem the functional $q \mapsto \int_e q \cdot n \, ds$, $e \in \partial T$, is continuous on $H^1(T)^2$. Moreover, we have $\rho_T q = q$ for $q \in \mathcal{P}_0^2$ since $\mathcal{P}_0^2 \subset RT_0$. Therefore, the Bramble–Hilbert lemma and a scaling argument yield

$$\|q - \rho_\Omega q\|_0 \le ch \, |q|_1 .$$

The bound for $\mathrm{div}(q - \rho_\Omega q)$ follows from the minnimal property 5.3, and the proof is complete. ☐

Now the error estimate of the finite element solution of (5.1)

$$\|\sigma - \sigma_h\|_{H(\mathrm{div},\Omega)} + \|u - u_h\|_0$$
$$\le c(h \, |\sigma|_1 + h \, \|u\|_1 + \inf_{f_h \in \mathcal{M}^0} \|f - f_h\|_0) \tag{5.13}$$

is a direct consequence of Theorem 4.3.

The error estimate for the u-component is weaker than that for standard finite elements. On the other hand, the Raviart–Thomas element is more robust than the standard method for a class of problems that we will encounter in Ch. VI. Moreover the above disadvantage can be eliminated by a postprocessing procedure.

Implementation and Postprocessing

In principle, the discretization leads to an indefinite system of equations. It can be turned into a positive definite system by a trick which was described by Arnold and Brezzi [1985].

Instead of initially choosing the gradients to lie in a subspace of $H(\mathrm{div}, \Omega)$, we first admit gradients in $L_2(\Omega)^2$, and later explicitly require that $\mathrm{div} \, \sigma_h \in L_2(\Omega)$. Equivalently, we require that the normal components $\sigma_h n$ do not have jumps on the edges. To achieve this, we enforce the continuity of $\sigma_h n$ on the edges as an explicit constraint. This introduces a further Lagrange multiplier.

The approximating functions for σ_h no longer involve continuity conditions, and each basis function has support on a single triangle. If we eliminate the associated variables by static condensation, the resulting equations are just as sparse as before the elimination process. In addition, we have avoided the costly construction of a basis of Raviart–Thomas elements.

A further advantage is that the Lagrange multiplier can be regarded as a finite element approximation of u on the edges. Arnold and Brezzi [1985] used them to improve the finite element solution.

Mesh-Dependent Norms for the Raviart–Thomas Element

Finite element computations with the Raviart–Thomas elements may also be analyzed in the framework of primal mixed methods, i.e., with the pairing $H^1(\Omega)$, $L_2(\Omega)$. Since the tangential components of the functions in (5.8) may have jumps on interelement boundaries, in this context the elements are nonconforming and we need mesh-dependent norms

$$\|\tau\|_{0,h} := \left(\|\tau\|_0^2 + h \sum_{e \in \Gamma_h} \|\tau n\|_{0,e}^2 \right)^{1/2},$$

$$|v|_{1,h} := \left(\sum_{T \in \mathcal{T}_h} |v|_{1,T}^2 + h^{-1} \sum_{e \in \Gamma_h} \|J(v)\|_{0,e}^2 \right)^{1/2}. \qquad (5.14)$$

Here, $\Gamma_h := \cup_T (\partial T \cap \Omega)$ is the set of interelement boundaries. On the edges of Γ_h the jump $J(v)$ of v and the normal component τn of τ are well defined. We note that both τn and $J(v)$ change sign if the orientation of an edge is reversed. Therefore, the product is independent of the orientation.

The continuity of the bilinear form $a(\cdot, \cdot)$ is obvious. Its coercivity follows from

$$\|\tau\|_{0,h} \le C\|\tau\|_0 \quad \text{for all } \tau \in RT_k$$

which in turn is obtained by a standard scaling argument. The bilinear form $b(\cdot, \cdot)$ is rewritten by the use of Green's formula

$$b(\tau, v) = - \sum_{T \in \mathcal{T}_(} \int_T \tau \cdot \operatorname{grad} v \, dx + \int_{\Gamma_h} J(v)\tau n \, ds. \qquad (5.15)$$

Now its continuity with respect to the norms (5.14) is immediate.

5.6 Lemma. *The inf-sup condition*

$$\sup_{\tau \in RT_k} \frac{b(\tau, v)}{\|\tau\|_{0,h}} \ge \beta |v|_{1,h} \quad \text{for all } v \in \mathcal{M}^k$$

holds with a constant $\beta > 0$ which depends only on k and the shape parameter of the triangulation $\mathcal{T}_h$.
Proof. We restrict ourselves to the case $k = 0$. Given $v \in \mathcal{M}^0$, we note that $J(v)$ is constant on each edge $e \in \Gamma_h$. Therefore, there exists $\tau \in RT_0$ such that

$$\tau n = h^{-1} J(v) \quad \text{on each edge } e \in \Gamma_h.$$

Since the area term in (5.15) vanishes on each T, it follows that

$$b(\tau, v) = h^{-1} \int_{\Gamma_h} |J(v)|^2 ds = |v|_{1,h}^2.$$

On the other hand we have $\|\tau\|_{0,h}^2 \le ch \sum_{e \in \Gamma_h} \|\tau\|_{0,e}^{1/2} = ch^{-1} \sum_{e \in \Gamma_h} \|J(v)\|_{0,e}^2 = c|v|_{1,h}^2$. Hence $b(\tau, v) \ge c^{-1} |v|_{1,h}^2 \|\tau\|_{0,h}$, and the proof of the inf-sup condition is complete.   □

The Softening Behaviour of Mixed Methods

The (primal) mixed method (5.2) provides a softening of the quadratic form $a(., .)$. We will study this phenomenon since an analogous procedure has become very popular in computational mechanics during recent years.

Let $u_h \in M_h \subset H_0^1(\Omega)$ and $\sigma_h \in X_h \subset L_2(\Omega)$ be the solution of the mixed method

$$\begin{aligned}
(\sigma_h, \tau)_{0,\Omega} - (\tau, \nabla u_h)_{0,\Omega} &= 0 && \text{for all } \tau \in X_h, \\
- (\sigma_h, \nabla v)_{0,\Omega} &= -(f, v)_{0,\Omega} && \text{for all } v \in M_h.
\end{aligned} \tag{5.2}_h$$

If $E_h := \nabla M_h \subset X_h$, then the first equation implies $\sigma_h = \nabla u_h$, and $(5.2)_h$ is equivalent to the classical treatment of the Poisson equation with the finite element space M_h. This is the uninteresting case.

More interesting is the case $E_h \not\subset X_h$. Let $P_h : L_2(\Omega) \to X_h$ be the orthogonal projector onto X_h. The first equation in $(5.2)_h$ reads

$$\sigma_h = P_h(\nabla u_h)$$

and the second one

$$(P_h \nabla u_h, \nabla v)_{0,\Omega} = (f, v) \quad \text{for all } v \in M_h.$$

This is the weak equation for the relaxed minimum problem

$$\frac{1}{2} \int_\Omega [P_h \nabla v_h]^2 dx - \int_\Omega f v_h \to \min_{v_h \in M_h}. \tag{5.16}$$

Only the part of the gradient that is projected onto X_h contributes to the energy in the variational formulation. The amount of the softening is fixed by the choice of the target space of the projection.

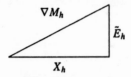

Fig. 36. Projection of the gradient onto X_h in the mixed method and the EAS method, resp.

There is another characterization. The variational equations $(5.2)_h$ can be rewritten in a form which leads to linear equations with a positive definite matrix. We may choose a subspace $\tilde{E}_h$ of the L_2-orthogonal complement of X_h such that

$$\nabla M_h \subset X_h \oplus \tilde{E}_h. \tag{5.17}$$

5.7 Remark. The mixed method $(5.2)_h$ is equivalent to the variational formulation

$$
\begin{aligned}
(\nabla u_h, \nabla v)_{0,\Omega} + (\tilde{\varepsilon}_h, \nabla v)_{0,\Omega} &= (f, v)_{0,\Omega} \quad \text{for all } v \in M_h, \\
(\nabla u_h, \tilde{\eta})_{0,\Omega} + (\tilde{\varepsilon}_h, \eta)_{0,\Omega} &= 0 \qquad\qquad \text{for all } \eta \in \tilde{E}_h.
\end{aligned}
\tag{5.18}
$$

if the space $\tilde{E}_h$ of *enhanced gradients* satisfies the decomposition rule (5.17). Here the relaxation of the variational form and the projector P_h are defined by $\tilde{E}_h$, i.e. by the orthogonal complement of the target space.

The proof of the equivalence follows Yeo and Lee [1996]. Let σ_h, u_h be a solution of $(5.2)_h$. From (5.17) we have a decomposition

$$\nabla u_h = \tilde{\sigma}_h - \tilde{\varepsilon}_h \quad \text{with } \tilde{\sigma}_h \in X_h \text{ and } \tilde{\varepsilon}_h \in \tilde{E}_h.$$

From the first equation in $(5.2)_h$ we conclude that $\nabla u_h - \sigma_h$ is orthogonal to X_h, and the uniqueness of the decomposition implies $\tilde{\sigma}_h = \sigma_h$. When we insert $\sigma_h = \nabla u_h + \tilde{\varepsilon}_h$ in $(5.2)_h$, we get the first equation of the system (5.18). The second one is a reformulation of $\nabla u_h + \tilde{\varepsilon}_h \in X_h$ and $X_h \perp \tilde{E}_h$.

The converse follows from the uniqueness of the solutions. The uniqueness of the solution of (5.18) follows from ellipticity which in turn is given by Problem 5.9. □

We note that in structural mechanics an equivalent concept was derived by Simo and Rifai [1998] and called *the method of enhanced assumed strains (EAS method)*.

The stability of the mixed method can be stated in terms of the enhanced elements; cf. Braess [1998]. It also shows that the stability is *not* independent of the choice of the space $\tilde{E}_h$.

5.8 Lemma. *The spaces X_h and M_h satisfy the inf-sup condition (4.16) with a constant $\beta \geq 0$ if and only if a strengthened Cauchy inequality*

$$(\nabla v_h, \eta_h)_{0,\Omega} \leq \sqrt{1 - \beta^2}\, \|\nabla v_h\|_0 \|\eta_h\|_0 \quad \text{for all } v_h \in M_h, \eta_h \in \tilde{E}_h \quad (5.19)$$

holds.

Proof. Given $v_h \in M_h$, by the inf-sup condition there is a $\sigma_h \in X_h$ such that $(\nabla v_h, \sigma_h)_0 \geq \beta\|\nabla v_h\|_0$ and $\|\sigma_h\| = 1$. Now for any $\eta_h \in \tilde{E}_h$ we conclude from the orthogonality of X_h and $\tilde{E}_h$ that

$$\|\nabla v_h - \eta_h\|_0 \geq (\nabla v_h - \eta_h, \sigma_h)_0 = (\nabla v_h, \sigma_h)_0 \geq \beta\|\nabla v_h\|_0. \quad (5.20)$$

Since the strengthened Cauchy inequality is homogeneous in its arguments, it is sufficient to verify it for the case $\|\eta_h\|_0 = (1 - \beta^2)^{1/2}\|\nabla v_h\|_0$,

$$\begin{aligned}
2(\nabla v_h, \eta_h)_0 &= \|\nabla v_h\|_0^2 + \|\eta_h\|_0^2 - \|\nabla v_h - \eta_h\|_0^2 \\
&\leq (1 - \beta^2)\|\nabla v_h\|_0^2 + \|\eta_h\|_0^2 = 2(1 - \beta^2)^{1/2}\|\nabla v_h\|_0^2 \|\eta_h\|_0^2,
\end{aligned}$$

and the proof of (5.19) is complete.

The converse is easily proved by using the decomposition of ∇v_h. $\qquad\square$

Problems

5.9 Show that the strengthened Cauchy inequality (5.19) is equivalent to the ellipticity property

$$\int_\Omega (\nabla v_h + \eta_h)^2 dx \geq (1 - \beta)(|v_h|_1^2 + \|\eta_h\|_0^2) \quad \text{for } v_h \in X_h, \eta_h \in \tilde{E}_h.$$

Further equivalent properties are presented in Problem V.5.7.

5.10 Define the vectors a_i and b_i in ℓ_2 by

$$(a_i)_j := \begin{cases} 1 & \text{if } j = 2i, \\ 0 & \text{otherwise,} \end{cases} \qquad (b_i)_j := \begin{cases} 1 & \text{if } j = 2i, \\ 2^{-i} & \text{if } j = 2i + 1, \\ 0 & \text{otherwise,} \end{cases}$$

and the subspaces $A := \text{span}\{a_i; \ i > 0\}$ and $B := \text{span}\{b_i; \ i > 0\}$. Show that A and B are closed, but that $A + B$ is not. Is there a nontrivial strengthened Cauchy inequality between the spaces A and B?

§ 6. The Stokes Equation

The *Stokes equation* describes the motion of an incompressible viscous fluid in an n-dimensional domain (with $n = 2$ or 3):

$$\begin{aligned} \Delta u + \operatorname{grad} p &= -f \quad &&\text{in } \Omega, \\ \operatorname{div} u \quad\quad &= 0 \quad &&\text{in } \Omega, \\ u &= u_0 \quad &&\text{on } \partial\Omega. \end{aligned} \tag{6.1}$$

Here $u : \Omega \longrightarrow \mathbb{R}^n$ is the velocity field and $p : \Omega \longrightarrow \mathbb{R}$ is the pressure. Since we are assuming that the fluid is incompressible, $\operatorname{div} u = 0$ when no sources or sinks are present.

In order for a divergence-free flow to exist with given boundary values u_0, by Gauss' integral theorem we must have

$$\int_{\partial\Omega} u_0 \cdot v\, ds = \int_{\partial\Omega} u \cdot v\, ds = \int_{\Omega} \operatorname{div} u\, dx = 0. \tag{6.2}$$

This *compatibility condition* on u_0 is obviously satisfied for *homogeneous* boundary values.

By an appropriate scaling we can assume that the viscosity is 1, which we have already done in writing (6.1).

The given external force field f causes an acceleration of the flow. The pressure gradient gives rise to an additional force which prevents a change in the density. In particular, a large pressure builds up at points where otherwise a source or sink would be created. From a mathematical point of view, the pressure can be regarded as a Lagrange multiplier.

If (6.1) is satisfied for some functions $u \in [C^2(\Omega) \cap C^0(\bar{\Omega})]^n$ and $p \in C^1(\Omega)$, then we call u and p a classical solution of the Stokes problem. – Note that (6.1) only determines the pressure p up to an additive constant, which is usually fixed by enforcing the normalization

$$\int_{\Omega} p\, dx = 0. \tag{6.3}$$

Variational Formulation

In view of the restriction $\operatorname{div} u = 0$, the weak formulation of the Stokes equation (6.1) leads to a saddle point problem. In order to make use of the general framework of §4, we set

$$X = H_0^1(\Omega)^n, \quad M = L_{2,0}(\Omega) := \{q \in L_2(\Omega); \int_\Omega q\, dx = 0\},$$

$$a(u, v) = \int_\Omega \operatorname{grad} u : \operatorname{grad} v\, dx,$$
$$b(v, q) = \int_\Omega \operatorname{div} v\, q dx. \tag{6.4}$$

Here $\operatorname{grad} u : \operatorname{grad} v := \sum_{ij} \frac{\partial u_i}{\partial x_j} \frac{\partial v_i}{\partial x_j}$.

As usual, we restrict our attention to homogeneous boundary conditions, i.e., we assume $u_0 = 0$. Then the saddle point problem becomes: *Find $(u, p) \in X \times M$ such that*

$$a(u, v) + b(v, p) = (f, v)_0 \quad \text{for all } v \in X,$$
$$b(u, q) \qquad\quad = 0 \qquad\quad \text{for all } q \in M. \tag{6.5}$$

A solution (u, p) of (6.5) is called a *classical solution* provided $u \in [C^2(\Omega) \cap C^0(\bar\Omega)]^n$ and $p \in C^1(\Omega)$.

6.1 Remark. For $v \in H_0^1$ and $q \in H^1$, Green's formula gives

$$b(v, q) = \int_\Omega \operatorname{div} v\, q\, dx = -\int_\Omega v \cdot \operatorname{grad} q\, dx + \int_\Gamma v \cdot q\, v\, ds$$
$$= -\int_\Omega v \cdot \operatorname{grad} q\, dx. \tag{6.6}$$

Thus, we can regard div and $-$ grad as adjoint operators. Moreover, from (6.6) we see that $b(v, q)$ does not change if we add a constant function to q. Thus, we can identify M with $L_2(\Omega)/\mathbb{R}$. In this quotient space we consider functions in L_2 to be equivalent whenever they differ only by a constant.

6.2 Remark. Every classical solution of the saddle point equation (6.5) is a solution of (6.1).

Proof. Let (u, p) be a classical solution. We split $\phi := \operatorname{div} u \in L_2$ into $\phi = q_0 + const$ with $q_0 \in M$. Since $u \in H_0^1$, combining the formula (6.6) with $v = u$ and $q = 1$ implies $\int_\Omega \operatorname{div} u\, dx = 0$. Substituting q_0 in (6.5), we get

$$\int_\Omega (\operatorname{div} u)^2 dx = b(u, q_0) + const \int_\Omega \operatorname{div} u\, dx = 0.$$

Thus, the flow is divergence-free.

By Remark 6.1, the first equation in (6.5) can be written in the form

$$(\operatorname{grad} u, \operatorname{grad} v)_{0,\Omega} = (f - \operatorname{grad} p, v)_{0,\Omega} \quad \text{for all } v \in H_0^1(\Omega)^n.$$

Since $u \in C^2(\Omega)^n$, by the theory of scalar equations in Ch. II, §2, it follows that u is a classical solution of

$$-\Delta u = f - \operatorname{grad} p \quad \text{in } \Omega,$$
$$u = 0 \quad \text{on } \partial\Omega,$$

and the proof is complete. $\qquad\qquad\square$

The inf-sup Condition

In order to apply the general theory described in the previous section, let

$$V := \{v \in X;\ (\operatorname{div} v, q)_{0,\Omega} = 0 \quad \text{for all } q \in L_2(\Omega)\}.$$

By Friedrichs' inequality, $|u|_{1,\Omega} = \|\operatorname{grad} u\|_{0,\Omega} = a(u,u)^{1/2}$ is a norm on X. Hence, the bilinear form a is H_0^1-elliptic. Thus it is elliptic not only on the subspace V, but also on the entire space X. This means that we could get by with an even simpler theory than in §4.

In order to ensure the existence and uniqueness of a solution of the Stokes problem, it remains to verify the Brezzi condition. To accomplish this, we need to use the following estimate. Its proof is beyond the scope of this book (see e.g. Duvaut and Lions [1976]). Recall that

$$V^\perp := \{u \in X;\ (\operatorname{grad} u, \operatorname{grad} v)_{0,\Omega} = 0 \quad \text{for all } v \in V\} \qquad (6.7)$$

is the H^1-orthogonal complement of V.

6.3 Theorem. *Let Ω be a bounded connected domain with Lipschitz continuous boundary.*
(1) The image of the linear mapping

$$\operatorname{grad} : L_2(\Omega) \longrightarrow H^{-1}(\Omega)^n \qquad (6.8)$$

is closed in $H^{-1}(\Omega)^n$.
(2) There exists a constant $c = c(\Omega)$ such that

$$\|p\|_{0,\Omega} \le c(\|\operatorname{grad} p\|_{-1,\Omega} + \|p\|_{-1,\Omega}) \quad \text{for all } p \in L_2(\Omega), \qquad (6.9)$$
$$\|p\|_{0,\Omega} \le c\,\|\operatorname{grad} p\|_{-1,\Omega} \quad \text{for all } p \in L_{2,0}(\Omega). \qquad (6.10)$$

This theorem immediately implies

6.4 Theorem. *Under the hypotheses of Theorem 6.3, the Stokes problem (6.5) satisfies the Brezzi condition (4.8).*

Proof. For $p \in L_{2,0}$, it follows from (6.10) that

$$\| \text{grad } p \|_{-1} \geq c^{-1} \| p \|_0.$$

By the definition of negative norms, there exists $v \in H_0^1(\Omega)^n$ with $\| v \|_1 = 1$ and

$$(v, \text{grad } p)_{0,\Omega} \geq \frac{1}{2} \| v \|_1 \| \text{grad } p \|_{-1} \geq \frac{1}{2c} \| p \|_0.$$

By (6.6),

$$\frac{b(-v, p)}{\| v \|_1} = (v, \text{grad } p)_{0,\Omega} \geq \frac{1}{2c} \| p \|_0,$$

which establishes the Brezzi condition. □

Remarks on the Brezzi Condition

By the abstract Lemma 4.2, the inf-sup condition can be expressed in terms of properties of the operators B and B'. In the concrete case of the Stokes equation with $b(v, q) = (\text{div } v, q)_{0,\Omega}$, the conditions are to be understood as properties of the operators div and grad.

6.5 Remark. Suppose $\Omega \subset \mathbb{R}^n$ is a bounded connected domain with Lipschitz continuous boundary.
(1) Let $f \in H^{-1}(\Omega)^n$. If

$$\langle f, v \rangle = 0 \quad \text{for all } v \in V, \tag{6.11}$$

then there exists a unique $q \in L_2(\Omega)$ with $\int_\Omega q dx = 0$ and $f = \text{grad } q$.
(2) The mapping

$$\text{div} : V^\perp \longrightarrow L_{2,0}(\Omega)$$

$$v \longmapsto \text{div } v$$

is an isomorphism. Moreover, for any $q \in L_2(\Omega)$ with $\int_\Omega q \, dx = 0$, there exists a function $v \in V^\perp \subset H_0^1(\Omega)^n$ with

$$\text{div } v = q \quad \text{and} \quad \| v \|_{1,\Omega} \leq c \| q \|_{0,\Omega}, \tag{6.12}$$

where $c = c(\Omega)$ is a constant.

Both 6.3(1) and (6.10) assert that the image of the mapping grad : $L_2(\Omega) \longrightarrow H^{-1}(\Omega)^n$ is closed. By (6.6), the divergence is the adjoint operator. Now Theorem 3.4 implies that the image is identical with the polar set of ker(div), and thus also with V^0. This is just 6.5(1). Moreover, (6.12) asserts that the divergence is an isomorphism. Thus, all of the statements in Theorem 6.3 and Remark 6.5 are equivalent.

Nearly Incompressible Flows

Instead of directly enforcing that the flow be divergence-free, sometimes a penalty term is added to the variational functional

$$\frac{1}{2}\int [(\nabla v)^2 + t^{-2}(\operatorname{div} v)^2 - 2fv]\, dx \longrightarrow \min!$$

Here t is a parameter. The smaller is t, the more weight is placed on the restriction. In this way a nearly incompressible flow is modeled.

The solution is characterized by the equation

$$a(u, v) + t^{-2}(\operatorname{div} u, \operatorname{div} v)_{0,\Omega} = (f, v)_{0,\Omega} \quad \text{for all } v \in H_0^1(\Omega)^n. \tag{6.13}$$

In order to establish a connection with the standard formulation (6.5), we set

$$p = t^{-2}\operatorname{div} u. \tag{6.14}$$

Now (6.13) together with the weak formulation of (6.14) leads to

$$
\begin{aligned}
a(u, v) + (\operatorname{div} v, p)_{0,\Omega} &= (f, v)_{0,\Omega} && \text{for all } v \in H_0^1(\Omega)^n, \\
(\operatorname{div} u, q)_{0,\Omega} - t^2(p, q)_{0,\Omega} &= 0 && \text{for all } q \in L_{2,0}(\Omega)^n.
\end{aligned}
\tag{6.15}
$$

Clearly, in comparison with (6.5), (6.15) contains a term which can be interpreted as a penalty term in the sense of §4. By the theory in §4, we know that the solution converges to the solution of the Stokes problem as $t \to 0$.

Problems

6.6 Show that among all representers of $q \in L_2(\Omega)/\mathbb{R}$, the one with the smallest L_2-norm $\|q\|_{0,\Omega} = \inf_{c\in\mathbb{R}} \|q + c\|_{0,\Omega}$ is characterized by $\int_\Omega q\, dx = 0$. [Consequently, $L_2(\Omega)/\mathbb{R}$ and $L_{2,0}(\Omega)$ are isometric.]

6.7 Find a Stokes problem with a suitable right-hand side to show that for every $q \in L_2(\Omega)$, there exists $u \in H_0^1(\Omega)$ with

$$\operatorname{div} u = q \quad \text{and} \quad \|u\|_1 \leq c\|q\|_0,$$

where as usual, c is a constant independent of q.

§ 7. Finite Elements for the Stokes Problem

In the study of convergence for saddle point problems we assumed that the finite element spaces for velocities and pressure satisfy the inf-sup condition. This raises the question of whether this condition is only needed to get a complete mathematical theory, or whether it plays an essential role in practice.

The answer to this question is given by a well-known finite element method for which the Brezzi condition is violated. Although instabilities had been observed in computations with this element in fluid mechanics, attempts to explain its instable behavior and to overcome it in a simple way mostly proved to be unsatisfactory. The Brezzi condition turned out to be the appropriate mathematical tool for understanding and removing this instability, and it also provided the essential breakthrough in practice. There are very few areas[7] where the mathematical theory is of as great importance for the development of algorithms as in fluid mechanics.

After discussing the instable element mentioned above, we present two commonly used stable elements and another one which is easier to implement. There is also a nonconforming divergence-free element which allows the elimination of the pressure.

An Instable Element

In the Stokes equation (6.1), Δu and grad p are the terms with derivatives of highest order for the velocity and pressure, respectively. Thus, the orders of the differential operators differ by 1. This suggests the rule of thumb: the degree of the polynomials used to approximate the velocities should be one larger than for the approximation of the pressure. However, this "rule" is not sufficient to guarantee stability – as we shall see.

Because of its simplicity, the so-called Q_1-P_0 element has been popular for a long time. It is a rectangular element which uses bilinear functions for the velocity and piecewise constants for the pressure:

$$X_h := \{v \in C^0(\bar{\Omega})^2; \ v|_T \in \mathcal{Q}_1 \text{ i.e., bilinear for } T \in \mathcal{T}_h\},$$
$$M_h := \{q \in L_{2,0}(\Omega); \ q|_T \in \mathcal{P}_0 \quad \text{for } T \in \mathcal{T}_h\}.$$

---

[7] There are two comparable situations where purely mathematical considerations have played a major role in the development of methods for differential equations. The approximation properties of the exponential function show that to solve stiff differential equations, we need to use implicit methods. (In particular, parabolic differential equations lead to stiff systems.) – For hyperbolic equations, we need to enforce the Courant–Levy condition in order to correctly model the domain of dependence in the discretization.

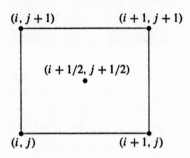

Fig. 37. Numbering of the nodes in the element T_{ij} for the Q_1-P_0 element

One indicator of the instability is the fact that the kernel of $B_h' : M_h \longrightarrow X_h'$ is nontrivial. In order to avoid unnecessary indices when showing this, we will denote the vector components of v by u and w, i.e.,

$$v = \begin{pmatrix} u \\ w \end{pmatrix}.$$

With the numbering shown in Fig. 37, the fact that q is constant and div v is linear implies

$$\int_{T_{ij}} q \, \mathrm{div}\, v \, dx = h^2 q_{i+1/2,j+1/2} \, \mathrm{div}\, v_{i+1/2,j+1/2}$$

$$= h^2 q_{i+1/2,j+1/2} \frac{1}{2h} [u_{i+1,j+1} + u_{i+1,j} - u_{i,j+1} - u_{i,j} \qquad (7.1)$$
$$+ w_{i+1,j+1} + w_{i,j+1} - w_{i+1,j} - w_{i,j}].$$

We now sum over the rectangles. Sorting the terms by grid points is equivalent to partial summation, and we get

$$\int_{\Omega} q \, \mathrm{div}\, v \, dx = h^2 \sum_{i,j} [u_{ij}(\nabla_1 q)_{ij} + w_{ij}(\nabla_2 q)_{ij}], \qquad (7.2)$$

where

$$(\nabla_1 q)_{i,j} = \frac{1}{2h} [q_{i+1/2,j+1/2} + q_{i+1/2,j-1/2} - q_{i-1/2,j+1/2} - q_{i-1/2,j-1/2}],$$

$$(\nabla_2 q)_{ij} = \frac{1}{2h} [q_{i+1/2,j+1/2} + q_{i-1/2,j+1/2} - q_{i+1/2,j-1/2} - q_{i-1/2,j-1/2}]$$

are the difference quotients. Since $v \in H_0^1(\Omega)^2$, the summation runs over all interior nodes. Now $q \in \ker(B_h')$ provided

$$\int_{\Omega} q \, \mathrm{div}\, v \, dx = 0 \quad \text{for all } v \in X_h,$$

and thus $\nabla_1 q$ and $\nabla_2 q$ vanish at all interior nodes. This happens if

$$q_{i+1/2,j+1/2} = q_{i-1/2,j-1/2}, \quad q_{i+1/2,j-1/2} = q_{i-1/2,j+1/2}.$$

These equations do not mean that q must be a constant. They only require that

$$q_{i+1/2,j+1/2} = \begin{cases} a & \text{for } i+j \text{ even,} \\ b & \text{for } i+j \text{ odd.} \end{cases}$$

Here the numbers a and b must be chosen so that (6.3) holds, and thus $q \in L_{2,0}(\Omega)$. In particular, a and b must have opposite signs, giving the *checkerboard pattern* shown in Fig. 38. In the following we use ρ to denote the corresponding pressure (up to a constant factor).

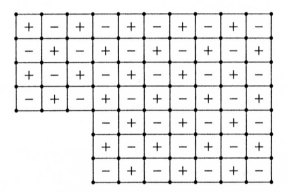

Fig. 38. Checkerboard instability

7.1 Remark. The inf-sup condition is an *analytic* property, and should not be interpreted just as a purely *algebraic* one. This fact becomes clear from the modification of Q_1-P_0 elements needed to achieve stability. We start with a reduction of the space M_h so that the kernel of B_h' becomes trivial. Since Ω is assumed to be connected, $\ker B_h' = \text{span}[\rho]$ has dimension 1. The mapping $B_h' : \mathcal{R}_h \longrightarrow X_h'$ is injective on the space

$$\mathcal{R}_h := \rho^\perp = \{q \in M_h; \ (q, \rho)_{0,\Omega} = 0\}.$$

Unfortunately this is not sufficient for full stability.

There is a constant $\beta_1 > 0$ such that

$$\sup_{v \in X_h} \frac{b(v, q)}{\|v\|_1} \geq \beta_1 h \|q\|_0 \quad \text{for } q \in \mathcal{R}_h \tag{7.3}$$

+3	−2	+1	0	−1	+2	−3
−3	+2	−1	0	+1	−2	+3
+3	−2	+1	0	−1	+2	−3
−3	+2	−1	0	+1	−2	+3

Fig. 39. Nearly instable pressure

for the pair X_h, $\mathcal{R}_h$ (see e.g. Girault and Raviart [1986]). However, the factor h in (7.3) cannot be avoided. Indeed, suppose Ω is a rectangle of width $B = (2n+1)h$ and height $2mh$ with $n \geq 4$. The pressure

$$q^*_{i+1/2,j+1/2} := i\,(-1)^{i+j} \quad \text{for } -n \leq i \leq +n,\ 1 \leq j \leq 2m \qquad (7.4)$$

(see Fig. 39) lies in $\mathcal{R}_h$.[8] Then

$$\|q^*\|^2_{0,\Omega} = h^2 \sum_{i=-n}^{+n} \sum_{j=1}^{2m} i^2 = h^2 \frac{1}{3} n(n+1)(2n+1)2m = \frac{1}{3}n(n+1)\mu(\Omega)$$

$$\geq \frac{1}{16} B^2 h^{-2} \mu(\Omega). \qquad (7.5)$$

In addition, obviously

$$(\nabla_1 q^*)_{ij} = 0, \quad (\nabla_2 q^*)_{ij} = (-1)^{i+j}\frac{1}{h}.$$

We now return to the node-oriented sum (7.2). We want to reorder it to get an element-oriented sum as in (7.1). To this end, we reassign one-quarter of each summand associated with an interior node to each of the four neighboring squares:

$$\int_\Omega q^* \operatorname{div} v\, dx = h \sum_{i,j}(-1)^{i+j} w_{ij}$$

$$= \frac{h}{4} \sum_{i,j}(-1)^{i+j}[w_{i+1,j} - w_{i,j+1} - w_{i+1,j+1} + w_{i,j}]. \qquad (7.6)$$

---

[8] Similarly, if the width is $B = 2nh$, we set

$$q^*_{i+1/2,j+1/2} = (-1)^{i+j}(i + \frac{1}{2}) \quad \text{for } -n \leq i \leq n-1,\ 1 \leq j \leq 2m.$$

For a bilinear function $\hat{w}$ on the reference square $[0, 1]^2$, the derivative $\partial_2 \hat{w}$ is linear in ξ. With $\hat{\phi}(\xi) = 2\xi - 1$, simple integration gives

$$\int_{[0,1]^2} \hat{\phi}(\xi)\partial_2\hat{w}\,d\xi\,d\eta = \frac{1}{6}[\hat{w}(1, 1) - \hat{w}(1, 0) - \hat{w}(0, 1) + \hat{w}(0, 0)].$$

For a bilinear function w, affine transformation to a square T with edges of length h and vertices a, b, c, d (in cyclic order) gives

$$\int_T \phi\partial_2 w\,dx\,dy = \frac{h}{6}[w(a) - w(b) - w(c) + w(d)].$$

Here ϕ is a function with $\|\phi\|_{0,T}^2 = \mu(T)/3$. Repeating this computation for each square of the partition of Ω and using (7.6), we get

$$\int_\Omega q^* \operatorname{div} v\,dx = \frac{3}{2}\int_\Omega \phi\partial_2 w\,dx. \tag{7.7}$$

Here $\|\phi\|_0^2 = \mu(\Omega)/3$. With the help of the Cauchy–Schwarz inequality, (7.6) and (7.7) imply

$$\left|\int_\Omega q^* \operatorname{div} v\,dx\right| \leq \frac{3}{2}\|\phi\|_{0,\Omega}\,\|\partial_2 w\|_{0,\Omega} \leq \mu(\Omega)^{1/2}\|v\|_{1,\Omega}$$

$$\leq 4B^{-1}h\|q^*\|_{0,\Omega}\,\|v\|_{1,\Omega}.$$

In fact,

$$\sup_{v\in X_h} \frac{b(v, q^*)}{\|v\|_{1,\Omega}} \leq 4B^{-1}h\|q^*\|_{0,\Omega}. \tag{7.8}$$

Thus, the inf-sup condition only holds for some constant depending on h. This clearly shows that *we cannot check the inf-sup condition by merely counting degrees of freedom and using dimensional arguments.*

In order to verify the Brezzi condition with a constant independent of h, we have to further restrict the space $\mathcal{R}_h$. This can be done by combining four neighboring squares into a macro-element. The functions sketched in Fig. 40 form a basis on the level of the macro-elements for the functions which are constant on every small square.

If we eliminate those functions in each macro-element which correspond to the pattern in Fig. 40d, we get the desired stability independent of h, see e.g. Girault and Raviart [1986], p. 167 or Johnson and Pitkäranta [1982]. However, in doing so, we lose much of the simplicity of the original approximations. Therefore, the stabilized Q_1-P_0 elements are not considered to be competitive.

.

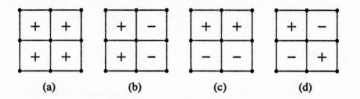

Fig. 40 a–d. Basis functions in M_h for the macro-element

The Taylor–Hood Element

The Taylor–Hood element is an often-used triangular element where the velocity polynomial has a higher degree than the pressure polynomial. The pressure is taken to be continuous:

$$X_h := (\mathcal{M}_{0,0}^2)^d \ = \{v_h \in C(\bar{\Omega})^d \cap H_0^1(\Omega)^d; \ v_h|_T \in \mathcal{P}_2 \quad \text{for } T \in \mathcal{T}_h\},$$
$$M_h := \mathcal{M}_0^1 \cap L_{2,0} = \{q_h \in C(\Omega) \cap L_{2,0}(\Omega); \ q_h|_T \in \mathcal{P}_1 \quad \text{for } T \in \mathcal{T}_h\}.$$

Here $\mathcal{T}_h$ is a partition of Ω into triangles. For a proof of the inf-sup condition, see Verfürth [1984] and the book of Girault and Raviart [1986].

Another stable element can be obtained by a simple modification. For the velocities we use piecewise linear functions on the triangulation obtained by dividing each triangle into four congruent subtriangles:

$$X_h := \mathcal{M}_{0,0}^1(\mathcal{T}_{h/2})^2 = \{v_h \in C(\bar{\Omega})^2 \cap H_0^1(\Omega)^2; \ v_h|_T \in \mathcal{P}_1 \quad \text{for } T \in \mathcal{T}_{h/2}\},$$
$$M_h := \mathcal{M}_0^1 \cap L_{2,0} = \{q_h \in C(\Omega) \cap L_{2,0}(\Omega); \ q_h|_T \in \mathcal{P}_1 \quad \text{for } T \in \mathcal{T}_h\}. \quad (7.9)$$

Thus, the number of degrees of freedom is the same as for the Taylor–Hood element. Often this variant is also called the *modified Taylor–Hood element* in the literature (see Fig. 41).

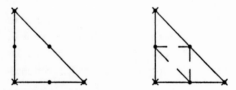

Fig. 41. The Taylor–Hood element and its variant. Here u is given at the nodes ($\bullet$) and p at the nodes ($\times$)

The approximation properties for the velocities can be obtained directly from results for piecewise quadratic functions. For the approximation of the pressure, we have to verify that the restriction (6.3) to functions with zero integral mean does not reduce the order. Let $\tilde{q}_h$ be an interpolant to $q \in L_{2,0}(\Omega)$. In general, $\int_\Omega \tilde{q}_h dx \neq 0$. By the Cauchy–Schwarz inequality,

$$\left| \int_\Omega \tilde{q}_h dx \right| = \left| \int_\Omega (q - \tilde{q}_h) dx \right| \leq \mu(\Omega)^{1/2} \| q - \tilde{q}_h \|_{0,\Omega}.$$

Thus adding a constant of order $\| q - \tilde{q}_h \|_{0,\Omega}$ gives an approximation in the desired subspace $L_{2,0}(\Omega)$ with the same approximation order.

The MINI Element

One disadvantage of the Taylor–Hood element is that the nodal values of velocity and pressure occur on different triangulations. This complication is avoided with the so-called *MINI element*; see Arnold, Brezzi, and Fortin [1984].

The key idea for the MINI element is to include a *bubble function* in the space X_h for the velocities. Let λ_1, λ_2, and λ_3 be the barycentric coordinates of a triangle (e.g., x_1, x_2, and $(1 - x_1 - x_2)$ in the unit triangle). Then

$$b(x) = \lambda_1 \lambda_2 \lambda_3 \tag{7.10}$$

vanishes on the edges of the triangle. The addition of such a *bubble function* does not affect the continuity of the elements:

$$X_h := [\mathcal{M}_{0,0}^1 \oplus B_3]^2, \qquad M_h := \mathcal{M}_0^1 \cap L_{2,0}(\Omega)$$
$$\text{with} \quad B_3 := \{v \in C^0(\bar{\Omega}); \ v|_T \in \text{span}[\lambda_1 \lambda_2 \lambda_3] \ \text{for } T \in \mathcal{T}_h\}. \tag{7.11}$$

Fig. 42. MINI element. u is given at the nodes ($\bullet$) and p at the nodes ($\times$)

Since the support of a bubble is restricted to the element, we can eliminate the associated variable from the resulting system of linear equations by static condensation. The MINI element requires less computation than the Taylor–Hood element and its variant, but according to many reports, it yields a poorer approximation of the pressure.

7.2 Theorem. *Assume that Ω is convex or has a smooth boundary. Then the MINI element (7.11) satisfies the inf-sup condition.*

Proof. In order to apply Fortin's criterion, we will use arguments introduced in II.7.8 in our treatment of the boundedness of the L_2-projector. We will restrict

ourselves to uniform meshes, and note that the extension to shape-regular triangulations is possible by the use of Clément's approximation process.

Let $\pi_h^0 : H_0^1(\Omega) \to \mathcal{M}_{0,0}^1$ be the L_2-projector. From Corollary II.7.8 we
know that $\|\pi_h^0 v\|_1 \leq c_1 \|v\|_1$ and $\|v - \pi_h^0 v\|_0 \leq c_2 h \|v\|_1$. Moreover, we fix a
linear mapping $\pi_h^1 : L_2(\Omega) \to B_3$ such that

$$\int_T (\pi_h^1 v - v) dx = 0 \quad \text{for each } T \in \mathcal{T}_h. \tag{7.12}$$

We may interpret the map π_h^1 as a process with two steps. First, we apply the
L_2-projection onto the space of piecewise constant functions. Afterwards, in each
triangle the constant is replaced by a bubble function with the same integral. In
this way we get $\|\pi_h^1 v\|_0 \leq c_3 \|v\|_0$.

Now we set

$$\Pi_h v := \pi_h^0 v + \pi_h^1 (v - \pi_h^0 v). \tag{7.13}$$

By construction,

$$\int_T (\Pi_h v - v) dx = \int_T (\pi_h^1 - \text{id})(v - \pi_h^1 v) dx = 0 \quad \text{for each } T \in \mathcal{T}_h. \tag{7.14}$$

The definition of the mapping Π_h is now extended to vector-valued functions.
Specifically, each component is to be treated as specified in (7.13).

Since p is continuous, we can apply Green's formula. We recall (7.14), and
that the gradient of the pressure is piecewise constant:

$$b(v - \Pi_h v, q_h) = \int_\Omega \text{div}(v - \Pi_h v) q_h dx$$

$$= \int_{\partial\Omega} (v - \Pi_h v) \cdot n q_h ds - \int_\Omega (v - \Pi_h v) \cdot \text{grad } q_h dx = 0.$$

The boundedness of Π_h now follows from (7.12) and an inverse estimate for bubble
functions

$$\|\Pi_h v\|_1 \leq \|\pi_h^0 v\|_1 + \|\pi_h^1 (v - \pi_h^0 v)\|_1$$
$$\leq c_1 \|v\|_1 + c_4 h^{-1} \|\pi_h^1 (v - \pi_h^0 v)\|_0$$
$$\leq c_1 \|v\|_1 + c_4 h^{-1} c_3 \|v - \pi_h^0 v\|_0$$
$$\leq c_1 \|v\|_1 + c_4 c_3 c_2 \|v\|_1.$$

Now by Fortin's criterion an inf-sup condition holds.   □

The Divergence-Free Nonconforming P_1 Element

The Crouzeix–Raviart element plays a special role. We can select from the non-conforming P_1 elements those functions which are piecewise divergence-free, and we can get by without the pressure. We choose

$$X_h := \{v \in L_2(\Omega)^2; \ v|_T \text{ is linear and divergence-free for every } T \in \mathcal{T}_h,$$

$$v \text{ is continuous at the midpoints of the triangle edges,}$$

$$v = 0 \text{ at the midpoints of the triangle edges in } \partial\Omega\},$$

i.e., $X_h := \{v \in (\mathcal{M}^1_{*,0})^2; \ \text{div } v = 0 \text{ on every } T \in \mathcal{T}_h\}$. As in the scalar case in §1, we set

$$a_h(u, v) := \sum_{T \in \mathcal{T}_h} \int_T \nabla u \cdot \nabla v \, dx.$$

We seek $u_h \in X_h$ with

$$a_h(u_h, v) = (f, v)_0 \quad \text{for all } v \in X_h.$$

For a convergence proof, see Crouzeix and Raviart [1973].

It is easy to construct a basis for X_h by geometric means. By the Gauss integral theorem, for $v \in X_h$

$$0 = \int_T \text{div } v \, dx = \int_{\partial T} v \cdot n \, ds = \sum_{e \in \partial T} v(e_m) n \, \ell(e), \qquad (7.15)$$

for every triangle T. Here e_m is the midpoint of the edge e, and $\ell(e)$ is its length.

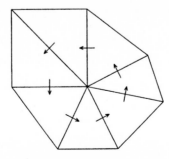

Fig. 43. Basis functions of the nonconforming P_1 element associated with one node. The normal components indicated by arrows have nonzero values

Since the tangential components do not enter into (7.15), we can prescribe them at the midpoint of each edge. For every interior edge e, we get one basis function $v = v_e$ in X_h with

$$v(e_m) \cdot t = 1,$$
$$v(e_m) \cdot n = 0, \qquad\qquad\qquad (7.16)$$
$$v(e'_m) \quad = 0 \quad \text{for } e' \neq e.$$

Let p be an arbitrary vertex of a triangle. Suppose the edges connected to p are oriented as follows: If we move around the point p in the mathematically positive direction, we cross the edges in the directions of the normals. Clearly, (7.15) holds if

$$v(e_m) \cdot n \;\; = \frac{1}{\ell(e)} \quad \text{for all edges connected to } p,$$
$$v(e_m) \cdot n \;\; = 0 \qquad \text{for all other edges,} \qquad (7.17)$$
$$v(e_m) \cdot t \;\; = 0 \qquad \text{for all edges,}$$

see Fig. 43. The functions in (7.16) and (7.17) are linearly independent, and a dimension count shows that they form a basis.

An analogous quadrilateral element was developed and studied by Rannacher and Turek [1992].

Problems

7.3 For Q_1-P_0 elements, the pressure is

$$q_{i+1/2, j+1/2} = \begin{cases} +(-1)^{i+j} & \text{for } i < i_0, \\ -(-1)^{i+j} & \text{for } i \geq i_0, \end{cases}$$

using the same notation as in (7.5). Here $-n < i_0 < +n$. This gives a checkerboard pattern – up to a shift. Show that

$$\left| \int_\Omega q \operatorname{div} v \, dx \right| \leq c\sqrt{h} \|v\|_{1,\Omega}.$$

Note that in order to get a constant in the inf-sup condition which is independent of h, we have to put increasingly stronger restrictions on M_h as $h \to 0$.

7.4 If Ω is convex or sufficiently smooth, then one has for the Stokes problem the regularity result

$$\|u\|_2 + \|p\|_1 \leq c\|f\|_0; \qquad\qquad (7.18)$$

see Girault and Raviart [1986]. Show by a duality argument the L_2 error estimate

$$\|u - u_h\|_0 \leq ch(\|u - u_h\|_1 + \|p - p_h\|_0). \qquad\qquad (7.19)$$

§ 8. A Posteriori Error Estimates

It frequently happens in practical problems that due to the nature of the data in certain subdomains, a solution of a boundary-value problem is less regular. In this case we would like to increase the accuracy of the finite element approximation without using too many additional degrees of freedom. One way to do this is to *adaptively* perform *local grid refinement* in those subdomains where it is needed. We first carry out the finite element calculations on a provisional grid, and then compute an *a posteriori estimate* for the error whose purpose is to indicate what part of the grid induces large errors. Using this information, we then locally refine the grid, and repeat the finite element computation. If necessary, the process can be repeated several times.

To simplify our discussion, we restrict ourselves to the case of the Poisson equation

$$-\Delta u = f \tag{8.1}$$

with homogeneous Dirichlet boundary conditions. Moreover, we consider only conforming elements, although this still involves arguments which are usually associated with the analysis of nonconforming elements. This is why we have not presented a posteriori estimates earlier.

Let $\mathcal{T}_h$ be a shape-regular triangulation. In addition, suppose u_h is a finite element solution lying in $S_h := \mathcal{M}_{0,0}^2$ (or in $\mathcal{M}_{0,0}^1$). Suppose Γ_h is the set of all *inter-element boundaries*, i.e., edges of the triangles $T \in \mathcal{T}_h$ which lie in the interior of Ω.

If we insert u_h into the differential equation in its classical form, we get a residual. Moreover, u_h differs from the classical solution in that grad u has jumps on the edges of elements. Both the area-based residuals

$$R_T := R_T(u_h) := \Delta u_h + f \quad \text{for } T \in \mathcal{T}_h \tag{8.2}$$

and the edge-based jumps

$$R_e := R_e(u_h) := \lceil \frac{\partial u_h}{\partial n} \rceil \quad \text{for } e \in \Gamma_h \tag{8.3}$$

enter either directly or indirectly into many estimators. [Note that both the jump $\lceil \nabla u_h \rceil$ and the normal direction change if we reverse the orientation of the edge e, but that the product $\lceil \frac{\partial u_h}{\partial n} \rceil = \lceil \nabla u_h \rceil \cdot n$ remains fixed.] Moreover we need the following notation for the neighborhoods of elements and edges:

$$\omega_T := \bigcup \{T' \in \mathcal{T}_h; \ T \text{ and } T' \text{ have a common edge or } T' = T\},$$

$$\omega_e := \bigcup \{T' \in \mathcal{T}_h; \ e \in \partial T'\}. \tag{8.4}$$

There are essentially five different approaches to building a posteriori estimators.

1. Residual Estimators.

We bound the error on an element T in terms of the size of the residual R_T and the jumps R_e on the edges $e \in \partial T$. These estimators are due to Babuška and Rheinboldt [1978a].

2. Estimators based on a local Neumann Problem

On every triangle T we solve a local variational problem which is a discrete analog of

$$\begin{aligned} -\Delta z &= R_T \quad \text{in } T, \\ \frac{\partial z}{\partial n} &= R_e \quad \text{on } e \in \partial T. \end{aligned} \tag{8.5}$$

We choose the approximating space to contain polynomials whose degrees are one higher than those in the underlying finite element space. These estimators are obtained using the energy norm $\|z\|_{1,T}$, and are due to Bank and Weiser [1985].

3. Estimators based on a local Dirichlet Problem.

For every element T, we solve a variational problem on the set ω_T:

$$\begin{aligned} -\Delta z &= f \quad \text{in } \omega_T, \\ z &= u_h \quad \text{on } \partial \omega_T. \end{aligned} \tag{8.6}$$

Again, we expand the approximating space to include polynomials of higher degree than in the actual finite element space. Following Babuška and Rheinboldt [1978b], the norm of the difference $\|z - u_h\|_{1,\omega_T}$ provides an estimator.

4. Estimators based on Averaging.

We construct a continuous approximation σ_h of Δu by a two-step process. At every node of the triangulation, let σ_h be a weighted average of the gradients ∇u_h on the neighboring triangles, where the weight is proportional to the areas of the triangles. We then extend σ_h to the whole element by linear interpolation. Then following Zienkiewicz and Zhu [1987], we use the difference between ∇u_h and σ_h as an estimator.

5. Hierarchical Estimators.

In principle the difference from a finite element approximation on an expanded space is estimated. The difference can be estimated by using a strengthened Cauchy inequality, see Deuflhard, Leinen, and Yserentant [1989].

To get started, following Dörfler [1996] we first show how to extract a part of the expression (8.2) which can be determined already before computing the finite element solution. Let

$$f_h := P_h f \in S_h \tag{8.7}$$

be the L_2-projection of f onto S_h. Since $(f - f_h, v_h)_{0,\Omega} = 0$ for $v_h \in S_h$, the variational problems corresponding to f and f_h lead to the same finite element approximation in S_h. Thus, the a priori computable quantity

$$h_T \| f - f_h \|_{0,T} \tag{8.8}$$

appears in many estimates. In particular, clearly

$$\| \Delta u_h + f \|_{0,T} \leq \| \Delta u_h + f_h \|_{0,T} + \| f - P_h f \|_{0,T}. \tag{8.9}$$

As usual, h_T denotes the diameter of T. Similarly, h_e is the length of e. [As an alternative to (8.7), we can compute f_h as the projection onto piecewise constant functions.]

Residual Estimators

To get residual estimators, we use the functions introduced in (8.2) and (8.3) to compute the local quantities

$$\eta_{T,R} := \left\{ h_T^2 \| R_T \|_{0,T}^2 + \frac{1}{2} \sum_{e \in \partial T} h_e \| R_e \|_{0,e}^2 \right\}^{1/2} \quad \text{for } T \in \mathcal{T}_h. \tag{8.10}$$

Summing the squares over all triangles, we get a global quantity:

$$\eta_R := \left\{ \sum_{T \in \mathcal{T}_h} h_T^2 \| R_T \|_{0,T}^2 + \sum_{e \in \Gamma_h} h_e \| R_e \|_{0,e}^2 \right\}^{1/2}. \tag{8.11}$$

8.1 Theorem. *Let $\mathcal{T}_h$ be a shape-regular triangulation with shape parameter κ. Then there exists a constant $c = c(\Omega, \kappa)$ such that*

$$\| u - u_h \|_{1,\Omega} \leq c \left\{ \sum_{T \in \mathcal{T}_h} \eta_{T,R}^2 \right\}^{1/2} \tag{8.12}$$

and

$$\eta_{T,R} \leq c \left\{ \| u - u_h \|_{1,\omega_T}^2 + \sum_{T' \subset \omega_T} h_T^2 \| f - f_h \|_{0,T'}^2 \right\}^{1/2} \tag{8.13}$$

for all $T \in \mathcal{T}_h$.

Proof of the upper estimate (8.12). We start by using a duality argument to find

$$|u - u_h|_1 = \sup_{|w|_1 = 1, w \in H_0^1} (\nabla(u - u_h), \nabla w)_0. \tag{8.14}$$

We shall make use of the following formula which also appeared in establishing the Céa Lemma:

$$(\nabla(u - u_h), \nabla v_h)_0 = 0 \quad \text{for } v_h \in S_h. \tag{8.15}$$

We now consider the functional ℓ corresponding to (8.15), apply Green's formula, and insert the residuals (8.3) and (8.4):

$$\begin{aligned}
\langle \ell, w \rangle &:= (\nabla(u - u_h), \nabla w)_{0,\Omega} \\
&= (f, w)_{0,\Omega} - \sum_T (\nabla u_h, \nabla w)_{0,T} \\
&= (f, w)_{0,\Omega} - \sum_T \left\{ (-\Delta u_h, w)_{0,T} + \sum_{e \in \partial T} (\nabla u_h \cdot n, w)_{0,e} \right\} \\
&= \sum_T (\Delta u_h + f, w)_{0,T} + \sum_{e \in \Gamma_h} (\lceil \frac{\partial u_h}{\partial n} \rceil, w)_{0,e} \\
&= \sum_T (R_T, w)_{0,T} + \sum_{e \in \Gamma_h} (R_e, w)_{0,e}.
\end{aligned} \tag{8.16}$$

By Clément's results on approximation, cf. II.6.9, for given $w \in H_0^1(\Omega)$ there exists an element $I_h w \in S_h$ with

$$\|w - I_h w\|_{0,T} \le c h_T \|\nabla w\|_{0,\tilde{\omega}_T} \quad \text{for all } T \in \mathcal{T}_h, \tag{8.17}$$

$$\|w - I_h w\|_{0,e} \le c h_e^{1/2} \|\nabla w\|_{0,\tilde{\omega}_T} \quad \text{for all } e \in \Gamma_h. \tag{8.18}$$

Here $\tilde{\omega}_T$ is the neighborhood of T specified in (II.6.14) which is larger than ω_T. Since the triangulations are assumed to be shape regular, $\bigcup \{\tilde{\omega}_T; \ T \in \mathcal{T}_h\}$ covers Ω only a finite number of times. Hence, (8.15) implies

$$\begin{aligned}
\langle \ell, w \rangle &= \langle \ell, w - I_h w \rangle \\
&\le \sum_T \|R_T\|_{0,T} \|w - I_h w\|_{0,T} + \sum_{e \in \Gamma_h} \|R_e\|_{0,e} \|w - I_h w\|_{0,e} \\
&\le c \sum_T h_T \|R\|_{0,T} |w|_{1,T} + c \sum_{e \in \Gamma_h} h_e^{1/2} \|R_e\|_{0,e} |w|_{1,\omega_e} \\
&\le c \sum_T \eta_{T,R} |w|_{1,T} \le c \eta_R |w|_{1,\Omega}.
\end{aligned} \tag{8.19}$$

The last inequality follows from the Cauchy–Schwarz inequality for finite sums. Combining (8.18) and (8.19) with Friedrichs' inequality and the duality argument (8.14), we get the global upper error bound (8.12). $\qquad \qquad \square$

8.2 Remark. For P_1 elements, $\Delta u_h = 0$ piecewise, and we have the special case where the complete area-based estimator R_T is a priori computable. As noted by Verfürth [1997], this term is dominated by the other term and can be neglected for grids which are nearly uniform. In this case, we can choose the operator I_h so that in addition to (8.17) and (8.18), we also have that $w - I_h w$ is L_2-orthogonal to $\mathcal{M}_{0,0}^1$.

Lower Estimates

The lower estimate (8.13) provides information on local properties of the discretization. It can be obtained using test functions with local support. The following cutoff functions ψ_T and ψ_e provide an essential tool: ψ_T is the well-known bubble function associated with the triangle T, so that

$$\psi_T \in B_3, \ \operatorname{supp} \psi_T = T, \ 0 \le \psi_T \le 1 = \max \psi_T. \tag{8.20}$$

ψ_e has support on a pair of neighboring triangles sharing the edge e, and consists of quadratic polynomials joined together continuously so that

$$\psi_e \in \mathcal{M}_0^2, \ \operatorname{supp} \psi_e = \omega_e; \ 0 \le \psi_e \le 1 = \max \psi_e. \tag{8.21}$$

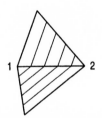

Fig. 44. Level curves for the extension of a function from e to ω_e. The level curves in the triangle lying below [1,2] are given by $\lambda_1 = const$, where λ_1 is the barycentric coordinate w.r.t. the point 1. The level curves in the triangle lying on the opposite side [1,2] are similarly described using λ_2.

We also need a mapping $E : L_2(e) \to L_2(\omega_e)$ which extends any function defined on an edge e to the pair of neighboring triangles making up ω_e. We take

$$E\sigma(x) := \sigma(x') \text{ in } T, \text{ if } x' \in e \text{ is the point in } e \text{ with } \lambda_j(x') = \lambda_j(x).$$

Here λ_j is one of the two barycentric coordinates in T which is not constant on e; see Fig. 44.

8.3 Lemma. *Let $\mathcal{T}_h$ be a shape-regular triangulation. Then there exists a constant c which depends only on the shape parameter κ such that*

$$\|\psi_T v\|_{0,T} \le \|v\|_{0,T} \qquad\qquad \text{for all } v \in L_2(T), \quad (8.22)$$

$$\|\psi_T^{1/2} p\|_{0,T} \ge c\|p\|_{0,T} \qquad\qquad \text{for all } p \in \mathcal{P}_2, \qquad (8.23)$$

$$\|\nabla(\psi_T p)\|_{0,T} \le ch_T^{-1}\|\psi_T p\|_{0,T} \qquad \text{for all } p \in \mathcal{P}_2, \qquad (8.24)$$

$$\|\psi_e^{1/2} \sigma\|_{0,e} \ge c\|\sigma\|_{0,e} \qquad\qquad \text{for all } \sigma \in \mathcal{P}_2, \qquad (8.25)$$

$$ch^{1/2}\|\sigma\|_{0,e} \le \|\psi_e E\sigma\|_{0,T} \le ch_e^{1/2}\|\sigma\|_{0,e} \qquad \text{for all } \sigma \in \mathcal{P}_2, \qquad (8.26)$$

$$\|\nabla(\psi_e E\sigma)\|_{0,T} \le ch_T^{-1}\|\psi_e E\sigma\|_{0,T} \qquad \text{for all } \sigma \in \mathcal{P}_2, \qquad (8.27)$$

for all $T \in \mathcal{T}_h$ and all $e \in \partial T$.

The estimate (8.22) follows directly from $0 \le \psi_T \le 1$. For a fixed reference triangle, the others are obvious because of the finite dimensionality of $\mathcal{P}_2$. The assertions for arbitrary triangles then follow from the usual scaling argument. Details have been elaborated by Verfürth [1994]. □

Proof of (8.13). Let $T \in \mathcal{T}_h$. In view of (8.9), in analogy with (8.2) we introduce

$$R_{T,\text{red}} := \Delta u_h + f_h. \qquad\qquad (8.28)$$

By construction, $R_{T,\text{red}} \in \mathcal{P}_2$. Let

$$w := w_T := \psi_T \cdot R_{T,\text{red}}.$$

Then (8.16), (8.23), and the fact that supp $w = T$ imply

$$
\begin{aligned}
c^{-1}\|R_{T,\text{red}}\|_{0,T}^2 &\le \|\psi_T^{1/2} R_{T,\text{red}}\|_{0,T}^2 \\
&= (R_{T,\text{red}}, w)_{0,T} \\
&= (R_T, w)_{0,T} + (f - f_h, w)_{0,T} \\
&= \langle \ell, w \rangle + (f - f_h, w)_{0,T} \\
&\le |u - u_h|_{1,T} \cdot |w|_{1,T} + \|f - f_h\|_{0,T}\|w\|_{0,T}.
\end{aligned}
$$

Note that because of (8.22), $\|w\|_{0,T} \le \|R_{T,\text{red}}\|_{0,T}$. Now using Friedrichs' inequality and the inverse inequality (8.24), after dividing by $\|R_{T,\text{red}}\|_{0,T}$ we get

$$\|R_{T,\text{red}}\|_{0,T} \le c(h_T^{-1}\|u - u_h\|_{1,T} + \|f - f_h\|_{0,T}).$$

By (8.9), this gives

$$h_T\|R_T\|_{0,T} \le c(\|u - u_h\|_{1,T} + h_T\|f - f_h\|_{0,T}). \qquad (8.29)$$

We now show that the edge-based terms in the error estimator can be treated in a similar way. Let $e \in \Gamma_h$. With R_e as in (8.3), we define

$$w := w_e := \psi_e \cdot E(R_e).$$

In particular, supp $w = \omega_e$, and $R_e \in \mathcal{P}_2(e)$. Using (8.16), (8.25), we have

$$
\begin{aligned}
c\|R_e\|_{0,e}^2 &\leq \|\psi_e^{1/2} R_e\|_{0,e}^2 \\
&= (R_e, w)_{0,e} \\
&= \langle \ell, w \rangle - \sum_{T' \in \omega_e} (R_{T'}, w)_{0,T'} \\
&\leq |u - u_h|_{1,e} |w|_{1,\omega_e} + \sum_{T' \in \omega_e} \|R_{T'}\|_{0,T'} \|w\|_{0,T'}.
\end{aligned}
\tag{8.30}
$$

Now (8.27) implies that $|w|_{1,T'} \leq ch_{T'}^{-1} \|w\|_{0,T'}$, while (8.26) yields the bound $\|w\|_{0,T'} \leq h_e^{1/2} \|R_e\|_{0,e}$. But then (8.30) leads to

$$\|R_e\|_{0,e} \leq ch_e^{-1/2} |u - u_h|_{1,\omega_e} + ch_e^{1/2} \sum_{T' \in \omega_e} \|R_{T'}\|_{0,T'},$$

which combined with (8.29) gives

$$h_e^{1/2} \|R_e\|_{0,e} \leq c|u - u_h|_{1,\omega_e} + \sum_{T' \in \omega_e} h_{T'} \|f - f_h\|_{0,T}.
\tag{8.31}$$

Combining (8.29) and (8.31) and taking into account the fact that $\omega_T = \bigcup\{\omega_e; e \in \partial T\}$, we get the desired assertion (8.13). $\square$

Other Estimators

The limitations of the estimators in Theorem 8.1 are easier to recognize in a more general framework. Assuming we want to solve the variational problem $a(u, v) = (f, v)_0$ for $v \in H^1(\Omega)$ (or a subspace), then because of the coercivity we may estimate $\|u - u_h\|_1$ in terms of $\sup_v a(u - u_h, v)$. This expression can be expressed in terms of residuals and jump terms. Now if we regard the variational problem as an operator equation $\mathcal{L}u = f$ with $\mathcal{L} : H^1(\Omega) \to H^{-1}(\Omega)$, then $\mathcal{L}(u - u_h) = \ell$, where ℓ is defined by (8.16). Thus in effect we have determined $\|\mathcal{L}(u - u_h)\|_{-1}$. Since $\mathcal{L}$ is an isomorphism, this gives upper and lower bounds for $\|u - u_h\|_1$.

There may be a big gap between the lower and upper bounds when the condition number of $\mathcal{L}$ is large, i.e. when

$$\frac{\sup\{\|\mathcal{L}u\|_{-1}; \|u\|_1 = 1\}}{\inf\{\|\mathcal{L}u\|_{-1}; \|u\|_1 = 1\}} \gg 1.$$

In this case it is better to compute $\|\tilde{\mathcal{L}}^{-1}\ell\|_1$ instead of $\|\ell\|_{-1}$, where $\tilde{\mathcal{L}}^{-1}$ stands for an approximate inverse. Such an inverse is provided by the variational problem (8.5). Note that the corresponding linear functional is just the restriction of the functional ℓ to the contributions of T and its boundaries. In this sense, the estimator (8.12) is equivalent to the one obtained by solving the Neumann problem.

This and other equivalences are discussed in Verfürth [1996]. Babuška, Durán, and Rodríguez [1992] show that the efficiency of estimators can be much worse for unstructured grids than for regular ones.

To get other estimators, a different idea is used: to compare the current solution with another one which gives a better error order. This is why the auxiliary problem is solved using an enlarged space of approximating functions. The fact that local problems suffice can be shown using the same arguments as employed in the proof of Theorem 8.1.

Local Mesh Refinement

In finite element computations using local grid refinement, we generally start with a coarse grid and continue to refine it successively until the estimator $\eta_{T,R}$ is smaller than a prescribed bound for all elements T. In particular, those elements where the estimators give large values are the ones which are refined. The geometrical aspects have already been discussed in Ch. II, §8.

This leads to a triangulation for which the estimators have approximately equal values in all triangles. Numerical results obtained using this simple idea are quite good, although there is no strict proof of the optimality.

The above approach can be justified heuristically. Suppose the domain Ω in d-space is divided into m (equally large) subdomains where the derivatives of the solution have different size. Suppose an element with mesh size h_i in the i-th subdomain contributes $c_i h_i^\alpha$ to the error, where $\alpha > d$. The subdomains involve different factors c_i, but are all associated with the same exponent α. Now if the i-th subdomain is divided into n_i parts, then $h_i = n_i^{1/d}$, and the total error is of order

$$\sum_i n_i c_i h_i^\alpha = \sum_i c_i h_i^{\alpha-d}. \tag{8.32}$$

Our aim is to minimize the expression (8.32) subject to $\sum_i n_i = \sum_i h_i^{-d} = \text{const}$. The optimum is a stationary point of the Lagrange function

$$L(h,\lambda) := \sum_i c_i h_i^{\alpha-d} + \lambda(\sum_i h_i^{-d} - const). \tag{8.33}$$

If we relax the requirement that the n_i be integers, then we can find the optimum by differentiating (8.33), which leads to $c_i h_i^\alpha = \frac{d\lambda}{\alpha-d}$. This is just the condition that the contributions of all elements be equal. $\qquad\qquad \Box$

Chapter IV

The Conjugate Gradient Method

The discretization of boundary-value problems leads to very large systems of equations which often involve several thousand unknowns. The systems are particularly large for three-dimensional problems and for problems of higher order. Often the bandwidth of the matrices is so large that the classical Gauss elimination algorithm and its modern variants are not efficient methods. This suggests that even for linear problems, we should use iterative methods.

Iterative methods first became popular at the end of the fifties, primarily as a means for solving large problems using computers with a small memory. The methods developed then are no longer competitive, but they still provide useful ingredients for modern iterative methods, and so we review them in §1. The bulk of this chapter is devoted to the conjugate gradient method which is particularly useful for the solution of variational problems and saddle point problems. Since the CG methods discussed here can be applied to a broad spectrum of problems, they are competitive with the still faster multigrid methods to be discussed later (whose implementation is generally more complicated and requires more individual programming).

We begin by classifying problems according to the number n of unknowns:

1. *Small problems*: For linear problems we can use a direct method. For nonlinear problems (e.g. using the Newton method), all elements of the Jacobi matrices should be computed (at least approximately).

2. *Midsized problems*: If the matrices are sparse, we should make use of this fact. For nonlinear problems (e.g. for quasi-Newton methods), the Jacobi matrices should be approximated. Iterative methods can still be used even when the number of steps in the iteration exceeds n.

3. *Very large problems*: Here the only choice is to use iterative methods which require fewer than n steps to compute a solution.

For very large problems, we have to deal with completely different aspects of the method of conjugate gradients as compared with midsized problems. For example, the fact that in exact arithmetic CG methods produce a solution in n steps plays no role for very large problems. In this case it is more important that the accuracy of the approximate solution depends on the condition number of the matrix.

§ 1. Classical Iterative Methods
for Solving Linear Systems

For a small step size h, the finite element method leads to very large systems of equations. Although the associated matrices are sparse, unfortunately their structure is such that after a few (approximately $\sqrt{n}$) steps of Gauss elimination, the computational effort will grow significantly because many zero elements will be replaced by nonzero ones.

This is the reason why iterative methods were studied in the fifties. The methods introduced then, which we shall call *classical iterative methods*, converge very slowly. Nowadays they are rarely used as stand-alone iterative methods. Nevertheless, they have not lost their importance. Indeed they are often used in conjunction with modern iterative methods, for example in the CG method for preconditioning, and in the multigrid method for smoothing.

We now review classical iterative methods. For more detailed treatments, see the monographs of Varga [1962], Young [1971], and Hackbusch [1991].

Stationary Linear Processes

Many iterative methods for the solution of the system $Ax = b$ are based on a decomposition of the matrix

$$A = M - N.$$

Here M is assumed to be an easily invertible matrix, and the given system can be rewritten in the form

$$Mx = Nx + b.$$

This leads to the iteration

$$Mx^{k+1} = Nx^k + b,$$

or equivalently $x^{k+1} = M^{-1}(Nx^k + b)$, or

$$x^{k+1} = x^k + M^{-1}(b - Ax^k), \quad k = 0, 1, 2, \dots . \tag{1.1}$$

Clearly, (1.1) is an iteration of the form

$$x^{k+1} = Gx^k + d \tag{1.2}$$

with $G = M^{-1}N = I - M^{-1}A, d = M^{-1}b$.

The solution x^* of the equation $Ax = b$ is a *fixed point* of the process (1.2), i.e.,

$$x^* = Gx^* + d.$$

Subtracting the last two equations gives

$$x^{k+1} - x^* = G(x^k - x^*),$$

and by induction we see that

$$x^k - x^* = G^k(x^0 - x^*). \tag{1.3}$$

In view of (1.1) we may consider M^{-1} as an *approximate inverse* of A. The associated error reduction per step is described by the matrix $G = I - M^{-1}A$.

The iteration (1.2) is called *convergent* if for every arbitrary $x^0 \in \mathbb{R}^n$, we have $\lim_{k \to \infty} x^k = x^*$. By (1.3), this is equivalent to

$$\lim_{k \to \infty} G^k = 0. \tag{1.4}$$

1.1 Definition. Let A be a (real or complex) $n \times n$ matrix with eigenvalues λ_i, $i = 1, 2, \ldots, m, \, m \leq n$. Then

$$\rho(A) = \max_{1 \leq i \leq m} |\lambda_i|$$

is called the *spectral radius* of A.

The following characterization of (1.4) via the spectral radius is well known in linear algebra (see e.g. Varga [1962], p. 64, or Hackbusch [1991]).

1.2 Theorem. *Let G be an $n \times n$ matrix. Then the following assertions are equivalent:*

(i) The iteration (1.2) converges for every $x^0 \in \mathbb{R}^n$.

(ii) $\lim_{k \to \infty} G^k = 0$.

(iii) $\rho(G) < 1$.

The spectral radius also provides a quantitative measure of the rate of convergence. Note that $\|x^k - x^*\| \leq \|G^k\| \, \|x^0 - x^*\|$, and thus

$$\frac{\|x^k - x^*\|}{\|x^0 - x^*\|} \leq \|G^k\|. \tag{1.5}$$

1.3 Theorem. *For every $n \times n$ matrix G,*

$$\lim_{k \to \infty} \|G^k\|^{1/k} = \rho(G). \tag{1.6}$$

Proof. Let $\varepsilon > 0$, $\rho = \rho(G)$, and

$$B = \frac{1}{\rho + \varepsilon} G.$$

Then $\rho(B) = \rho/(\rho+\varepsilon) < 1$, and $\lim_{k\to\infty} B^k = 0$. Hence, $\sup_{k\geq 0} \|B^k\| \leq a < \infty$. This implies $\|G^k\| \leq a(\rho+\varepsilon)^k$ and $\lim_{k\to\infty} \|G^k\|^{1/k} \leq \rho+\varepsilon$. Thus, the left-hand side of (1.6) is at most $\rho(G)$.

On the other hand, the left-hand side of (1.6) cannot be smaller than $\rho(G)$ since G has an eigenvalue of modulus ρ. ☐

Here we have made use of the usual notation. In particular, M^{-1} can be thought of as an approximate inverse of A. In the framework of the CG method, the matrix M is called a *preconditioner*, and is usually denoted by the symbol C.

The Jacobi and Gauss–Seidel Methods

One way to get an iterative method is to start with the decomposition

$$A = D - L - U.$$

Here D is diagonal, L lower-triangular, and U upper-triangular:

$$D_{ik} = \begin{cases} a_{ik} & \text{if } i = k, \\ 0 & \text{otherwise,} \end{cases}$$

$$L_{ik} = \begin{cases} -a_{ik} & \text{if } i > k, \\ 0 & \text{otherwise,} \end{cases} \qquad U_{ik} = \begin{cases} -a_{ik} & \text{if } i < k, \\ 0 & \text{otherwise.} \end{cases}$$

The *Jacobi method* corresponds to the iteration

$$Dx^{k+1} = (L + U)x^k + b, \tag{1.7}$$

which can be written componentwise as

$$a_{ii}x_i^{k+1} = -\sum_{j \neq i} a_{ij}x_j^k + b_i, \quad i = 1, 2, \ldots, n. \tag{1.8}$$

The associated iteration matrix is $G = G_J = D^{-1}(L + U)$, and thus

$$G_{ik} = \begin{cases} -\dfrac{a_{ik}}{a_{ii}} & \text{for } k \neq i, \\ 0 & \text{otherwise.} \end{cases}$$

Normally we compute the components of the new vector x^{k+1} sequentially, i.e., we compute $x_1^{k+1}, x_2^{k+1}, \ldots, x_n^{k+1}$. During the computation of x_i^{k+1}, we already have the components of the new vector up to the index $j = i - 1$. If we use this information, we are led to *the method of successive relaxation*, also called the *Gauss–Seidel method*:

$$a_{ii}x_i^{k+1} = -\sum_{j<i} a_{ij}x_j^{k+1} - \sum_{j>i} a_{ij}x_j^k + b_i. \qquad (1.9)$$

Here the decomposition in matrix-vector form reads

$$Dx^{k+1} = Lx^{k+1} + Ux^k + b,$$

and the associated iteration matrix is $G = G_{GS} = (D - L)^{-1}U$.

1.4 Example. Consider the matrix

$$A = \begin{pmatrix} 1 & 0.1 \\ 4 & 1 \end{pmatrix}.$$

Then the iteration matrix for the Jacobi method is

$$G_J = \begin{pmatrix} 0 & -0.1 \\ -4 & 0 \end{pmatrix},$$

and the matrix for the Gauss–Seidel method is

$$G_{GS} = -\begin{pmatrix} 1 & 0 \\ 4 & 1 \end{pmatrix}^{-1}\begin{pmatrix} 0 & 0.1 \\ 0 & 0 \end{pmatrix} = -\begin{pmatrix} 1 & 0 \\ -4 & 1 \end{pmatrix}\begin{pmatrix} 0 & 0.1 \\ 0 & 0 \end{pmatrix} = \begin{pmatrix} 0 & -0.1 \\ 0 & 0.4 \end{pmatrix}.$$

A simple calculation shows that $\rho(G_{GS}) = \rho^2(G_J) = 0.4$, and hence both iterations converge.

The convergence of the above methods can often be established with the help of the following simple concept.

1.5 Definition. An $n \times n$ matrix A is said to be *strictly diagonally dominant* provided that

$$|a_{ii}| > \sum_{j \neq i} |a_{ij}|, \quad 1 \leq i \leq n.$$

A is said to be *diagonally dominant* if

$$|a_{ii}| \geq \sum_{j \neq i} |a_{ij}|, \quad 1 \leq i \leq n,$$

and strict inequality holds for at least one i.

1.6 Definition. An $n \times n$ matrix A is called *reducible* if there exists a proper subset $J \subset \{1, 2, \ldots, n\}$ such that

$$a_{ij} = 0 \quad \text{for } i \in J, \quad j \notin J.$$

After reordering, a reducible matrix can be written in block form as

$$\begin{pmatrix} A_{11} & 0 \\ A_{21} & A_{22} \end{pmatrix}.$$

In this case the linear system $Ax = b$ splits into two smaller ones. Thus, there is no loss of generality in restricting our attention to irreducible matrices.

1.7 Diagonal Dominance. *Suppose the $n \times n$ matrix A is diagonally dominant and irreducible. Then both the Jacobi and Gauss–Seidel methods converge.*

Supplement: If A is strictly diagonally dominant, then the hypothesis that A be irreducible is not needed, and the spectral radius of the iteration matrix is

$$\rho \le \max_{1 \le i \le n} \frac{1}{|a_{ii}|} \sum_{j \ne i} |a_{ik}| < 1.$$

Proof. Let $x \ne 0$. For the Jacobi method the iteration matrix is $G = D^{-1}(L + U)$, and with the maximum norm we have

$$\begin{aligned}
|(Gx)_i| &\le \frac{1}{|a_{ii}|} \sum_{j \ne i} |a_{ij}||x_j| \\
&\le \frac{1}{|a_{ii}|} \sum_{j \ne i} |a_{ij}| \cdot \|x\|_\infty \le \|x\|_\infty.
\end{aligned} \tag{1.10}$$

In particular, $\|Gx\|_\infty \le \beta \|x\|_\infty$ with $\beta < 1$ if the stronger condition holds. Thus, $\|G^k x\|_\infty \le \beta^k \|x\|_\infty$, and we have established the convergence.

The diagonal dominance implies $\|Gx\|_\infty \le \|x\|_\infty$, and so $\rho(G) \le 1$. Suppose that $\rho = 1$. Then there exists a vector x with $\|x\|_\infty = \|Gx\|_\infty = 1$, and equality holds in all components of (1.10). Set

$$J = \{j \in \mathbb{N}; \ 1 \le j \le n, \ |x_j| = 1\}.$$

In order for equality to hold in each component of (1.10), we must have

$$a_{ij} = 0 \quad \text{for } i \in J, j \notin J.$$

Moreover, by hypothesis $|a_{ii}| > \sum_{j \ne i} |a_{ij}|$ for some $i = i_0$, and it follows that $|(Gx)_{i_0}| < 1$. By definition, $i_0 \notin J$, and J is a proper subset of $\{1, 2 \ldots, n\}$, contradicting the assumption that A is irreducible.

For the Gauss–Seidel method, the iteration matrix G is given implicitly by

$$Gx = D^{-1}(LGx + Ux).$$

First we prove that $|(Gx)_i| \leq \|x\|_\infty$ for $i = 1, 2, \ldots, n$. By induction on i we can verify that

$$|(Gx)_i| \leq \frac{1}{|a_{ii}|}\{\sum_{j<i}|a_{ij}|\,|(Gx)_j| + \sum_{j>i}|a_{ij}|\,|x_j|\}$$

$$\leq \frac{1}{|a_{ii}|}\sum_{j\neq i}|a_{ij}|\,\|x\|_\infty \leq \|x\|_\infty.$$

The rest of the proof is similar to that for the Jacobi method.   □

The Model Problem

The system of equations which arises in using the five-point stencil (II.4.9) to discretize the Poisson equation on a square of lenth L has been frequently studied. With $h = L/m$ and $(m - 1) \times (m - 1)$ interior points, we have

$$4x_{i,j} - x_{i+1,j} - x_{i-1,j} - x_{i,j+1} - x_{i,j-1} = b_{i,j}, \quad 1 \leq i, j \leq m - 1, \quad (1.11)$$

where variables with index 0 or m are assumed to be zero. Clearly, the associated matrix is diagonally dominant, and since the system is irreducible, this suffices for the convergence.

Here we have the special case where the diagonal D is a multiple of the unit matrix. Hence, the iteration matrix for the Jacobi method can be obtained immediately from the original matrix:

$$G = I - \frac{1}{4}A.$$

In particular, A and G have the same eigenvectors $z^{k,\ell}$, as can be seen by simple substitution using elementary properties of trigonometric functions:

$$\left.\begin{array}{l} (z^{k,\ell})_{i,j} = \sin\dfrac{ik\pi}{m}\sin\dfrac{j\ell\pi}{m}, \\[2mm] Az^{k,\ell} = (4 - 2\cos\dfrac{k\pi}{m} - 2\cos\dfrac{\ell\pi}{m})z^{k,\ell}, \\[2mm] Gz^{k,\ell} = (\dfrac{1}{2}\cos\dfrac{k\pi}{m} + \dfrac{1}{2}\cos\dfrac{\ell\pi}{m})z^{k,\ell}, \end{array}\right\} \quad 1 \leq k, \ell \leq m - 1. \quad (1.12)$$

The eigenvalue of G with largest modulus corresponds to $k = \ell = 1$, and we have

$$\rho(G) = \frac{1}{2}(1 + \cos\frac{\pi}{m}) = 1 - \frac{\pi^2}{4m^2} + \mathcal{O}(m^{-4}). \quad (1.13)$$

For large m, the quantity $\rho(G)$ tends to 1, and so the rate of convergence is very poor for grids with small mesh size h.

The situation is slightly better for the Gauss–Seidel method, but even then we have $\rho(G) = 1 - \mathcal{O}(m^{-2})$.

Overrelaxation

Overrelaxation was the first acceleration technique. In updating via (1.9), the change in the value of x_i is immediately multiplied by a factor ω:

$$a_{ii}x_i^{k+1} = \omega[-\sum_{j<i} a_{ij}x_j^{k+1} - \sum_{j>i} a_{ij}x_j^k + b_i] - (\omega - 1)a_{ii}x_i^k, \qquad (1.14)$$

or in matrix form

$$Dx^{k+1} = \omega[Lx^{k+1} + Ux^k + b] - (\omega - 1)Dx^k. \qquad (1.15)$$

If the factor ω is larger than 1, this is called *overrelaxation*, otherwise it is called *underrelaxation*. The process of *successive overrelaxation* ($\omega > 1$) is referred to as the SOR method.

Let A be a symmetric positive definite matrix. Then the Gauss–Seidel method (with or without relaxation) can be viewed as the process of minimizing the expression

$$f(x) = \frac{1}{2}x'Ax - b'x. \qquad (1.16)$$

Clearly,

$$\frac{\partial f}{\partial x_i} = (Ax - b)_i.$$

If we fix $x_1, x_2, \ldots, x_{i-1}, x_{i+1} \ldots, x_n$ and vary only x_i, then a minimum of $f(x)$ as a function of x_i is characterized by the equation $\partial f/\partial x_i = 0$, i.e., by $(Ax-b)_i = 0$. This gives the formula (1.9). In particular, the improvement in the value of f after one step is given by $\frac{1}{2}a_{ii}(x_i^{k+1} - x_i^k)^2$. The value of f can be reduced using relaxation for any choice of $\omega \in (0, 2)$. The improvement is then $\frac{1}{2}\omega(2 - \omega)$ $[(x_i^{k+1} - x_i^k)/\omega]^2$. These quantities also arise in the proof of the following theorem.

1.8 Ostrowski–Reich Theorem. *Suppose A is a symmetric $n \times n$ matrix with positive diagonal elements, and let $0 < \omega < 2$. Then the SOR method (1.14) converges if and only if A is positive definite.*

Proof. After an elementary manipulation, it follows from (1.15) that

$$(1 - \frac{\omega}{2})D(x^{k+1} - x^k) = \omega[(L - \frac{1}{2}D)x^{k+1} + (U - \frac{1}{2}D)x^k + b].$$

Multiplying on the left by $(x^{k+1} - x^k)$ and taking into account $y'Lz = z'Uy$, we have

$$(1 - \frac{\omega}{2})(x^{k+1} - x^k)'D(x^{k+1} - x^k)$$
$$= \frac{\omega}{2}[-x^{k+1'}Ax^{k+1} + x^{k'}Ax^k + 2b'(x^{k+1} - x^k)] \qquad (1.17)$$
$$= \omega[f(x^k) - f(x^{k+1})].$$

Since the convergence only depends on the spectral radius of the iteration matrix, without loss of generality we can assume that $b = 0$. Moreover, observe that $x^{k+1} \neq x^k$ since $Ax^k - b \neq 0$.

(1) Let A be positive definite. For every x^0 in the unit sphere

$$S^{n-1} := \{x \in \mathbb{R}^n; \ \|x\| = 1\},$$

(1.17) implies

$$\frac{x^{1'}Ax^1}{x^{0'}Ax^0} < 1.$$

In view of the continuity of the mapping $x^0 \mapsto x^1$ and the compactness of the sphere S^{n-1}, there exists $\beta < 1$ with

$$\frac{x^{1'}Ax^1}{x^{0'}Ax^0} \leq \beta < 1, \tag{1.18}$$

for all $x^0 \in S^{n-1}$. Since the quotient on the left-hand side of (1.18) does not change if x^0 is multiplied by a factor $\neq 0$, (1.18) holds for all $x^0 \neq 0$. It follows by induction that

$$x^{k'}Ax^k \leq \beta^k x^{0'}Ax^0,$$

and thus $x^{k'}Ax^k \to 0$. By the definiteness of A, we have $\lim_{k\to\infty} x^k = 0$.

(2) Let A be indefinite. Without loss of generality, consider the iteration (1.9) for the homogeneous equation with $b = 0$. Then there exists $x^0 \neq 0$ with $\alpha := f(x^0) < 0$. Now (1.17) implies $f(x^k) \leq f(x^{k-1})$ and

$$f(x^k) \leq \alpha < 0, \quad k = 0, 1 \dots,$$

and we conclude that $x^k \nrightarrow 0$. □

In carrying out the relaxation methods (1.9) and (1.14), the components of the vectors are recomputed in the order from $i = 1$ to $i = n$. Obviously, we can also run through the indices in the reverse order. In this case, the roles of the submatrices L and U are reversed. In the *symmetric* SOR *method*, SSOR method for short, the iteration is performed alternately in the forward and backward directions. Thus, each iteration step consists of two half steps:

$$\begin{aligned} Dx^{k+1/2} &= \omega[Lx^{k+1/2} + Ux^k + b] - (\omega - 1)Dx^k, \\ Dx^{k+1} &= \omega[Lx^{k+1/2} + Ux^{k+1} + b] - (\omega - 1)Dx^{k+1/2}. \end{aligned} \tag{1.19}$$

By the remark following (1.16), we conclude that the assertion of the theorem also holds for the symmetric iteration process.

The SSOR method has two advantages. The computational effort in performing k SSOR cycles is not the same as that for $2k$ cycles of the SOR method, but only to $k + 1/2$ cycles. Moreover, the associated iteration matrix

$$M^{-1} = \omega(2 - \omega)(D - \omega U)^{-1}D(D - \omega L)^{-1} \tag{1.20}$$

in the sense of (1.1) is symmetric; cf. Problem 1.10.

Problems

1.9 Write the SSOR method (1.19) in componentwise form similar to (1.14).

1.10 Verify (1.20). – Hint: Under iteration the matrix A corresponds to the approximate inverse M^{-1}. In particular, $x^1 = M^{-1}b$ for $x^0 = 0$.

1.11 Consider the matrices

$$A_1 = \begin{pmatrix} 1 & 2 & -2 \\ 1 & 1 & 1 \\ 2 & 2 & 1 \end{pmatrix} \quad \text{and} \quad A_2 = \frac{1}{2}\begin{pmatrix} 2 & -1 & 1 \\ 2 & 2 & 2 \\ -1 & -1 & 2 \end{pmatrix}.$$

For which of the matrices do the Jacobi method and Gauss–Seidel methods converge?

1.12 Let G be an $n \times n$ matrix with $\lim_{k\to\infty} G^k = 0$. In addition, suppose $\|\cdot\|$ is an arbitrary vector norm on $\mathbb{R}^n$. Show that

$$\|\|x\|\| := \sum_{k=0}^{\infty} \|G^k x\|$$

defines a norm on $\mathbb{R}^n$, and that $\|\|G\|\| < 1$ for the associated matrix norm. – Give an example to show that $\rho(G) < 1$ does not imply $\|G\| < 1$ for every arbitrary norm.

1.13 If we choose $\omega = 2$ in the SSOR method, then $x^{k+1} = x^k$. How can this be shown without using any formulae?

1.14 Suppose the Jacobi and Gauss–Seidel methods converge for the equation $Ax = b$. In addition, suppose D is a nonsingular diagonal matrix. Do we still get convergence if AD (or DA) is substituted for A?

1.15 A matrix B is called *nonnegative*, written as $B \geq 0$, if all matrix elements are nonnegative. Let $D, L, U \geq 0$, and suppose the Jacobi method converges for $A = D - L - U$. Show that this implies $A^{-1} \geq 0$. – What is the connection with the discrete maximum principle?

1.16 Let M^{-1} be an approximate inverse of A in the sense of (1.1). Determine the approximate inverse that corresponds to k steps of the iteration.

§ 2. Gradient Methods

In developing iterative methods for the solution of systems of equations associated with a positive definite matrix A, it is very useful to observe that the solution of $Ax = b$ is also the minimum of

$$f(x) = \frac{1}{2}x'Ax - b'x. \tag{2.1}$$

The simplest method for finding a minimum is the *gradient method*. In its classical form it is a stable method, but converges very slowly if the condition number $\kappa(A)$ is large. Unfortunately, this is usually the case for the systems which arise in the discretization of elliptic boundary-value problems. For problems of order $2m$, the condition number typically grows like h^{-2m}.

The so-called *PCG method* (see §§3 and 4) was developed from the gradient method by making two modifications to it, and is one of the most effective system solvers.

The General Gradient Method

For completeness, we formulate the gradient method in a more general form as a method for the minimization of a C^1 function f defined on an open set $M \subset \mathbb{R}^n$.

2.1 Gradient Method with Complete Line Search.

Choose $x_0 \in M$. For $k = 0, 1, 2, \ldots$, perform the following calculations:

1. Determine the direction: Compute the negative gradient

$$d_k = -\nabla f(x_k). \tag{2.2}$$

2. Line search: Find a point $t = \alpha_k$ along the line $\{x_k + td_k : t \geq 0\} \cap M$ where f attains a (local) minimum. Set

$$x_{k+1} = x_k + \alpha_k d_k. \tag{2.3}$$

□

Clearly, the method generates a sequence (x_k) with

$$f(x_0) \geq f(x_1) \geq f(x_2) \geq \cdots,$$

where equality holds only at points where the gradient vanishes.

For the special case of the quadratic function (2.1), we have

$$d_k = b - Ax_k, \tag{2.4}$$

$$\alpha_k = \frac{d_k'd_k}{d_k'Ad_k}. \tag{2.5}$$

2.2 Remark. In practice the line search is done only approximately. There are two possibilities:

(1) Find a point for which the directional derivative is small, say

$$|d_k' \nabla f(x_k + t d_k)| < \frac{1}{4} |d_k' \nabla f(x_k)|.$$

(2) First choose a reasonable step size t, and continue halving it until the corresponding improvement in the function value is at least one quarter of what we would get by linearization, i.e., until

$$f(x_k + t d_k) \leq f(x_k) - \frac{t}{4} |d_k' \nabla f(x_k)|.$$

This version of the gradient method leads to global convergence: *either there is some subsequence which converges to a point $x \in M$ with $\nabla f(x) = 0$, or the sequence tends to the boundary of M.* The latter cannot happen if f is greater than $f(x_0)$ as we approach ∂M (or as $x \to \infty$, respectively).

Gradient Methods and Quadratic Functions

Rather than presenting a general proof of convergence, we instead focus on a more detailed examination of the case of quadratic functions. In this case the rate of convergence is determined by the size of $\kappa(A)$.

As a measure of distance, we use the *energy norm*

$$\|x\|_A := \sqrt{x'Ax}. \tag{2.6}$$

If x^* is a solution of the equation $Ax = b$, then as in II.2.4

$$f(x) = f(x^*) + \frac{1}{2} \|x - x^*\|_A^2. \tag{2.7}$$

Now (2.4) and (2.5) imply

$$\begin{aligned}
f(x_{k+1}) &= f(x_k + \alpha_k d_k) \\
&= \frac{1}{2}(x_k + \alpha_k d_k)' A(x_k + \alpha_k d_k) - b'(x_k + \alpha_k d_k) \\
&= f(x_k) + \alpha_k d_k'(Ax_k - b) + \frac{1}{2}\alpha_k^2 \, d_k' A d_k \\
&= f(x_k) - \frac{1}{2} \frac{(d_k' d_k)^2}{d_k' A d_k}.
\end{aligned}$$

Combining this with (2.7), we have

$$\|x_{k+1} - x^*\|_A^2 = \|x_k - x^*\|_A^2 - \frac{(d_k' d_k)^2}{d_k' A d_k}.$$

Since $d_k = -A(x_k - x^*)$, we have $\|x_k - x^*\|_A^2 = (A^{-1}d_k)'A(A^{-1}d_k) = d_k'A^{-1}d_k$ and

$$\|x_{k+1} - x^*\|_A^2 = \|x_k - x^*\|_A^2 \left[1 - \frac{(d_k'd_k)^2}{d_k'Ad_k \, d_k'A^{-1}d_k}\right]. \qquad (2.8)$$

We now estimate the quantity in the square brackets. If we compute the condition number of a matrix with respect to the Euclidean vector norm, then for positive definite matrices, it coincides with the *spectral condition number*

$$\kappa(A) = \frac{\lambda_{\max}(A)}{\lambda_{\min}(A)}.$$

2.3 The Kantorovitch Inequality. *Let A be a symmetric, positive definite matrix with spectral condition number κ. Then*

$$\frac{(x'Ax)(x'A^{-1}x)}{(x'x)^2} \leq (\frac{1}{2}\sqrt{\kappa} + \frac{1}{2}\sqrt{\kappa^{-1}})^2 \qquad (2.9)$$

for every vector $x \neq 0$.

Proof. Set $\mu := [\lambda_{\max}(A)\,\lambda_{\min}(A)]^{1/2}$, and let λ_i be an eigenvalue of A. It follows that $\kappa^{-1/2} \leq \lambda_i/\mu \leq \kappa^{1/2}$. Since the function $z \mapsto z + z^{-1}$ is monotone on the interval $(0, 1)$ and for $z > 1$, we know that

$$\lambda_i/\mu + \mu/\lambda_i \leq \kappa^{1/2} + \kappa^{-1/2}.$$

The eigenvectors of A are also eigenvectors of the matrix $\frac{1}{\mu}A + \mu A^{-1}$, and the eigenvalues of the latter are bounded by $\kappa^{1/2} + \kappa^{-1/2}$. Hence, it follows from Courant's maximum principle that

$$\frac{1}{\mu}(x'Ax) + \mu(x'A^{-1}x) \leq (\kappa^{1/2} + \kappa^{-1/2})\,(x'x)$$

holds for all $x \in \mathbb{R}^n$. The variant of Young's inequality $ab \leq \frac{1}{4}(|a| + |b|)^2$ now yields

$$(x'Ax)(x'A^{-1}x) \leq \frac{1}{4}\left[\frac{1}{\mu}(x'Ax) + \mu(x'A^{-1}x)\right]^2 \leq \frac{1}{4}(\kappa^{1/2} + \kappa^{-1/2})^2\,(x'x)^2,$$

and the proof is complete. $\qquad\qquad\square$

The left-hand side of (2.9) attains its maximum when the vector x contains only components from the eigenvectors that are associated to the largest and the smallest eigenvalue and if the two contributions have the same size. Furthermore, Example 2.5 below shows that the bound is sharp.

Now combining (2.8), (2.9), and the identity $1 - 4/(\sqrt{\kappa} + \sqrt{\kappa^{-1}})^2 = (\kappa - 1)^2/(\kappa + 1)^2$, we get

2.4 Theorem. *Suppose A is a symmetric positive definite matrix with spectral condition number κ. Then applying the gradient method to the function (2.1) generates a sequence with*

$$\|x_k - x^*\|_A \le \left(\frac{\kappa - 1}{\kappa + 1}\right)^k \|x_0 - x^*\|_A. \tag{2.11}$$

Convergence Behavior in the Case of Large Condition Numbers

If we solve a system of equations with a very large condition number κ, the convergence rate

$$\frac{\kappa - 1}{\kappa + 1} \approx 1 - \frac{2}{\kappa}$$

is very close to 1. We can show that this unfavorable rate dominates the iteration with a simple example involving two unknowns.

2.5 Example. Let $a \gg 1$. We seek the minimum of

$$f(x, y) = \frac{1}{2}(x^2 + ay^2), \quad \text{i.e.,} \quad A = \begin{pmatrix} 1 & \\ & a \end{pmatrix}. \tag{2.12}$$

Here $\kappa(A) = a$. Suppose we choose

$$(x_0, y_0) = (a, 1)$$

as the starting vector. Then the direction of steepest descent is given by $(-1, -1)$. We claim that

$$x_{k+1} = \rho x_k, \quad y_{k+1} = -\rho y_k, \quad k = 0, 1, \dots, \tag{2.13}$$

where $\rho = (a - 1)/(a + 1)$. This is easily checked for $k = 0$, and it follows for all k by symmetry arguments. Thus, the convergence rate is exactly as described in Theorem 2.4.

The contour lines of the function (2.12) are strongly elongated ellipses (see Fig. 45), and the angle between the gradient and the direction leading to a minimum can be very large (cf. Problem 2.8).

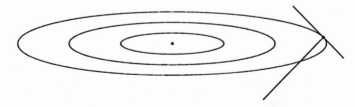

Fig. 45. Contour lines and the gradient for Example 2.5

For two-dimensional problems, there is one exceptional case. Since $d_k' \nabla f(x_{k+1}) = 0$, we have $d_k' d_{k+1} = 0$. Thus, d_k and d_{k+2} are parallel in $\mathbb{R}^2$. This cycling does not occur in n-space for $n \geq 3$. If we start with an arbitrary vector in $\mathbb{R}^n$ ($n \geq 3$), after a few iterations we get a vector which lies in a position similar to that in Example 2.5. This agrees with the observation that the iteration nearly stops after a few steps, and any further approach to the solution will be extremely slow.

Problems

2.6 By (2.5), $\alpha_k \geq \alpha^* := 1/\lambda_{\max}(A)$. Show that convergence is guaranteed for every fixed step size α with $0 < \alpha < 2\alpha^*$.

2.7 Establish the recurrence (2.13) in Example 2.5.

2.8 (a) In Example 2.5 the directions d_k and d_{k+1} are orthogonal w.r.t. the Euclidean metric. Show that they are nearly parallel in the metric defined by (2.6). Here the angle between two vectors x and y is given by

$$\cos \varphi = \frac{x'Ay}{\|x\|_A \|y\|_A}. \tag{2.14}$$

(b) Show that $|\cos \varphi| \leq \frac{\kappa-1}{\kappa+1}$ for the angle defined in (2.14), provided $x'y = 0$.
Hint: For the unit vector,

$$\frac{1 - \cos \varphi}{1 + \cos \varphi} = \frac{\|x - y\|_A}{\|x + y\|_A}.$$

2.9 Consider the unbounded quadratic form

$$f(x) = \sum_{j=1}^{\infty} j |x^{(j)}|^2$$

in ℓ_2. Let $x_0 = (1, 1/8, 1/27, \ldots)$, or more generally let x_0 be any element with $f(x_0) < \infty$, $\nabla f(x_0) \in \ell_2$, such that the rate of decay of the components is polynomial. Show that the gradient method stops after finitely many steps.

2.10 Use the Kantorovitch inequality to give upper and lower bounds for the quotients of

$$(x'Ax)^2 \quad \text{and} \quad (x'A^2x)(x'x)$$

and also for those of

$$(x'Ax)^n \quad \text{and} \quad (x'A^nx)(x'x)^{n-1}.$$

§ 3. Conjugate Gradients
and the Minimal Residual Method

The conjugate gradient method was developed in 1952 by Hestenes and Stiefel, and first became of importance in 1971 when it was combined with a simple method for preconditioning by Reid [1971]. The big advantage of the conjugate gradient method became apparent when faster computers became available and larger problems were attacked.

For the systems of equations which arise in the discretization of two-dimensional second order elliptic boundary-value problems, the method is more efficient than Gauss elimination as soon as 400 to 800 unknowns are involved, and also uses much less memory. For three-dimensional problems, the advantages are even more significant than for two dimensions. The advantages are less clear for problems of fourth order.

In this section we restrict our discussion to linear problems. We discuss further details in the next section, where we also present a generalization to nonlinear minimization problems.

The basic idea behind the *conjugate gradient method* (CG method) is to make sure that successive directions are not nearly parallel (in the sense of Problem 2.8). The idea is to use orthogonal directions, where orthogonality is measured in a metric more suited to problem (2.1) than the *Euclidean metric*.

3.1 Definition. Let A be a symmetric nonsingular matrix. Then two vectors x and y are called *conjugate* or *A-orthogonal* provided $x'Ay = 0$.

In the following we will assume that A is positive definite. In this case, any set of k pairwise conjugate vectors $x_1, x_2, \ldots, x_k$ are linearly independent provided that none of them is the zero vector.

In particular, suppose $d_0, d_1, \ldots, d_{n-1}$ are conjugate directions, and that the desired solution $x^* = A^{-1}b$ has the expansion

$$x^* = \sum_{k=0}^{n-1} \alpha_k d_k$$

in terms of this basis. Then in view of the orthogonality relations, we have $d_i' A x^* = \sum_k d_i' A \alpha_k d_k = \alpha_i d_i' A d_i$, and

$$\alpha_i = \frac{d_i' A x^*}{d_i' A d_i} = \frac{d_i' b}{d_i' A d_i}. \tag{3.1}$$

Thus, when using conjugate vectors – in contrast to using an arbitrary basis – we can compute the coefficients α_i in the expansion of x^* directly from the given vector b.

Let x_0 be an arbitrary vector in $\mathbb{R}^n$. Then an expansion of the desired correction vector $x^* - x_0$ in terms of conjugate directions can be computed recursively directly from the gradients $g_k := \nabla f(x_k)$.

3.2 Lemma on Conjugate Directions. *Let $d_0, d_1, \ldots, d_{n-1}$ be conjugate directions, and let $x_0 \in \mathbb{R}^n$. Then the sequence generated by the recursion*

$$x_{k+1} = x_k + \alpha_k d_k$$

with
$$\alpha_k = -\frac{d_k' g_k}{d_k' A d_k}, \quad g_k := A x_k - b,$$

gives a solution $x_n = A^{-1} b$ after (at most) n steps.

Proof. Writing $x^* - x_0 = \sum_i \alpha_i d_i$, from (3.1) we immediately have

$$\alpha_k = \frac{d_k' A (x^* - x_0)}{d_k' A d_k} = -\frac{d_k' (A x_0 - b)}{d_k' A d_k}.$$

Since the vector d_k is assumed to be conjugate to the other directions, we have $d_k' A (x_k - x_0) = d_k' A \sum_i^{k-1} \alpha_i d_i = 0$. Hence,

$$\alpha_k = -\frac{d_k' (A x_k - b)}{d_k' A d_k} = -\frac{d_k' g_k}{d_k' A d_k}. \qquad \square$$

3.3 Corollary. *Under the hypotheses of 3.2, x_k minimizes the function f not only on the line $\{x_{k-1} + \alpha d_{k-1}; \ \alpha \in \mathbb{R}\}$, but also over $x_0 + V_k$, where $V_k := \text{span}[d_0, d_1, \ldots, d_{k-1}]$. In particular,*

$$d_i' g_k = 0 \quad \text{for } i < k. \tag{3.2}$$

Proof. It suffices to establish the relation (3.2). The choice of α_i ensures that

$$d_i' g_{i+1} = 0. \tag{3.3}$$

Thus, (3.2) holds for $k = 1$. Assume that the assertion has been proved for $k - 1$. From $x_k - x_{k-1} = \alpha_{k-1} d_{k-1}$ it follows that $g_k - g_{k-1} = A(x_k - x_{k-1}) = \alpha_{k-1} A d_{k-1}$. Since the directions $d_0, d_1, \ldots, d_{k-1}$ are conjugate, we have $d_i'(g_k - g_{k-1}) = 0$ for $i < k - 1$. Combining this with the induction hypothesis, i.e. (3.2) for $k - 1$, we conclude that the formula is also correct for k and $i \le k - 2$. The remaining equation for k and $i = k - 1$ is a direct consequence of (3.3). $\qquad \square$

The CG Algorithm

In the conjugate gradient method, the directions $d_0, d_1, \ldots, d_{n-1}$ are not selected in advance, but are computed from the current gradients g_k by addition of a correction factor. From an algorithmic point of view, this means that we do not need a complicated orthogonalization process, but instead can use a simple three-term recurrence formula.[9] We show later that this approach makes sense from an analytic point of view.

3.4 The Conjugate Gradient Method.
Let $x_0 \in \mathbb{R}^n$, and set $d_0 = -g_0 = b - Ax_0$.
For $k = 0, 1, 2, \ldots$, compute

$$
\begin{aligned}
\alpha_k &= \frac{g_k' g_k}{d_k' A d_k}, \\
x_{k+1} &= x_k + \alpha_k d_k, \\
g_{k+1} &= g_k + \alpha_k A d_k, \\
\beta_k &= \frac{g_{k+1}' g_{k+1}}{g_k' g_k}, \\
d_{k+1} &= -g_{k+1} + \beta_k d_k,
\end{aligned}
\tag{3.4}
$$

while $g_k \neq 0$.   □

We note that in the derivation of the above, we initially get

$$
\alpha_k = -\frac{g_k' d_k}{d_k' A d_k}, \quad \beta_k = \frac{g_{k+1}' A d_k}{d_k' A d_k}.
\tag{3.5}
$$

For quadratic problems, the expressions in (3.4) are equivalent, but are more stable from a numerical point of view, and require less memory.

3.5 Properties of the CG Method. *While $g_{k-1} \neq 0$, we have the following:*
(1) $d_{k-1} \neq 0$.
(2)
$$
\begin{aligned}
V_k :&= \operatorname{span}[g_0, Ag_0, \ldots, A^{k-1} g_0] \\
&= \operatorname{span}[g_0, g_1, \ldots, g_{k-1}] = \operatorname{span}[d_0, d_1, \ldots, d_{k-1}].
\end{aligned}
$$

(3) The vectors $d_0, d_1, \ldots, d_{k-1}$ are pairwise conjugate.
(4)
$$
f(x_k) = \min_{z \in V_k} f(x_0 + z).
\tag{3.6}
$$

---

[9] Three-term recurrence formulae are well known for orthogonal polynomials. The connection is explored in Problem 3.9.

Proof. The assertions are obvious for $k = 1$. Suppose that they hold for some $k \geq 1$. Then

$$g_k = g_{k-1} + A(x_k - x_{k-1}) = g_{k-1} + \alpha_{k-1}Ad_{k-1},$$

and thus $g_k \in V_{k+1}$ and span$[g_i]_{i=0}^{k} \subset V_{k+1}$. By the induction hypothesis, $d_0, d_1, \ldots, d_{k-1}$ are conjugate, and by the optimality of x_k, we have

$$d_i'g_k = 0 \quad \text{for } i < k. \tag{3.7}$$

Thus, g_k is linearly dependent on $d_0, d_1, \ldots, d_{k-1}$ only if $g_k = 0$, and so $g_k \neq 0$ implies $g_k \notin V_k$. It follows that span$[g_i]_{i=0}^{k}$ is a $(k + 1)$-dimensional space, and cannot be a proper subspace of V_{k+1}. This establishes the first equality in (2) for $k + 1$. Moreover, V_{k+1} coincides with span$[d_i]_{i=0}^{k}$, since in view of $g_k + d_k \in V_k$, we could just as well have included d_k.

Now $d_k \neq 0$ immediately implies $g_k + d_k \in V_k$ provided $g_k \neq 0$, and thus (1) holds.

For the proof of (3), we compute

$$d_i'Ad_k = -d_i'Ag_k + \beta_{k-1}d_i'Ad_{k-1}. \tag{3.8}$$

For $i \leq k-2$ the first term on the right-hand side vanishes because $Ad_i \in AV_{k-1} \subset V_k$ and (3.7) holds. Moreover, by assumption the second term is zero. Finally, with β_k as in (3.5), the right-hand side of (3.8) vanishes for $i = k - 1$.

The last assertion follows from Corollary 3.3, and the inductive proof is complete.   □

The use of conjugate directions prevents the iteration from proceeding in nearly parallel directions (cf. Problem 2.8). In order to get conjugate directions from nearly parallel directions, we need to use very large factors β_k. Thus in principle, we have to keep in mind the possibility that small denominators will be encountered due to roundoff errors, and the iteration will have to be restarted. This doesn't happen as often as might be expected, however, since by Problem 3.13, the denominator can be reduced by at most a factor κ; see e.g. Powell [1977].

In this connection we should note that the accumulated roundoff errors do not lead to a loss of stability. Since d_k is a linear combination of g_k and d_{k-1}, x_{k+1} provides a minimum for f over the two-dimensional manifold

$$x_k + \text{span}[g_k, d_{k-1}].$$

This problem is still well posed if the variables with roundoff errors are used in the calculation instead of the exact vectors g_k and d_{k-1}. The choice of the coefficients α_k and β_k according to (3.4) ensures that the *perturbed* two-dimensional minimum problem is well posed; see 3.13 and 4.11. This is the reason for the numerical stability of the CG method.

Analysis of the CG method as an Optimal Method

The CG method finds a minimum of a quadratic function on $\mathbb{R}^n$ in (at most) n steps. However, at the time of its discovery, the fact that this iterative method finds a solution in n steps was overemphasized. Indeed, for problems with 1000 or more unknowns, this property is of little significance. What is much more important is the fact that a very good approximation can already be found with many fewer than n steps.

First we note that for all iterative methods of the form

$$x_{k+1} = x_k + \alpha_k(b - Ax_k), \quad \text{with } \alpha_k \in \mathbb{R}$$

– and thus also for the simplest gradient method – the approximation x_k lies in $x_0 + V_k$, where V_k is defined as in 3.5(2). It follows from 3.5(2) that among all of these methods, the CG method is the one which gives the smallest error $\|x_k - x^*\|_A$.

As usual, the *spectrum* $\sigma(A)$ of a matrix A is the set of its eigenvalues.

3.6 Lemma. *Suppose there exists a polynomial $p \in \mathcal{P}_k$ with*

$$p(0) = 1 \quad and \quad |p(z)| \le r \text{ for all } z \in \sigma(A). \tag{3.9}$$

Then for arbitrary $x_0 \in \mathbb{R}^n$, the CG method satisfies

$$\|x_k - x^*\|_A \le r\|x_0 - x^*\|_A. \tag{3.10}$$

Proof. Set $q(z) = (p(z) - 1)/z$. Using the same notation as in 3.5, we have $y := x_0 + q(A)g_0 \in x_0 + V_k$, and $g_0 = A(x_0 - x^*)$ implies

$$y - x^* = x_0 - x^* + y - x_0 = x_0 - x^* + q(A)g_0$$
$$= p(A)(x_0 - x^*).$$

Now let $\{z_j\}_{j=1}^n$ be a complete system of orthonormal eigenvectors with $Az_j = \lambda_j z_j$ and $x_0 - x^* = \sum_j c_j z_j$. Then

$$y - x^* = \sum_j c_j p(A) z_j = \sum_j c_j p(\lambda_j) z_j. \tag{3.11}$$

The orthogonality of the eigenvectors implies

$$\|x_0 - x^*\|_A^2 = \sum_j \lambda_j |c_j|^2 \tag{3.12}$$

and

$$\|y - x^*\|_A^2 = \sum_j \lambda_j |c_j p(\lambda_j)|^2 \le r^2 \sum_j \lambda_j |c_j|^2.$$

Thus, $\|y - x^*\|_A \le r\|x_0 - x^*\|_A$. Combining this with $y \in x_0 + V_k$ and the minimal property 3.5(4) for x_k, we get (3.10). $\qquad\square$

If we only know that the spectrum lies in the interval $[a, b]$, where $b/a = \kappa$, then we get optimal estimates using the so-called *Chebyshev polynomials*[10]

$$T_k(x) := \frac{1}{2}[(x + \sqrt{x^2 - 1})^k + (x - \sqrt{x^2 - 1})^k], \quad k = 0, 1, \ldots \quad (3.13)$$

The formulae (3.13) define polynomials with real coefficients since after multiplying out and using the binomial formula, the terms with odd powers of the roots cancel each other. Moreover, $|x + \sqrt{x^2 - 1}| = |x + i\sqrt{1 - x^2}| = 1$ for real $|x| \le 1$, and thus

$$T_k(1) = 1 \quad \text{and} \quad |T_k(x)| \le 1 \quad \text{for} \; -1 \le x \le 1. \quad (3.14)$$

The special polynomial $p(z) := T([b + a - 2z]/[b - a])/T([b + a]/[b - a])$ satisfies $p(0) = 1$, and so Lemma 3.6 implies the main result:

3.7 Theorem. *For any starting vector $x_0 \in \mathbb{R}^n$, the CG method generates a sequence x_k satisfying*

$$\|x_k - x^*\|_A \le \frac{1}{T_k\left(\dfrac{\kappa + 1}{\kappa - 1}\right)} \|x_0 - x^*\|_A$$

$$(3.15)$$

$$\le 2\left(\frac{\sqrt{\kappa} - 1}{\sqrt{\kappa} + 1}\right)^k \|x_0 - x^*\|_A .$$

Proof. Since $\sigma(A) \subset [\lambda_{\min}, \lambda_{\max}]$ and $\kappa = \lambda_{\max}/\lambda_{\min}$, the first inequality follows with the special polynomial above. Recalling (3.13) we clearly have $T_k(z) \ge \frac{1}{2}(z + \sqrt{z^2 - 1})^k$ for $z \in [1, \infty)$. We evaluate $z + \sqrt{z^2 - 1}$ for $z := (\kappa + 1)/(\kappa - 1)$ and use $\kappa - 1 = (\sqrt{\kappa} + 1)(\sqrt{\kappa} - 1)$:

$$\frac{\kappa + 1}{\kappa - 1} + \sqrt{\left(\frac{\kappa + 1}{\kappa - 1}\right)^2 - 1} = \frac{\kappa + 1 + \sqrt{4\kappa}}{\kappa - 1} = \frac{\sqrt{\kappa} + 1}{\sqrt{\kappa} - 1}.$$

This establishes the second assertion. $\qquad\qquad\qquad\qquad\qquad\qquad\qquad\square$

A comparison with Theorem 2.4 and Example 2.5 shows that the computation of conjugate directions has the same positive effect as replacing the condition number by its square root. In practice, the improvement in each iteration is usually even better than theoretically predicted by Theorem 3.7. The inequality (3.15) also covers the pessimistic case where the eigenvalues are uniformly distributed between $\lambda_{\min}$ and $\lambda_{\max}$. Frequently, the eigenvalues appear in groups, and because of the gaps in the spectrum, Theorem 3.7 is too pessimistic.

Here we should point out the connection with the so-called *semi-iterative methods* [Varga 1962]. They are based on modifying the relaxation method

---

[10] The usual definition $T_k(x) := \cos(k \arccos x)$ is equivalent to (3.13).

so that using the comparison polynomial q_k constructed in Theorem 3.7, we actually get $x_k = x_0 + q_k(A)(x_0 - x^*)$ for certain k. The process is optimal if $\lambda_{\min}$ and $\lambda_{\max}$ are known. – The situation is better for the conjugate gradient method since for it, even this information is not needed, and since the gaps in the spectrum automatically lead to an acceleration in the convergence.

In practice, the errors decrease nonuniformly in the course of the iteration. After a clear reduction of the error in the first few steps, there is often a phase with only small improvements. Then the convergence speeds up again. Often, in terms of computational time, it is not important whether we choose a relative accuracy of 10^{-4} or 10^{-5}, see the numerical example in §4. – Here we note that the factors α_k and β_k in (3.4) vary from step to step, and the iteration is *not* a stationary linear process.

The Minimal Residual Method

There is an easy modification of Method 3.4 for which x_k in the linear manifold $x_0 + V_k$ minimizes the error in the norm

$$\|x_k - x^*\|_{A^\mu}$$

for some $\mu \geq 1$, rather than in the energy norm. To achieve this, we replace the scalar products $u'v$ in the quotients in (3.4) by $u'A^{\mu-1}v$. The Euclidean norm $\|x_k - x^*\| = \|x_k - x^*\|_{A^0}$ can also be easily minimized by using the space $x_0 + AV_k$ instead of $x_0 + V_k$.

The case $\mu = 2$ is of some practical importance. Since

$$\|x - x^*\|_{A^2}^2 = \|Ax - b\|^2 = x'A^2x - 2b'Ax + const, \tag{3.16}$$

it is called the *minimal residual method*. The method is also applicable for indefinite or unsymmetric matrices. We shall use this method to illustrate that the strength of the CG method is due more to its *analytic* properties than to its simple *algebraic* properties.

For $\mu > 1$, Lemma 3.6 and Theorem 3.7 hold for positive definite matrices. It follows from (3.11) that the vector y constructed in the proof of the corollary satisfies

$$\|y - x^*\|_{A^\mu} \leq r\|x_0 - x^*\|_{A^\mu}. \tag{3.17}$$

Although the leading term in the quadratic form (3.16) is determined by the matrix A^2, the rate of convergence depends on $\kappa(A)$, rather than $\kappa(A^2)$.

Indefinite and Unsymmetric Matrices

We now turn to indefinite problems. First, we recall that an indefinite system $Ax = b$ cannot be converted into a system $A^2 x = Ab$ with positive definite matrix simply by multiplication by A. Indeed, since $\kappa(A^2) = [\kappa(A)]^2$, the condition number increases significantly under the transformation. This raises the question of whether we can avoid this shortcoming by applying the minimal residual method.

This is in fact the case for problems with only a few negative eigenvalues, and also for unsymmetric spectra. However, if the spectrum is symmetric w.r.t. zero, then unfortunately we are faced with the same effect as squaring the matrix.

3.8 Example. Suppose we double the system $Ay = b$,

$$\begin{pmatrix} A & \\ & -A \end{pmatrix} \begin{pmatrix} y \\ z \end{pmatrix} = \begin{pmatrix} b \\ -b \end{pmatrix},$$

where A is positive definite. Let $y_0 = z_0$. Then the residual has the form $(h_0, -h_0)'$. Since the expression $\|A(y_0 + \alpha h_0) - b\|^2 + \| - A(z_0 - \alpha h_0) + b\|^2$ assumes its minimum at $\alpha = 0$, it follows that $y_1 = y_0 = z_1 = z_0$. This shows that in general, we get an improvement only for an even number of steps, and the minimum for $x_0 + \text{span}[Ag_0, A^3 g_0, A^5 g_0, \ldots]$ will be computed. Unfortunately, this again corresponds to the calculation with the squared matrix. $\quad\square$

Since in contrast to 3.5(2), here the gradients $g_1, g_2, g_3, \ldots$ are not linearly independent, the formal extension of the minimal residual algorithm breaks down. More importantly, as we shall see later in Remark 4.3, for minimal residuals, the preconditioning generally can no longer be built into a three-term recurrence.

To treat problems with indefinite or unsymmetric matrices, we need to make modifications; cf. Paige and Saunders [1975], Stoer [1983], Golub and van Loan [1983]. They are still relatively simple for symmetric indefinite matrices. The Cholesky decomposition hidden in the CG method is replaced by a QR decomposition; cf. Paige and Saunders [1975]. Otherwise, we have to distinguish between an incomplete minimization and a very short recurrence with stabilization. QM-RES and its variants belong to the first group; see Saad and Schultz [1985] and Saad [1993]. It generates directions which are conjugate only to the last few difference vectors. The other methods use two systems of biorthogonal vectors; see van der Vorst [1992]. In order to avoid degeneracies as in Example 3.8, several steps are carried out together. This is called the "look ahead strategy"; cf. Freund, Gutknecht, and Nachtigal [1993]. Various studies have shown that no optimal algorithm exists for indefinite and unsymmetric problems.

Because of this phenomenon, completely different methods based on the Uzawa algorithm have been developed for the class of indefinite problems involving saddle point problems. They are described in §5 below.

Problems

3.9 Suppose $z \in \mathbb{R}^n$ and $k \geq 1$. In addition, let A, B and C be positive definite $n \times n$ matrices. Suppose that the matrices A and B commute. Using as few arithmetic operations as possible, compute A-orthogonal directions $d_0, d_1, \ldots, d_k$, which span the same space as

(a) $z, Az, A^2 z, \ldots, A^k z,$

(b) $z, CAz, (CA)^2 z, \ldots, (CA)^k z,$

(c) $z, Bz, B^2 z, \ldots, B^k z.$

How many matrix-vector multiplications, and how many scalar products are needed? When can A^2-orthogonal directions be computed simply?

3.10 Let $S = \{a_0\} \cup [a, b]$ with $0 < a_0 < a < b$ and $\kappa = b/a$. Show that there exists a polynomial p of degree k such that

$$p(0) = 1 \quad \text{and} \quad |p(x)| \leq \frac{2b}{a_0} \left(\frac{\sqrt{\kappa} - 1}{\sqrt{\kappa} + 1} \right)^{k-1} \quad \text{for } x \in S.$$

3.11 How does the iteration in Problem 2.9 perform using conjugate directions? Does the difficulty discussed in Problem 2.9 disappear if we use conjugate directions?

3.12 Let $\kappa(A) = 1000$. How many iteration steps are needed in the gradient and the conjugate gradient methods in order to reduce the error by 0.01 in the worst case?

3.13 The choice of α_k guarantees that $d_k' g_{k+1} = 0$, independent of the roundoff errors in the previous steps. Thus, $\|d_{k+1}\|_A$ is just the distance of the vector g_{k+1} from the one-dimensional linear space $\text{span}[d_k]$. Show that

$$\|d_{k+1}\|_A \geq \frac{1}{\kappa(A)^{1/2}} \|g_{k+1}\|_A .$$

Hint: First compare the Euclidean norm of the vectors d_{k+1} and g_{k+1}.

§ 4. Preconditioning

The conjugate gradient method becomes an especially efficient method when it is coupled with preconditioning. The combination is called the *preconditioned conjugate gradient method* or, for short, the *PCG method*. We describe two standard preconditioning methods which suffice for the solution of many systems of equations arising from second order elliptic problems. These methods do not need to be tailored to the problem at hand, and can even be built into a black box.

Given the equation $Ax = b$, suppose we have an easily invertible positive definite matrix C which approximates the matrix A. We discuss later how to measure the quality of the approximation. Given $x_0 \in \mathbb{R}^n$, consider

$$x_1 = x_0 - \alpha C^{-1} g_0, \tag{4.1}$$

where $g_0 = Ax_0 - b$. If $C = A$, then we already get the solution in the first step. Thus, it is to be expected that choosing C to be any (reasonable) approximation to A will get us to the solution faster than the trivial choice $C = I$.

This idea leads to the following algorithm:

4.1 The Conjugate Gradient Method with Preconditioning.
Let $x_0 \in \mathbb{R}^n$. Set $g_0 = Ax_0 - b$, $d_0 = -h_0 = -C^{-1} g_0$, and compute

$$
\begin{aligned}
x_{k+1} &= x_k + \alpha_k d_k, \\
\alpha_k &= \frac{g_k' h_k}{d_k' A d_k}, \\
g_{k+1} &= g_k + \alpha_k A d_k, \\
h_{k+1} &= C^{-1} g_{k+1}, \\
d_{k+1} &= -h_{k+1} + \beta_k d_k, \\
\beta_k &= \frac{g_{k+1}' h_{k+1}}{g_k' h_k},
\end{aligned}
\tag{4.2}
$$

for $k \geq 0$. $\qquad\qquad\qquad\qquad\qquad\qquad\qquad\qquad\qquad\qquad\qquad$ □

If C is positive definite, then in analogy with 3.5 we have

4.2 Properties of the PCG Method. *As long as $g_{k-1} \neq 0$, we have*
(1) $d_{k-1} \neq 0$.
(2) $V_k := \mathrm{span}[g_0, AC^{-1} g_0, \dots, (AC^{-1})^{k-1} g_0] = \mathrm{span}[g_0, g_1, \dots, g_{k-1}]$
and $\mathrm{span}[d_0, d_1, \dots, d_{k-1}] = C^{-1} \mathrm{span}[g_0, g_1, \dots, g_{k-1}]$.

(3) The vectors $d_0, d_1, \ldots, d_{k-1}$ are pairwise conjugate.
(4)

$$f(x_k) = \min_{z \in V_k} f(x_0 + C^{-1}z). \qquad (4.3)$$

The proof of these algebraic properties proceeds in exactly the same way as the proof of 3.5, and can be left to the reader.

4.3 Remark. The matrices C and A do not have to commute for the method to work, see Problem 3.9. Indeed, we only need to compute scalar products of the form

$$((AC^{-1})^j u)' A (C^{-1}A)^k v,$$

and the matrix $((AC^{-1})^j)' A(C^{-1}A)^k$ depends only on the sum $k + j$. Unfortunately, for $((AC^{-1})^j)' A^2 (C^{-1}A)^k$ this holds only in exceptional cases, and combining preconditioning with the minimal residual method is not so simple; see Axelsson [1980], Young and Kang [1980], Saad and Schultz [1985]. Conjugate directions can no longer be determined by three-term recurrence relations. Therefore, we do not attempt to find a complete orthogonalization, and instead make sure that the new direction is conjugate to the last five directions, say.

The convergence theory for the CG method can be generalized as follows.

4.4 Theorem. *(1) Suppose there exists a polynomial $p \in \mathcal{P}_k$ with*

$$p(0) = 1 \quad and \quad |p(z)| \le r \quad for \ all \ z \in \sigma(C^{-1}A).$$

Then for arbitrary $x_0 \in \mathbb{R}^n$, the PCG method satisfies

$$\|x_k - x^*\|_A \le r \|x_0 - x^*\|_A.$$

(2) With $\kappa = \kappa(C^{-1}A)$,

$$\|x_k - x^*\|_A \le 2 \left(\frac{\sqrt{\kappa} - 1}{\sqrt{\kappa} + 1} \right)^k \|x_0 - x^*\|_A.$$

Proof. Consider $q(z) := (p(z) - 1)/z$. Then $y := x_0 + q(C^{-1}A)C^{-1}g_0 \in x_0 + C^{-1}V_k$, and so $y - x^* = p(C^{-1}A)(x_0 - x^*)$. Now let $\{z_j\}_{j=1}^n$ be a complete system of eigenvectors for the problem

$$Az_j = \lambda_j C z_j, \quad j = 1, 2, \ldots, n. \qquad (4.4)$$

In particular, suppose the vectors are normalized so that

$$z_i' C z_j = \delta_{ij} \quad for \ i, j = 1, 2, \ldots, n.$$

0	12.95	14	3.93	28	3.13_{-2}
2	12.31	16	1.76	30	1.33_{-2}
4	11.99	18	0.519	32	5.79_{-3}
6	11.64	20	0.273	34	1.82_{-3}
8	10.55	22	0.175	36	6.21_{-4}
10	7.47	24	0.130	38	1.51_{-4}
12	4.76	26	0.086	40	3.35_{-5}

Table 6 and Fig. 46. Reduction in the energy norm of the error when the PCG method is applied to a cantilever problem with 544 unknowns. The slow decrease at the beginning and again in the middle is typical for the CG method

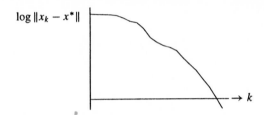

Then $z_i' A z_j = \lambda_j \delta_{ij}$, and (3.12) again follows. Moreover, $(C^{-1}A)^\ell z_j = \lambda_j^\ell z_j$ for all ℓ. The rest of the proof of (1) follows as in Lemma 3.5.

Since the numbers λ_j in (4.4) are actually the eigenvalues of $C^{-1}A$, the assertion (2) follows by the arguments used in Theorem 3.7. $\qquad\qquad\Box$

Preconditioning also helps to reduce the effect of the following difficulty. In principle, in using gradient methods we want to choose the *direction of steepest descent*. Which direction gives the steepest descent depends on the metric of the space. For the simple gradient method, we implicitly use the Euclidean metric. But if $\|x\|_C := \sqrt{x'Cx}$ is a better approximation to the metric $\|x\|_A$ than the Euclidean metric $\|x\| = \sqrt{x'x}$, then C is a good choice for preconditioning. By Theorem 4.4, the *oscillation* of the quotient

$$\frac{x'Ax}{x'Cx} \tag{4.5}$$

is the main determining factor for the rate of convergence. We shall make use of similar arguments in the following section.

Although the widely applicable methods described below can be used for the solution of second order boundary-value problems, for large problems of fourth order, we usually have to tailor the preconditioning to the problem. This is due to the strong growth of order h^{-4} of κ. There are three common approaches to constructing special methods:

1. Subdivide the domain. The solution of the much smaller systems corresponding to the partial domains serves as a preconditioning, see e.g. Widlund [1988].

2. Alter the boundary conditions to give a simpler problem. (For example, a modification of the boundary conditions for the biharmonic equation leads to two decoupled Laplace equations, see Braess and Peisker [1986].) The approximate solution so obtained is then used for the preconditioning.

3. Use so-called *hierarchical bases*, i.e., choose basis functions consisting of low and high frequency functions, respectively; see Yserentant [1986], Xu [1992], or Bramble, Pasciak, and Xu [1990]. The condition number can be significantly reduced by a suitable scaling of the different parts.

Preconditioning by SSOR

A simple but effective preconditioning can be obtained from the Gauss–Seidel method, despite its slow convergence when it is used as stand-alone iteration. We decompose the given symmetric matrix A as

$$A = D - L - L',$$

where L is a lower triangular matrix and D is a diagonal matrix. Then for $1 < \omega < 2$,

$$x \mapsto (D - \omega L)^{-1}(\omega b + \omega L' x - (\omega - 1)Dx)$$

defines an iteration step in the forward direction; cf. (1.19). Similarly, the relaxation in the backwards direction is defined by

$$x \mapsto (D - \omega L')^{-1}(\omega b + \omega L x - (\omega - 1)Dx).$$

Then the first half step gives

$$x^{1/2} = \omega(D - \omega L)^{-1} g_k,$$

where $x = 0$ and $b = g_k$, and the second half step gives

$$h_k = \omega(2 - \omega)(D - \omega L')^{-1} D(D - \omega L)^{-1} g_k,$$

since $\omega g_k + \omega L x^{1/2} - D x^{1/2} = 0$. In particular, $h_k = C^{-1} g_k$ with $C := [\omega(2 - \omega)]^{-1}(D - \omega L)D^{-1}(D - \omega L')$. Clearly, the matrix C is symmetric and positive definite.

We point out that multiplying the preconditioning matrix C by a positive factor has no influence on the iteration. Thus, the factor $\omega(2 - \omega)$ can be ignored in the calculation.

Experience shows that the quality of the preconditioning is not very sensitive to the choice of the parameter ω. Calculation with the fixed value $\omega = 1.3$ is only slightly worse than using the corresponding optimal values, which in practice lie between 1.2 and 1.6 [Axelsson and Barker 1984].

On the other hand, the numbering of the variables has a major influence on the performance of the method. The differences are very evident for the equations arising from five-point stencils on a rectangular mesh. We recommend that the *lexicographical* ordering $x_{11}, x_{12}, \ldots, x_{1n}, x_{21}, x_{22}, \ldots, x_{nn}$ be used. The *checkerboard ordering*, where all variables x_{ij} with $i + j$ even appear first, followed by all those with $i + j$ odd (or conversely), reduces the efficiency of the SSOR preconditioner dramatically. Thus it cannot be recommended. The disadvantages of this numbering are so great that they cannot even be compensated by vectorization or parallelization.

Preconditioning by ILU

Another preconditioning method can be developed from a variant of the Cholesky decomposition. For symmetric matrices of the type which appear in the finite element method, the Cholesky decomposition $A = LDL^t$ or $A = LL^t$ leads to a triangular matrix L which is significantly less sparse than A. Using an approximate inverse leads to the so-called *incomplete Cholesky decomposition (ICC)* or *incomplete LU decomposition (ILU)*; see Varga [1960]. In the simplest case, we simply avoid calculation with all matrix elements which vanish in the given matrix. This leads to a decomposition

$$A = LL^t + R \tag{4.6}$$

with an error matrix in which $R_{ij} \neq 0$ only appears if $A_{ij} = 0$.

This preconditioning method is often faster than the one using SSOR relaxation. However, there does not seem to be a general rule for deciding in which cases SSOR or ICC is more effective.

There are many variants of the method, and often filling in of elements in the neighborhood of the diagonal is allowed. In the so-called *modified incomplete decomposition* due to Meijerink and van Vorst [1977], instead of suppressing matrix elements, they are moved onto the main diagonal.

Gustafsson [1978] developed a preconditioning method for the standard five-point stencil for the Laplace equation. While in general there is only empirical evidence for the improvement of the conditioning, in this case he proved that the condition number is reduced from $\mathcal{O}(h^{-2})$ to $\mathcal{O}(h^{-1})$.

It is crucial for the proof that the diagonal elements can be increased by a small amount. Let $\zeta > 0$. In view of Friedrichs' inequality, we can estimate the quadratic forms $a(u, u) = |u|_1^2$ and $|u|_1^2 + \zeta \|u\|_0^2$ in terms of each other. The discretization

$$\begin{bmatrix} & 0 & \\ b_{i-1} & a_i & 0 \\ & c_{i-m} & \end{bmatrix}_* \qquad \begin{bmatrix} & c_i & \\ 0 & a_i & b_i \\ & 0 & \end{bmatrix}_*$$

$$\begin{bmatrix} b_{i-1}c_{i-1} & & a_i c_i & \\ a_{i-1}b_{i-1} & a_i^2 + b_{i-1}^2 + c_{i-m}^2 & & a_i b_i \\ & c_{i-m}a_{i-m} & & b_{i-m}c_{i-m} \end{bmatrix}_*$$

$$\begin{bmatrix} & -\gamma_i & \\ -\beta_{i-1} & \alpha_i & -\beta_i \\ & -\gamma_{i-m} & \end{bmatrix}_* \qquad \begin{bmatrix} -r_i & & \\ & r_i + r_{i-m+1} & \\ & & -r_{i-m+1} \end{bmatrix}_*$$

Difference stencils for L, L^t (top), LL^t (middle) and A, R (bottom) for the incomplete Cholesky decomposition

of $\|u\|_0^2$ leads to the so-called *mass matrix*. Since its condition number is bounded independent of h, $\|u\|_0$ and $h^2\|u\|_{\ell_2}$ are equivalent norms. Thus, for the design of a preconditioning matrix, instead of the standard five-point stencil we can consider the following modified stencil:

$$\begin{bmatrix} & -1 & \\ -1 & 4 + \zeta_i h^2 & -1 \\ & -1 & \end{bmatrix}_* \qquad (4.7)$$

where $0 < \zeta_i < \zeta$.

For simplicity, we now assume that the same number m of nodes lie on each horizontal grid line. Suppose that the neighbors of the node i to the South and West have the indices $i - m$ and $i - 1$, respectively.

The incomplete Cholesky decomposition leads to triangular matrices with at most three nonzero elements in every row. The general form of the difference stencils can be seen in the schemes above. The error matrix (4.6) can only have nonzero elements on the diagonal and at positions to the Northwest and Southeast. Combining this with $\sum_j R_{ij} = 0$, we see that the matrix R must have the form shown in the scheme. The coefficients a_i, b_i and c_i can be found recursively by

$$\begin{aligned} a_i^2 &= \alpha_i - b_{i-1}^2 - c_{i-m}^2 - r_i - r_{i-m+1}, \\ b_i &= -\beta_i/a_i, \\ c_i &= -\gamma_i/a_i, \\ r_i &= b_{i-1}c_{i-1}. \end{aligned} \qquad (4.8)$$

Recalling (4.7) we set $\alpha_i := 4 + 8h^2$ and $\beta_i := \gamma_i := 1$ with the usual convention that β_i and γ_i are set to 0 at points next to the boundary. It follows by induction that

$$a_i \geq \sqrt{2}(1 + h), \quad 0 < b_i, c_i \leq 1/a_i, \quad r_i \leq \frac{1}{2}(1 + h)^{-2}.$$

We can now estimate $x'Rx$ with the help of the formula $(x + y)^2 \leq 2(x^2 + y^2)$:

$$\begin{aligned}
0 \leq x'Rx &= \sum_i r_i(x_i - x_{i+m-1})^2 \\
&\leq \sum_i \frac{1}{(1+h)^2}\{(x_i - x_{i-1})^2 + (x_{i-1} - x_{i+m-1})^2\} \\
&\leq \frac{1}{1+h}x'Ax.
\end{aligned}$$

Combining this with (4.6), we have $x'LL'x \geq h/(1 + h)x'Ax$ and

$$x'LL'x \leq x'Ax \leq \frac{1+h}{h}x'LL'x$$

and thus

$$\kappa([LL']^{-1}A) \leq \frac{1+h}{h} = \mathcal{O}(h^{-1}). \tag{4.9}$$

Since $\kappa(A) = \mathcal{O}(h^{-2})$, the preconditioning has the effect that the effective condition number κ is reduced by one power of h.

The equation (4.6) also clearly shows that multiplication by $(LL')^{-1}$ would be equivalent to an SSOR step if the overrelaxation factors were point-independent and approximately of size $2 - \mathcal{O}(h^{-2})$. Thus, with small modifications in the argument, it is possible to show that in applying preconditioning with the SSOR method using a (fixed) factor $\omega = 2 - \mathcal{O}(h^{-2})$, the condition number is also reduced by one power of h, see Axelsson and Barker [1984].

Remarks on Parallelization

SSOR relaxation and multiplication by L^{-1} and R^{-1}, where L and R are associated with an ILU decomposition, are recursive processes. Nevertheless, both parallelization and vectorization are possible. [We recall that we should not choose a checkerboard order for simple parallelization.] The implementation depends heavily on the computer architecture. There is intense activity surrounding the use of parallel and vector machines, and so we would like to give a first impression of how to treat the kinds of banded matrices which arise for finite element problems.

We restrict our considerations to the equations arising from the use of the five-point stencil on a square domain with m^2 unknowns. We write the unknowns with double indices. Then in the first phase (the preconditioning), to determine the current variable x_{ij}, we need to know the values x_{kj} for $k < i$ and $x_{i\ell}$ for $\ell < j$.

In a vector machine we can collect the calculations of all x_{ij} with $i +$ $j = const$. The calculation then proceeds in $2m$ groups. It is well known that we can save time in a vector machine by overlapping about eight arithmetic operations. The time saved is proportional to m^2, and is worthwhile if it exceeds the time required for the initialization of the $2m$ groups. Normally, this is the case when $m >$ approx. 40.

A different approach can be used on a parallel machine. We sketch the case of two processors [Wittum 1989a]. First we divide the domain into two parts. The first processor takes care of the nodes (i, j) with $j \leq n/2$, and the second one takes care of those with $j > n/2$. Once the first processor has dealt with $(1, 1)$, $(1, 2)$, ..., $(1, n/2)$, the second one is signaled. While the second processor works on its assigned values in the row $i = 1$, the first can do row $i + 1 = 2$. The two processors continue to work in parallel on succeeding rows.

We have to provide memory with access by two processors only at the boundaries, i.e. for $j = n/2$ and $j = n/2 + 1$. It is clear how memory can be freed for access by the other processor.

Without considering the memory restriction, there is also another approach which we could take. The first processor works on the entire first row. With the delay of one node, it signals the second processor to begin work on the second row. The remaining rows are then dealt with alternately by the two processors. In particular, with several processors, we get a complete parallelization after a short initialization time.

For more on parallelization, see e.g. Hughes, Ferencz and Hallquist [1987], Meier and Sameh [1988], Ortega and Voigt [1985] and Ortega [1988].

Nonlinear Problems

The CG method can be carried over to nonlinear problems for which the function f to be minimized is not necessarily a quadratic function. This avoids iterating with the Newton method, where the solution of the corresponding linear system of equations would again require an iterative method.

As in §2, let f be a C^1 function defined on an open set $M \subset \mathbb{R}^n$. Very often f has the form

$$f(x) = \frac{1}{2} x' A x - \sum_{i=1}^{n} d_i \phi(x_i) - b'x$$

with $\phi \in C^1(\mathbb{R})$. The first term has a more significant effect on poor conditioning than the second [Glowinski 1984]. Suppose we have a matrix C which is appropriate for preconditioning A (otherwise we choose $C = I$).

The minimization of

$$\frac{x'Ax}{x'x}$$

for the determination of the smallest eigenvalue of A also involves a non-quadratic problem.

4.5 The Conjugate Direction Method for Nonlinear Problems *following Fletcher and Reeves.*

Let $x_0 \in M$. Set $g_0 = \nabla f(x_0)$ and $d_0 = -h_0 = -C^{-1}g_0$.

For $k = 0, 1, 2, \ldots$, perform the following calculations:

1. Line search: Find the minimum of f on the line $\{x_k + td_k : t \geq 0\} \cap M$. Suppose the minimum (or a local minimum) is assumed at $t = \alpha_k$. Set

$$x_{k+1} = x_k + \alpha_k d_k.$$

2. Determination of the direction:

$$g_{k+1} = \nabla f(x_{k+1}),$$
$$h_{k+1} = C^{-1}g_{k+1},$$
$$d_{k+1} = -h_{k+1} + \beta_k d_k, \tag{4.10}$$
$$\beta_k = \frac{g'_{k+1}h_{k+1}}{g'_k h_k}.$$

$\square$

4.6 Remark. In the method of Polak and Ribière, which is a variant of the Fletcher and Reeves method, β_k is not chosen as in (4.10), but instead we compute

$$\beta_k = \frac{g'_{k+1}(h_{k+1} - h_k)}{g'_k h_k}. \tag{4.11}$$

Problems

The following three exercises deal with the inversion of the so-called mass matrix.

4.7 Let $A_1, A_2, \ldots, A_k, C_1, C_2, \ldots, C_k$ be positive semidefinite matrices with

$$a\,x'C_i x \leq x'A_i x \leq b\,x'C_i x \quad \text{for } x \in \mathbb{R}^n \text{ and } i = 1, 2, \ldots, k.$$

In addition, let $0 < a \leq b$. Suppose that the matrices $A = \sum_i A_i$ and $C = \sum_i C_i$ are positive definite. Show that $\kappa(C^{-1}A) \leq b/a$.

4.8 Show that the matrix

$$A = \begin{pmatrix} 2 & 1 & 1 \\ 1 & 2 & 1 \\ 1 & 1 & 2 \end{pmatrix}$$

is positive definite, and that its condition number is 4.

Hint: The quadratic form associated with the matrix A is $x^2 + y^2 + z^2 + (x+y+z)^2$.

4.9 The computation of the mass matrix $\int \psi_i \psi_j dx$ for linear triangular elements on the element level leads to the matrix in Problem 4.8 (w.r.t. the nodes which are involved). Show that we can get $\kappa \leq 4$ using preconditioning with an easily computable diagonal matrix. How much is the error reduced after three steps of the PCG method?

4.10 Consider Problem 3.12. How does the answer change if we replace κ by $2\sqrt{\kappa}$?

4.11 The equation $Ax = b$ implies

$$By = c \quad \text{with } B = C^{-1/2}AC^{-1/2}, \; c = C^{-1/2}b, \tag{4.12}$$

where $y = C^{1/2}x$, since $C^{-1/2}AC^{-1/2}C^{1/2}x = C^{-1/2}b$. Show that applying the PCG method 4.1 (with preconditioning by the matrix C) to the original equation is equivalent to applying the CG method 3.4 to (4.12). Use this to derive the properties 4.2.

4.12 For preconditioning we often use a change of basis, e.g. in the method of hierarchical bases; see Yserentant [1986]. Let

$$x = Sy,$$

where S is a nonsingular matrix. Show that carrying out Algorithm 3.4 with the variables y is equivalent to Algorithm 4.1 with preconditioning based on

$$C^{-1} = SS^t.$$

4.13 For preconditioning we often use a matrix C which is not exactly symmetric. (In particular this is the case if multiplication by C^{-1} is only done approximately.) This means that we are not requiring that all of the d_k be pairwise conjugate. But we still want x_{k+1} to be the minimum of the function (2.1) over the two-dimensional manifold $x_k + \mathrm{span}[h_k, d_{k-1}]$. Hence d_{k+1} and d_k (and d_k and d_{k-1}, respectively) should be conjugate. Which of the following formulae for β_k can be used for unsymmetric C?

(1) $\beta_k = \dfrac{g'_{k+1}h_{k+1}}{g'_k h_k}$,   (2) $\beta_k = \dfrac{(g'_{k+1} - g_k)h_{k+1}}{g'_k h_k}$,   (3) $\beta_k = \dfrac{h'_{k+1}Ad_k}{d'_k Ad_k}$.

The following problems are useful not only for constructing preconditioners but also as preparation for a multigrid theory.

4.14 Let $A \leq B$ denote that $B - A$ is positive semidefinite. Show that $A \leq B$ implies $B^{-1} \leq A^{-1}$, but it does not imply $A^2 \leq B^2$. — To prove the first part note that $(x, B^{-1}x) = (A^{-1/2}x, A^{1/2}B^{-1}x)$ and apply Cauchy's inequality. Next consider the matrices

$$A := \begin{pmatrix} 1 & a \\ a & 2a^2 \end{pmatrix} \quad \text{and} \quad B := \begin{pmatrix} 2 & 0 \\ 0 & 3a^2 \end{pmatrix}$$

for establishing the negative result. From the latter it follows that we cannot derive good preconditioners for the biharmonic equation by applying those for the poisson equation twice.

Note: The converse is more favorable, i.e., $A^2 \leq B^2$ implies $A \leq B$. Indeed, the Rayleigh quotient $\lambda = \max\{(x, Ax)/(x, Bx)$ is an eigenvalue, and the maximum is attained at an eigenvector, i.e., $Ax = \lambda Bx$. On the other hand, by assumption

$$0 \leq (x, B^2x) - (x, A^2x) = (1 - \lambda^2) \|Bx\|^2.$$

Hence, $\lambda \leq 1$ and the proof is complete.

4.15 Show that $A \leq B$ implies $B^{-1}AB^{-1} \leq B^{-1}$.

4.16 Let A and B be symmetric positive definite matrices with $A \leq B$. Show that

$$(I - B^{-1}A)^m B^{-1}$$

is positive definite for $m = 1, 2, \ldots$. To this end note that

$$q(XY)X = Xq(YX)$$

holds for any matrices X and Y if q is a polynomial. Which assumption may be relaxed if m is even?

4.17 Let B^{-1} be an approximate inverse of A. Moreover, assume that A and B are symmetric positive definite matrices and that

$$A \leq B.$$

Let B_m^{-1} be the approximate inverse for m steps of the iteration (1.1); cf. Problem 1.16. Show that

$$A \leq B_{m+1} \leq B_m \leq B \quad \text{for } m \geq 1$$

by making use of the preceding problems.

§ 5. Saddle Point Problems

The determination of a minimum of

$$J(u) = \tfrac{1}{2}u'Au - f'u$$

with the constraint   (5.1)

$$Bu = g$$

leads to an indefinite system of equations of the form

$$
\begin{aligned}
Au + B'\lambda &= f, \\
Bu &= g.
\end{aligned}
\tag{5.2}
$$

If B is an $m \times n$ matrix, then the Lagrange multiplier λ is an m-dimensional vector. Clearly, we can restrict our attention to the case where the restrictions are linearly independent.

In most cases, A is invertible. After multiplying the first equation in (5.2) by A^{-1}, we can eliminate u from the second equation:

$$BA^{-1}B'\lambda = BA^{-1}f - g. \tag{5.3}$$

The matrix $BA^{-1}B'$ for this so-called *reduced equation* is positive definite, although it is given only implicitly. It is called the *Schur complement* of A; cf. Problem 5.9.

The Uzawa Algorithm and its Variants

A widely known iterative method for saddle point problems is connected with the name Uzawa.

5.1 The Uzawa Algorithm. Let $\lambda_0 \in \mathbb{R}^m$. Find u_k and λ_k so that

$$
\left.
\begin{aligned}
Au_k &= f - B'\lambda_{k-1}, \\
\lambda_k &= \lambda_{k-1} + \alpha(Bu_k - g),
\end{aligned}
\right\} \quad k = 1, 2, \ldots
\tag{5.4}
$$

Here we assume that the step size parameter α is sufficiently small.   □

For the analysis of the Uzawa algorithm, we define the *residue*

$$q_k := g - Bu_k. \tag{5.5}$$

In addition, suppose the solution of the saddle point problem is denoted by (u^*, λ^*). Now substituting the iteration formula for u_k into (5.5) and using (5.3), we get

$$q_k = g - BA^{-1}(f - B^t\lambda_{k-1}) = BA^{-1}B^t(\lambda_{k-1} - \lambda^*).$$

This means that

$$\lambda_k - \lambda_{k-1} = -\alpha q_k = \alpha BA^{-1}B^t(\lambda^* - \lambda_{k-1}).$$

Thus the Uzawa algorithm is equivalent to applying the gradient method to the reduced equation using a fixed step size (cf. Problem 2.6). In particular, the iteration converges for

$$\alpha < 2\|BA^{-1}B^t\|^{-1}.$$

The convergence results of §§2 and 3 can be carried over directly. We need to use a little trick in order to get an efficient algorithm. The formula (2.5) gives the step size

$$\alpha_k = \frac{q_k' q_k}{(B^t q_k)' A^{-1} B^t q_k}.$$

However, if we were to use this rule formally, we would need an additional multiplication by A^{-1} in every step of the iteration. This can be avoided by storing an auxiliary vector. – Here we have to pay attention to the differences in the sign.

5.2 Uzawa Algorithm *(the variant equivalent to the gradient method).*
Let $\lambda_0 \in \mathbb{R}^m$ and $Au_1 = f - B^t\lambda_0$.
For $k = 1, 2, \ldots$, compute

$$\begin{aligned}
q_k &= g - Bu_k, \\
p_k &= B^t q_k, \\
h_k &= A^{-1} p_k, \\
\lambda_k &= \lambda_{k-1} - \alpha_k q_k, \qquad \alpha_k = \frac{q_k' q_k}{p_k' h_k}, \\
u_{k+1} &= u_k + \alpha_k h_k.
\end{aligned}$$

□

Because of the size of the condition number $\kappa(BA^{-1}B^t)$, it is often more effective to use conjugate directions. Since the corresponding factor β_k in (3.4) is already independent of the matrix of the system, the extension is immediately possible.

5.3 Uzawa Algorithm with Conjugate Directions.

Let $\lambda_0 \in \mathbb{R}^m$ and $Au_1 = f - B^t\lambda_0$. Set $d_1 = -q_1 = Bu_1 - g$.
For $k = 1, 2, \ldots$, find

$$
\begin{aligned}
p_k &= B^t d_k, \\
h_k &= A^{-1} p_k, \\
\lambda_k &= \lambda_{k-1} + \alpha_k d_k, & \alpha_k &= \frac{q_k' q_k}{p_k' h_k}, \\
u_{k+1} &= u_k - \alpha_k h_k, \\
q_{k+1} &= g - Bu_{k+1}, \\
d_{k+1} &= -q_{k+1} + \beta_k d_k, & \beta_k &= \frac{q_{k+1}' q_{k+1}}{q_k' q_k}.
\end{aligned}
$$

An Alternative

In performing k steps of Algorithms 5.2 and 5.3, $k + 1$ multiplications by A^{-1} are required. Thus, there are $k + 1$ equations to be solved. In practice, we do this only approximately. In particular, we approximate A^{-1} by C^{-1}, where C is considered as a preconditioner for A and is again assumed to be a symmetric positive definite matrix.

We can go one step further and replace the matrix A in the initial problem (5.1) by a matrix C which is understood to be a preconditioner. This leads to the modified minimum problem

$$
\frac{1}{2}u'Cu - f'u \to \min!
$$

$$
Bu = g.
$$
(5.6)

As can be seen by carrying over (5.2), the matrix corresponding to this problem is

$$
\begin{pmatrix} C & B^t \\ B & \end{pmatrix}.
$$

Inserting this matrix in (1.1) in place of the matrix M, we get the iteration

$$
\begin{pmatrix} u_{k+1} \\ \lambda_{k+1} \end{pmatrix} = \begin{pmatrix} u_k \\ \lambda_k \end{pmatrix} + \begin{pmatrix} C & B^t \\ B & \end{pmatrix}^{-1} \begin{pmatrix} f - Au_k - B^t\lambda_k \\ g - Bu_k \end{pmatrix}.
$$
(5.7)

The rate of convergence of this iteration is determined by the gap between the upper and lower bounds on the quotients $\frac{u'Au}{u'Cu}$, $u \neq 0$; cf. (4.5). In fact, it suffices to examine the bounds for the subspace $V = \{u \in \mathbb{R}^n; \; Bu = 0\}$. (They can of course be estimated by the coarser bounds for $\mathbb{R}^n$.)

In view of the following (cf. Braess and Sarazin [1997]), the iteration (5.4) of Uzawa and the iteration (5.7) are extreme cases. If the iteration (5.7) is built into a cg-iteration, then u-variables and the Lagrangian multipliers have to be treated in a different way; see Braess, Deuflhard, and Lipnikov [1999].

5.4 Remark. In the Uzawa algorithm (5.4), u_{k+1} and λ_{k+1} are independent of u_k. In the iteration (5.7), u_{k+1} and λ_{k+1} are independent of λ_k.

The assertion about the Uzawa algorithm follows directly from the definition (5.4) of the algorithm. The other assertion is a consequence of the following formula which is equivalent to (5.7):

$$\begin{pmatrix} u_{k+1} \\ \lambda_{k+1} \end{pmatrix} = \begin{pmatrix} C & B^t \\ B & \end{pmatrix}^{-1} \begin{pmatrix} f - (A - C)u_k \\ g \end{pmatrix}. \tag{5.8}$$

Bramble and Pasciak [1988] took a completely different approach. By employing a different metric for the indefinite problem, they were able to get a preconditioning in almost the same way as in the positive definite case.

Problems

5.5 Consider the special case $A = I$, and compare the condition number of $BA^{-1}B^t$ with that of the squared matrix. In particular, show that the Uzawa algorithm is better than the gradient method for the squared matrix.

5.6 For the case $m \ll n$, the restriction can be elimininated indirectly. Let F be an $m \times m$ matrix with $FF^t = BB^t$, e.g. say F stems from the Cholesky decomposition of BB^t. In the special case $A = I$, we have the triangular decomposition:

$$\begin{pmatrix} I & \\ B & F \end{pmatrix} \begin{pmatrix} I & B^t \\ & -F^t \end{pmatrix} = \begin{pmatrix} I & B^t \\ B & \end{pmatrix}.$$

How can we construct a corresponding triangular decomposition for the matrix in (5.2) if a decomposition $A = L^t L$ is known?

5.7 Show that $\kappa(BA^{-1}B^t) \le \kappa(A)\kappa(BB^t)$.

5.8 For the saddle point problem (5.2), the norm $\| \cdot \|_A$ is obviously the natural norm for the u components. Show that the norm $\| \cdot \|_{BA^{-1}B^t}$ is then the natural one for the λ components in the following sense: the inf-sup condition holds for the mapping $B^t : \mathbb{R}^m \mapsto \mathbb{R}^n$ with the constant $\beta = 1$.

5.9 Verify the block Cholesky decomposition for the matrix

$$\begin{pmatrix} A & B^t \\ B & 0 \end{pmatrix} = \begin{pmatrix} A & 0 \\ B & I \end{pmatrix} \begin{pmatrix} A^{-1} & 0 \\ 0 & -BA^{-1}B^t \end{pmatrix} \begin{pmatrix} A & B^t \\ 0 & I \end{pmatrix}$$

appearing in the saddle point problem. What is the connection between this factorization and the computation of the reduced equation (5.3)? In addition, prove that the inverse has the following decomposition:

$$\begin{pmatrix} A & B^t \\ B & 0 \end{pmatrix}^{-1} = \begin{pmatrix} A^{-1} - A^{-1}B^t S^{-1}BA^{-1} & A^{-1}B^t S^{-1} \\ S^{-1}BA^{-1} & -S^{-1} \end{pmatrix},$$

where $S = BA^{-1}B^t$ is the Schur complement.

Chapter V
Multigrid Methods

The multigrid method is one of the fastest methods for solving systems of equations involving a large number of unknowns. The method is due to Fedorenko, who formulated it first as a two-grid method [1961], and then later as a multigrid method [1964]. He showed that the algorithm requires only $\mathcal{O}(n)$ operations, where n is the number of unknowns. Bachvalov [1966] continued the study for difference equations, and allowed nonconstant coefficients. A. Brandt was the first to discover in the mid-seventies that the multigrid method is considerably better than other known methods, even for values of n which occur in actual problems. Nearly at the same time, the multigrid method was discovered independently by Hackbusch [1976], whose approach also led to a simplification of the concepts involved.

Our starting point is the observation that in solving a system of equations, we should use different methods for the high frequency (oscillating) and low frequency (smooth) parts. The idea of the multigrid method is to combine two different methods to get an algorithm which will be effective on the entire spectrum.

Classical iterative methods work essentially by smoothing, i.e. they quickly eliminate the high-frequency parts of the error function. The low frequency parts of the functions can then be computed relatively well on a coarser grid. Although we cannot strictly separate the low and high frequency parts, we are able to get iterative methods whose rate of convergence (i.e. error reduction factors) are in the range from $\frac{1}{20}$ to $\frac{1}{4}$; see Table 7. If we use the multigrid idea in combination with good starting values, we will get convergence in just one or two iterations, and the iterative character of the method almost disappears.

In developing a convergence theory, we have to take account of the smoothness as well as the absolute size of the error. This means that we need to work with (at least) two norms. Natural candidates are the Sobolev norms and their discrete analogs.

The algorithmic aspects discussed in §§1, 4 and 5 are essentially independent of the convergence theory presented in §§2 and 3. For more on multigrid methods, see the books of Hackbusch [1985], Hackbusch and Trottenberg [1982], Briggs [1987], McCormick [1989], and Wesseling [1992].

§ 1. Multigrid Methods for Variational Problems

Smoothing Properties of Classical Iterative Methods

Multigrid methods are based on the observation that the classical iterative methods result in smoothing. This is most easily seen by examining the model Example II.4.3 involving the Poisson equation on a rectangle. This example was also discussed in connection with the Gauss–Seidel and Jacobi methods; cf. (IV.1.11).

1.1 Example. The discretization of the Poisson equation on the unit square using the standard five-point stencil leads to the system of equations

$$4x_{i,j} - x_{i+1,j} - x_{i-1,j} - x_{i,j+1} - x_{i,j-1} = b_{ij} \quad \text{for } 1 \le i, j \le n - 1. \quad (1.1)$$

Here $x_{i,0} = x_{i,n} = x_{0,j} = x_{n,j} = 0$. We consider the iterative solution of (1.1) using the Jacobi method with relaxation parameter ω:

$$x_{i,j}^{\nu+1} = \frac{\omega}{4}(x_{i+1,j}^{\nu} + x_{i-1,j}^{\nu} + x_{i,j+1}^{\nu} + x_{i,j-1}^{\nu}) + \frac{\omega}{4}b_{ij} + (1 - \omega)x_{i,j}^{\nu}. \quad (1.2)$$

By (IV.1.12), the eigenvectors $z^{k,m}$ of the iteration matrix defined implicitly in (1.2) can be thought of as the discretizations of the eigenfunctions

$$(z^{k,m})_{i,j} = \sin\frac{ik\pi}{n}\sin\frac{jm\pi}{n}, \quad 1 \le i, j, k, m \le n - 1, \quad (1.3)$$

of the Laplace operator, with corresponding eigenvalues

$$\lambda^{km} = \frac{1}{2}\cos\frac{k\pi}{n} + \frac{1}{2}\cos\frac{m\pi}{n}, \quad \text{if } \omega = 1,$$

and

$$\lambda^{km} = \frac{1}{4}\cos\frac{k\pi}{n} + \frac{1}{4}\cos\frac{m\pi}{n} + \frac{1}{2}, \quad \text{if } \omega = \frac{1}{2}.$$

In each step of the iteration, the individual spectral parts of the errors are multiplied by the corresponding factors λ^{km}. Thus, those terms corresponding to eigenvalues whose moduli are near 1 are damped the least.

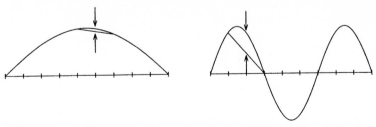

Fig. 47. Replacing values on a one-dimensional grid by the average values of their neighbors has only a minor effect on low frequency terms, but a substantial effect on high frequency ones.

1.2 Corollary. *After a few iterations of the Jacobi method (with $\omega = 1$), the error essentially contains only terms for which*

$\qquad$ *k and m are both small, or*

$\qquad$ *k and m are both close to n.*

Terms of the latter type are strongly reduced if we use $\omega = 1/2$ (instead of $\omega = 1$.) This leaves only the low frequency terms; see Fig. 47.

Accordingly, in carrying out the iteration we will get a clear reduction in the error as long as the error still contains highly oscillatory parts. However, as soon as the error becomes smooth, the iteration will essentially stop, and subsequently the error reduction per step will only be of order $1 - \mathcal{O}(n^{-2})$.

We will show later in connection with the convergence theory that the Jacobi method has a smoothing effect in general, and not just for the model problem, see Lemma 2.4. The following methods are used in practice for smoothing:

$\qquad$ the Jacobi method and the Richardson iteration,

$\qquad$ the successive overrelaxation method (SOR),

$\qquad$ the symmetric successive overrelaxation method (SSOR),

$\qquad$ iteration with the incomplete Cholesky decomposition (ICC).

We now list a few suggestions for the choice of the smoother. If the entire matrix of the system is to be stored, either the SOR or the SSOR method (with a small amount of underrelaxation) is useful in the standard case. For parallel computations, Jacobi relaxation is appropriate. Smoothing via ICC requires more work than the other methods, but turns out to be more robust for anisotropic problems; see Hemker [1980] and Wittum [1989b]. For example, we have a strongly anisotropic problem if one direction in space is preferred, such as for the differential equation

$$100\,u_{xx} + u_{yy} = f.$$

The Multigrid Idea

The above discussion suggests the following approach.

First we carry out several relaxation steps in order to strongly damp all oscillating components of the error. Then we go to a coarser grid, and approximate the remaining smooth part. This is possible because smooth functions can be approximated well on coarse grids.

We then alternately repeat the *smoothing step* on the fine grid and the *coarse-grid correction*. This results in an iterative method.

The system of equations corresponding to the problem on the coarse grid is usually simpler to solve than the original problem. In particular, in the planar case, if we go from a grid with mesh size h to one with mesh size $2h$, the number of unknowns is reduced by a factor of about 4. Moreover, the bandwidth of the matrix for the coarse grid is about half as large. This means that the number of operations needed for Gauss elimination is reduced by a factor of about 16. This leads to major savings after just two or three iterations.

In general, the number of operations needed to solve the system corresponding to mesh size $2h$ will still be too large, and so we repeat the process. We continue to double the size of the grid until we get a sufficiently small system of equations.

The Algorithm

For the sake of simplicity, we first describe the two-grid algorithm for conforming finite elements.

To formulate multigrid algorithms, we need some notation. It is standard to use the letter S for the smoothing operator. For example, when smoothing via Richardson relaxation, we have

$$x \longmapsto Sx := x - \omega(A_h x - b_h). \tag{1.4}$$

In order to avoid confusion with the finite element spaces S_h, we suppress the subscript h in our notation for smoothing operators.

Let $\{\psi_i\}$ be a basis for S_h. Each vector $x \in \mathbb{R}^N$ with $N = N_h = \dim S_h$ is associated with the function $u = \sum_i x_i \psi_i \in S_h$. The indices are inherited in the correspondence.

The iterates u_h^k as well as the intermediate values $u_h^{k,1}$ are associated with the fine grid, while the quantities v_{2h} (which actually depend on k) correspond to the coarser grid with double mesh size.

1.3 Two-Level Iteration (k-th cycle):

Let u_h^k be a given approximation in S_h.

1. *Smoothing Step.* Perform v smoothing steps:

$$u_h^{k,1} = S^v u_h^k.$$

2. *Coarse-Grid Correction.* Compute the solution $\hat{v}_{2h}$ of the variational problem at level $2h$:

$$J(u_h^{k,1} + v) \longrightarrow \min_{v \in S_{2h}}!$$

Set

$$u_h^{k+1} = u_h^{k,1} + \hat{v}_{2h}. \qquad\qquad \square$$

1.4 Remarks. (1) The parameter v controls the number of smoothing steps. In the standard case,

$$1 \le v \le 3.$$

The results of Ries, Trottenberg, and Winter [1983] presented in Table 7 show that for well-behaved problems such as the Poisson equation on a rectangle, it does not pay to do more than 2 smoothing steps. For more complicated problems (such as those involving nonconforming elements or mixed methods), it can happen that a much larger number of smoothing steps are necessary.

(2) In principle, the question of which bases to choose for the finite element spaces depends on the smoothing process. We can use the usual nodal basis functions, as will become clear from the convergence theory in §2.

Table 7. Bounds for the spectral radius ρ of the iteration matrix for the two-grid method for the Poisson equation. Here v steps of the Gauss–Seidel method with the checkerboard order are used for smoothing.

v	1	2	3	4
ρ	0.25	0.074	0.053	0.041

We turn now to the complete algorithm for several levels. It is easiest to give a precise formulation for conforming finite elements based on nested grids.

First choose a coarse triangulation $\mathcal{T}_{h_0}$. Let $\mathcal{T}_{h_1}$ be the triangulation which arises if we subdivide each triangle of $\mathcal{T}_{h_0}$ into four congruent subtriangles. Further subdivision leads to the grids[11] $\mathcal{T}_{h_2}, \mathcal{T}_{h_3}, \ldots, \mathcal{T}_{h_q}$ (see Fig. 48). We write $\mathcal{T}_\ell$ in place of $\mathcal{T}_{h_\ell}$ for $0 \le \ell \le \ell_{\max} =: q$. Suppose the finite element spaces corresponding to the triangulations $\mathcal{T}_\ell$ are S_{h_ℓ} (S_ℓ for short). Then

$$S_0 \subset S_1 \subset \ldots \subset S_{\ell_{\max}}. \tag{1.5}$$

Our goal is to compute the finite element solution of the boundary-value problem on the finest grid. The spaces S_ℓ corresponding to coarser grids arise only in the intermediate computations.

The variables in the multigrid algorithm involve up to three indices. Here
ℓ denotes the grid level,
k counts the iterations,
m counts the substeps inside each iteration.
We will usually write $u^{\ell,k}$ instead of $u^{\ell,k,0}$.

---

[11] Not all elements have to be refined as long as we take account of the rules discussed in Ch. II, §8 in connection with local mesh refinement.

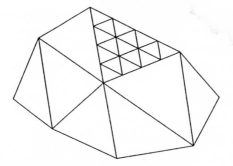

Fig. 48. A coarse triangulation for which one of the (coarse) triangles has been decomposed into 16 subtriangles in two steps. The other triangles should be decomposed in the analogous way.

We now define the multigrid method for conforming elements *recursively*.

1.5 Multigrid Iteration MGM$_\ell$ (k-th cycle at level $\ell \geq 1$):

Let $u^{\ell,k}$ be a given approximation in S_ℓ.

   1. *Pre-Smoothing.* Carry out ν_1 smoothing steps:

$$u^{\ell,k,1} = S^{\nu_1} u^{\ell,k}. \tag{1.6}$$

   2. *Coarse-Grid Correction.* Let $\hat{v}^{\ell-1}$ denote the solution of the variational problem at level $\ell - 1$,

$$J(u^{\ell,k,1} + v) \longrightarrow \min_{v \in S_{\ell-1}} ! \tag{1.7}$$

If $\ell = 1$, find the solution and set $v^{\ell-1} = \hat{v}^{\ell-1}$.
If $\ell > 1$, compute an approximation $v^{\ell-1}$ of $\hat{v}^{\ell-1}$ by carrying out μ steps of **MGM$_{\ell-1}$** with the starting value $u^{\ell-1,0} = 0$.
Set

$$u^{\ell,k,2} = u^{\ell,k,1} + v^{\ell-1}. \tag{1.8}$$

   3. *Post-Smoothing.* Carry out ν_2 smoothing steps,

$$u^{\ell,k,3} = S^{\nu_2} u^{\ell,k,2},$$

and set $u^{\ell,k+1} = u^{\ell,k,3}$.   □

1.6 Remarks. (1) If only two levels are being used, then we have only the case $\ell = 1$, and the coarse-grid correction will be done exactly. For more than two levels, we compute the solution on the coarse grid only approximately, and for the convergence theory we treat the multigrid iteration as a perturbed two-grid iteration.

(2) For more than two levels, it makes a difference whether we perform the smoothing before or after the coarse-grid correction. This is controlled by the parameters ν_1 and ν_2. For simplicity, frequently only the pre-smoothing is performed. However, for the V-cycle, it is better to do an equal amount of smoothing both before and after, i.e. to choose $\nu_1 = \nu_2$.

(3) Choosing $\mu = 1$ or $\mu = 2$ leads to either a *V-cycle* or a *W-cycle*. The reason for this terminology is clear from the shape of the corresponding schemes shown in Fig. 49. Obviously a W-cycle is more expensive than a V-cycle.

In order to ensure that in running through several levels the error does not build up too much, in the early use of the multigrid method most people chose W-cycles. However, most problems are so well-behaved that multigrid algorithms with the V-cycle are faster. (For more than four levels, it is better to insert one W-cycle after every three V-cycles; see Problem 3.12.)

(4) We solve the system of equations corresponding to the variational problem on the coarsest grid using Gauss elimination or some other direct method.

(5) In practice, an auxiliary grid can be so coarse that it would never be used as the final grid. For the Poisson equation on the unit square, it is even possible that the grid is coarsened so much that the coarsest grid contains only one (interior) point. This does not ruin the convergence rate of multigrid algorithms.

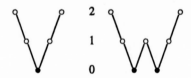

Fig. 49. V-cycle and W-cycle on three levels

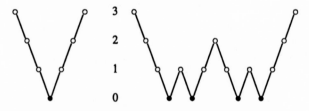

Fig. 50. V-cycle and W-cycle on four levels

Transfer Between Grids

The formulation (1.7) of the variational problem corresponding to a coarse-grid correction requires moving from the grid S_ℓ to $S_{\ell-1}$. The calculation uses the matrix-vector form

$$A_{\ell-1} y_{\ell-1} = b_{\ell-1}$$

of the system of equations which gives the solution of the auxiliary problem (1.7) on the coarser grid. The matrix $A_{\ell-1}$, which is of smaller dimension than A_ℓ, and the current right-hand side $b_{\ell-1}$ have to be computed.

All multigrid methods for linear systems of equations involve formulae of the form

$$\begin{aligned} A_{\ell-1} &:= r A_\ell p, \\ b_{\ell-1} &:= r d_\ell \quad \text{with } d_\ell := b_\ell - A_\ell u^{\ell,k,1}. \end{aligned} \tag{1.9}$$

The matrix $r = r_\ell$ is called the *restriction*, and the matrix $p = p_\ell$ is called the *prolongation*. The choice of p and r has a major influence on the rate of convergence.

There is a canonical choice for p and r when using conforming Lagrange elements. Then the spaces are nested, i.e., $S_{\ell-1} \subset S_\ell$. In these cases we take p to be the matrix representation of the injection $j : S_{\ell-1} \hookrightarrow S_\ell$, and $r := p^t$ to be the matrix of the adjoint operator $j^* : S'_\ell \hookrightarrow S'_{\ell-1}$.

To express the right-hand side, let $\{\psi_i^\ell\}_{i=1}^{N_\ell}$ be a basis for S_ℓ, and let $\{\psi_j^{\ell-1}\}_{j=1}^{N_{\ell-1}}$ be a basis for $S_{\ell-1}$. Since $S_{\ell-1} \subset S_\ell$, there exists an $N_{\ell-1} \times N_\ell$ matrix r with

$$\psi_j^{\ell-1} = \sum_i r_{ji} \psi_i^\ell, \quad j = 1, 2, \ldots, N_{\ell-1}. \tag{1.10}$$

Consider again the weak formulation of the variational problem (1.7):

$$a(u^{\ell,k,1} + v, w) = (f, w)_0 \quad \text{for } w \in S_{\ell-1}, \tag{1.7$'$}$$

or[12]

$$a(v, w) = (d, w)_0 \quad \text{for } w \in S_{\ell-1}.$$

---

[12] The connection with the following calculation will be somewhat clearer if we make use of a little (unneeded) formalism. It is important for nonconforming problems because the bilinear forms on S_ℓ and $S_{\ell-1}$ can be different. For the iteration at level ℓ, the bilinear form $a(\cdot, \cdot)$ is first defined on S_ℓ, and then passing to $S_{\ell-1}$, we can formally include the injection

$$a(jv, jw) = \langle j^*d, jw \rangle \quad \text{for } w \in S_{\ell-1}.$$

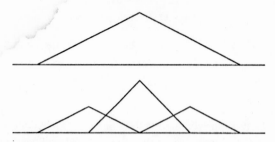

Fig. 51. Decomposition of a nodal basis function on the coarse grid (top) in terms of nodal basis functions on the fine grid (bottom)

Here d is defined by $(d, w)_0 := (f, w)_0 - a(u^{\ell,k,1}, w)$. In particular, $d = 0$ if $u^{\ell,k,1}$ is a solution at level ℓ. As in the derivation of (II.4.4), we successively insert $w = \psi_j^{\ell-1}$ for $j = 1, 2, \ldots, N_{\ell-1}$, and immediately take into account (1.10):

$$\sum_i r_{ji}\, a(u^{\ell,k,1} + v,\, \psi_i^\ell) = \sum_i r_{ji}\,(f, \psi_i^\ell)_0, \quad j = 1, 2, \ldots N_{\ell-1}.$$

We recall that $u^{\ell,k,1} = \sum_t x_t^{\ell,k,1} \psi_t^\ell$. Next we set $v = \sum_s y_s^{\ell-1} \psi_s^{\ell-1}$, and return to the basis of S_ℓ:

$$\sum_t \sum_{i,s} r_{ji}\, a(\psi_s^\ell, \psi_i^\ell)\, r_{st} y_t^{\ell-1} = \sum_i r_{ji}\Big[(f, \psi_i^\ell)_0 - \sum_t a(\psi_t^\ell, \psi_i^\ell) x_t^{\ell,k,1}\Big],$$
$$j = 1, 2, \ldots, N_{\ell-1}. \qquad (1.11)$$

The expression in the square brackets is just the i-th component of the residue d_ℓ defined in (1.9). Thus, (1.11) is the componentwise version of the equation $r A_\ell r^t\, y^{\ell-1} = r d_\ell$, and (1.9) follows with $p = r^t$. $\qquad \square$

For completeness we note that the vector representation of the approximate solution after the coarse-grid correction (1.8) is

$$x^{\ell,k,2} = x^{\ell,k,1} + p\, y^{\ell-1}. \qquad (1.12)$$

In practice we usually compute the prolongation and restriction matrices via interpolation. Let $\{\psi_i^\ell\}$ be a nodal basis for S_ℓ. Then we have N_ℓ points z_t^ℓ with

$$\psi_i^\ell(z_t^\ell) = \delta_{it}, \quad i, t = 1, 2, \ldots, N_\ell.$$

For every $v \in S_\ell$, $v = \sum_i v(z_i^\ell)\, \psi_i^\ell$, and so $\psi_j^{\ell-1} = \sum_i \psi_j^{\ell-1}(z_i^\ell)\, \psi_i^\ell$ for the basis functions of $S_{\ell-1}$. Comparing coefficients with (1.10), we get

$$r_{ji} = \psi_j^{\ell-1}(z_i^\ell). \qquad (1.13)$$

This is a convenient description of the restriction matrix.

The matrix r is simplest and easiest to find for variational problems in $\mathbb{R}^1$; see Fig. 51. For piecewise linear functions,

$$
r = \begin{pmatrix}
\frac{1}{2} & 1 & \frac{1}{2} & & & & \\
& & \frac{1}{2} & 1 & \frac{1}{2} & & \\
& & & & \frac{1}{2} & 1 & \frac{1}{2} \\
& & & & & \ddots & \ddots & \ddots \\
& & & & & & \frac{1}{2} & 1 & \frac{1}{2}
\end{pmatrix}.
\tag{1.14}
$$

For affine families of finite elements, we only need to compute the coefficients for one (reference) element. In particular, for piecewise linear triangular elements,

$$
r_{ji} = \begin{cases}
1 & \text{if } z_j^{\ell-1} = z_i^{\ell}, \\
\frac{1}{2} & \text{if } z_i^{\ell} \text{ is not a grid point of } \mathcal{T}_{\ell-1}, \\
& \text{but is a neighbor in } \mathcal{T}_{\ell} \text{ of } z_j^{\ell-1}, \\
0 & \text{otherwise.}
\end{cases}
$$

If the variables are numbered as in the model Example II.4.3 on a rectangular grid, then the operators can be expressed as stencils. For the example we have

$$
p = \begin{bmatrix}
& \frac{1}{2} & \frac{1}{2} \\
\frac{1}{2} & 1 & \frac{1}{2} \\
\frac{1}{2} & \frac{1}{2} &
\end{bmatrix}_*.
\tag{1.15}
$$

Note that r is always a sparse matrix, and thus there is never a need to store a full matrix. Frequently, it is given only in operator form, i.e., we have a procedure for computing the vector $rx^{\ell} \in \mathbb{R}^{N_{\ell-1}}$ for any given $x^{\ell} \in \mathbb{R}^{N_{\ell}}$. The way in which the nodes of the grid $\mathcal{T}_{\ell-1}$ are related to those of $\mathcal{T}_{\ell}$, and the way they are numbered, are critical to the efficiency of an algorithm.

For completeness, we describe the multigrid algorithm once again, paying more attention to the computational details. This formulation can also be used for difference methods. Suppose we are given the smoothing $\mathcal{S} = \mathcal{S}_{\ell}$, the restriction $r = r_{\ell}$, and the prolongation $p = p_{\ell}$.

1.7 Multigrid Iteration MGM$_{\ell}$ *(k-th cycle at level $\ell \geq 1$ in matrix-vector form):*
Let $x^{\ell,k}$ be a given approximation in S_{ℓ}.

1. *Pre-Smoothing.* Carry out ν_1 smoothing steps:

$$
x^{\ell,k,1} = \mathcal{S}^{\nu_1} x^{\ell,k}.
\tag{1.16}
$$

2. *Coarse-Grid Correction.* Compute the residue $d_{\ell} = b_{\ell} - A_{\ell} x^{\ell,k,1}$ and the restriction $b_{\ell-1} = r d_{\ell}$. Let

$$
A_{\ell-1} \hat{y}^{\ell-1} = b_{\ell-1}.
$$

If $\ell = 1$, find the solution, and set $y^{\ell-1} = \hat{y}^{\ell-1}$.
If $\ell > 1$, compute an approximation $y^{\ell-1}$ of $\hat{y}^{\ell-1}$ by carrying out μ steps of
$MGM_{\ell-1}$ with the starting value $x^{\ell-1,0} = 0$.
Set

$$x^{\ell,k,2} = x^{\ell,k,1} + p\, y^{\ell-1}.$$

3. *Post-Smoothing.* Carry out ν_2 smoothing steps

$$x^{\ell,k,3} = S^{\nu_2} x^{\ell,k,2},$$

and set $x^{\ell,k+1} = x^{\ell,k,3}$.   □

The development of the multigrid method for nonconforming elements and
for saddle point methods is more complicated. For nonconforming elements, usually $S_{2h} \not\subset S_h$. Then the prolongation and restriction operators have to be specially
computed. These elements have frequently been used as a model for the nonconforming P_1 element; see Brenner [1989] and Braess and Verfürth [1990]. On the
other hand, for mixed methods, it is not clear at the outset how to select suitable
smoothing operators. Previously, Jacobi smoothing was frequently applied to the
squared system, as e.g. in Verfürth [1988]. More recently, so-called transforming
smoothing has provided a completely different approach, see Brandt and Dinar
[1979], Wittum [1989], Braess and Sarazin [1997]. Moreover Bank, Welfert, and
Yserentant [1990] propose a smoother which is based on the Uzawa algorithm.

Problems

1.8 Suppose we apply the Jacobi method in Example 1.1 with $\omega = \frac{2}{3}$. With $n = 32$,
what frequencies are damped by less than a factor of 2 after ten iterations?

1.9 Suppose we are given a series of data points $\{y_i\}_{i=1}^{2n+1}$ which are the values of
a function at equally spaced grid points, but suppose we only need data on a grid
with double the mesh size. Suppose that in carrying out the elimination process we
want to eliminate as much measurement error as possible. Which of the following
three procedures does the best job of smoothing (here $i = 1, 2, \ldots, n$):

(a) $z_i = y_{2i}$,
(b) $z_i = \frac{1}{2}(y_{2i-1} + y_{2i+1})$,
(c) $z_i = \frac{1}{4}(y_{2i-1} + 2y_{2i} + y_{2i+1})$?

In performing the analysis, express the values cyclically, and compare the Fourier
coefficients of $\{z_i\}$ and $\{y_i\}$.

1.10 Write Algorithm 1.7 as a formal procedure $\mathbf{MGM}_\ell(A_\ell, b_\ell, x^{\ell,k})$ in PASCAL or some other programming language.

1.11 What is the (one-dimensional) stencil of the restriction operator with the matrix representation (1.14)?

1.12 Dividing the squares into triangles in the model Example II.4.3 results in an unsymmetric stencil for the prolongation in (1.15). The symmetric form

$$p = \begin{bmatrix} \frac{1}{4} & \frac{1}{2} & \frac{1}{4} \\ \frac{1}{2} & 1 & \frac{1}{2} \\ \frac{1}{4} & \frac{1}{2} & \frac{1}{4} \end{bmatrix}_* \tag{1.17}$$

appears more natural. What are the stencils for the restriction operators which arise from (1.15) and (1.17) via the matrix equation $r = p^t$?

§ 2. Convergence of Multigrid Methods

A multigrid method is said to have *multigrid convergence* if the error is reduced by a factor of at least $\rho < 1$ in each iteration cycle, where ρ is *independent of h*. In this case the convergence as $h \to 0$ cannot be arbitrarily slow, in contrast to classical iterative methods. The factor ρ is called the *convergence rate*. Clearly, it is a measure of the speed of convergence.

Independently from Fedorenko [1961], Hackbusch [1976] and Nicolaides [1977] also presented convergence proofs. Here we make use of an idea employed by Bank and Dupont [1981] in their proof. A general framework due to Hackbusch [1985] admits to break convergence proofs into two separate parts. In this way they become very transparent. A *smoothing property*

$$\|S^\nu v_h\|_X \le c\, h^{-\beta} \frac{1}{\nu^\gamma} \|v_h\|_Y \tag{2.1}$$

is combined with an *approximation property*

$$\|v_h - u_{2h}\|_Y \le c\, h^\beta \|v_h\|_X, \tag{2.2}$$

where u_{2h} is the coarse-grid approximation of v_h. Then for large ν, the product of the two factors is smaller than 1 and independent of h. In particcular, it follows that the convergence rate tends to zero for large numbers of smoothing steps.

The various proofs differ in the choice of the norms $\|\cdot\|_X$ and $\|\cdot\|_Y$, where $\|\cdot\|_X$ generates a stronger topology than $\|\cdot\|_Y$. The pair (2.1) and (2.2) have to fit together in exactly the same way as the approximation property (II.6.20) and the inverse estimate (II.6.21). It is clear that we need two norms, or more generally two measures, for specifying the error. In addition to measuring the *size* of the error (w.r.t. whichever norm), we also have to measure the *smoothness* of the error function.

It is the goal of this section to establish convergence of the two-level iteration under the following hypotheses:

2.1 Hypotheses.
(1) The boundary-value problem is H^1- or H_0^1-elliptic.
(2) The boundary-value problem is H^2-regular.
(3) The spaces S_ℓ belong to a family of conforming finite elements with uniform triangulations, and the spaces are nested, i.e., $S_{\ell-1} \subset S_\ell$.
(4) We use nodal bases.

For most of our discussion we can get by with weaker hypotheses, but in that case the proofs become more technical. For example, $H^{1+\alpha}$-regularity with $\alpha > 0$ would suffice for this section; see Problem 2.12.

There is a different theory due to Bramble, Pasciak, Wang, and Xu [1991] and Xu [1992]. Multigrid algorithms are connected with a decomposition of finite element spaces. The space decomposition method has the advantage that it does not require regularity assumptions. On the other hand, it does not model the fact that the convergence rate is improved by increasing the number of smoothing steps and it applies only to the energy norm. Since the arguments of the theory are far away from the finite element theory in this book, we restrict ourselves in §5 to the motivation and the proof of the central tool.

We will also not dicuss quantitative results on the convergence rate obtained by Fourier methods [Brandt 1977, Ries, Trottenberg, and Winter 1983].

Discrete Norms

So far, the quality of the approximation of a function by finite elements has been expressed in terms of higher Sobolev norms. This no longer works in dealing with the approximation of a function $v_h \in S_h$ by a function in the space S_{2h} on the coarse grid. In particular, if S_h consists of C^0 elements, then in general $S_h \not\subset H^2(\Omega)$, and we cannot employ estimates in the H^2 norm.

This leads us to assign another Hilbert scale to the N_h-dimensional space S_h. The new scale should be connected with the scale of the Sobolev spaces as closely as possible. – It is easiest to discuss how to do this in the following abstract form.

For a symmetric positive definite matrix A, the powers A^s are well defined for all values of s, and not just for integers. Thus, we go back to the spectral decomposition. The matrix A has a complete system of orthonormal eigenvectors $\{z_i\}_{i=1}^N$:

$$Az_i = \lambda_i z_i, \qquad i = 1, 2, \ldots, N,$$
$$(z_i, z_j) = \delta_{ij}, \qquad i, j = 1, 2, \ldots, N.$$

Every vector $x \in \mathbb{R}^N$ can be written in the form

$$x = \sum_{i=1}^N c_i z_i, \tag{2.3}$$

and

$$A^s x = \sum_i c_i \lambda_i^s z_i \tag{2.4}$$

is well defined.

2.2 Definition. Let A be a symmetric positive definite $N \times N$ matrix, and suppose $s \in \mathbb{R}$. Then

$$\||x|\|_s := (x, A^s x)^{1/2} \tag{2.5}$$

defines a norm, where $(\cdot, \cdot)$ is the Euclidean scalar product in $\mathbb{R}^N$.

Using (2.3), (2.4), and the orthogonality relation, we have

$$(x, A^s x) = \Big(\sum_k c_k z_k, \sum_i c_i \lambda_i^s z_i\Big) = \sum_i \lambda_i^s c_i^2.$$

Thus the norm (2.5) has the following alternative representation:

$$\||x|\|_s = \Big(\sum_{i=1}^N \lambda_i^s |c_i|^2\Big)^{1/2} = \|A^{s/2} x\|. \tag{2.6}$$

2.3 Properties of the Norm (2.5).

(1) *Connection with the Euclidean norm:* $\||x|\|_0 = \|x\|$, where $\|\cdot\|$ is the Euclidean norm.

(2) *Logarithmic convexity:* For $r, t \in \mathbb{R}$ and $s = \frac{1}{2}(r + t)$,

$$\||x|\|_s \leq \||x|\|_r^{1/2} \cdot \||x|\|_t^{1/2},$$
$$|(x, A^s y)| \leq \||x|\|_r \cdot \||y|\|_t.$$

Indeed, with the help of the Cauchy–Schwarz inequality, it follows that

$$|(x, A^s y)| = |(A^{r/2} x, A^{t/2} y)| \leq \|A^{r/2} x\| \, \|A^{t/2} y\| = \||x|\|_r \, \||y|\|_t .$$

This is the second inequality. The first follows if we choose $x = y$. Taking the logarithm of both sides and using the continuity, we see that the function

$$s \longmapsto \log \||x|\|_s$$

is convex provided that $x \neq 0$.

(3) *Monotonicity:* Let α be the constant of ellipticity, i.e., $(x, Ax) \geq \alpha(x, x)$. Then

$$\alpha^{-t/2} \||x|\|_t \geq \alpha^{-s/2} \||x|\|_s, \quad \text{for } t \geq s.$$

For the special case $\alpha = 1$, this follows immediately from (2.6) and $\lambda_i \geq \alpha = 1$. Otherwise we have the monotonicity property for the normalized matrix $\alpha^{-1} A$ which implies the monotonicity as stated for A.

(4) *Shift theorem.* The solution of $Ax = b$ satisfies

$$\||x|\|_{s+2} = \||b|\|_s$$

for all $s \in \mathbb{R}$. This follows from $(x, A^{s+2} x) = (Ax, A^s Ax) = (b, A^s b)$. $\square$

Using the scale defined with (2.5), we immediately get the following property of Richardson relaxation without any additional hypotheses. It can be thought of as a smoothing property, as we shall see later.

2.4 Lemma. *Let $\omega \geq \lambda_{\max}(A)$, $s \in \mathbb{R}$ and $t > 0$, and consider the iteration*

$$x^{\nu+1} = (1 - \frac{1}{\omega}A)x^{\nu}.$$

Then

$$|||x^{\nu}|||_{s+t} \leq c\nu^{-t/2}|||x^0|||_s,$$

where $c = \left(\frac{t\omega}{2e}\right)^{t/2}$.

Proof. If x^0 is expanded as in (2.3), then $x^{\nu} = \sum_i (1 - \lambda_i/\omega)^{\nu}c_i z_i$, and since $0 < \lambda_i/\omega \leq 1$ we have

$$|||x^{\nu}|||^2_{s+t} = \sum_i \lambda_i^{s+t}\left[(1 - \frac{\lambda_i}{\omega})^{\nu} c_i\right]^2$$

$$= \omega^t \sum_i (\frac{\lambda_i}{\omega})^t (1 - \frac{\lambda_i}{\omega})^{2\nu} \lambda_i^s c_i^2$$

$$\leq \omega^t \max_{0 \leq \zeta \leq 1}\{\zeta^t(1 - \zeta)^{2\nu}\} \sum_i \lambda_i^s c_i^2. \qquad (2.7)$$

To compute the maximum appearing in (2.7), we examine the function $\zeta(1 - \zeta)^p$ in the interval $[0,1]$ for $p > 0$. It attains its maximum at $\zeta = 1/(p + 1)$. Thus,

$$\zeta(1 - \zeta)^p \leq \frac{1}{p + 1}(\frac{p}{p + 1})^p = \frac{1}{p}\frac{1}{(1 + \frac{1}{p})^{p+1}} \leq \frac{1}{p}\frac{1}{e}$$

in $[0,1]$. With $p = 2\nu/t$, it follows that $\max\{\zeta^t(1 - t)^{2\nu}\} \leq (t/[2e\nu])^t$. Since the sum appearing in (2.7) is exactly $|||x^0|||^2_s$, the proof is complete. $\qquad \square$

The assertion of Lemma 2.4 is independent of the choice of basis. For the following results, this happens only under some additional hypotheses.

Connection with the Sobolev Norm

In Example 1.1, smooth eigenfunctions are associated with the small eigenvalues of A, while eigenfunctions with strongly oscillating terms are associated with the large eigenvalues. This fact naturally depends on the choice of basis. As we will see, this holds in general if we select the nodal basis for affine families, because in this case, the norm $||| \cdot |||_0$ is equivalent to the Sobolev norm $\| \cdot \|_{0,\Omega}$.

2.5 Lemma. *Let $\mathcal{T}_h$ be a family of uniform partitions of $\Omega \subset \mathbb{R}^n$, and suppose S_h belongs to an affine family of finite elements. Suppose we normalize the functions in the nodal basis so that*

$$\psi_i(z_j) = h^{-n/2}\delta_{ij}. \qquad (2.8)$$

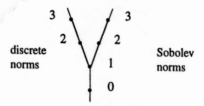

Fig. 52. Connection between the scales defined by discrete norms and Sobolev norms

Then for $v_h \in S_h$, $\||v_h\||_0 = [h^n \sum_i v_h(z_i)^2]^{1/2}$, which is just the Euclidean norm of the coefficient vector relative to the basis $\{\psi_i\}$. Moreover, there exists a constant c independent of h such that

$$c^{-1}\|v_h\|_{0,\Omega} \le \||v_h\||_0 \le c\|v_h\|_{0,\Omega}. \tag{2.9}$$

To establish (2.9), we make use of Problem II.6.12: On the reference element $T_{\text{ref}} \subset \mathbb{R}^n$, $\|p\|_{0,T_{\text{ref}}}^2$ and $\sum_i p(z_i)^2$ are equivalent. Now from the properties of affine transformations, we conclude that for every $T \in \mathcal{T}_h$, the quantities $\|v_h\|_{0,T}^2$ and $h^n \sum_i v_h(z_i)^2$ are equivalent, where the sum runs over the nodes belonging to T. The factor h^n enters through the transformation of the domain. Summing over the elements gives (2.9). □

In practice, we always normalize the basis functions so that $\psi_i(z_j) = \delta_{ij}$. We can ignore the difference in normalization factors since there is no essential change in the following results if the system matrix is multiplied by a constant factor.

We now further restrict ourselves to second order elliptic problems. For $v_h \in S_h$, we define $\||v_h\||_s$ by associating with v_h its coefficients relative to a nodal basis (2.8), and define the scale according to Definition 2.2 using the stiffness matrix A_h. Then in particular,

$$\||v_h\||_1^2 = (v_h, A_h v_h) = a(v_h, v_h),$$

and because of the ellipticity of the bilinear form a, it follows that

$$c^{-1}\|v_h\|_{1,\Omega} \le \||v_h\||_1 \le c\|v_h\|_{1,\Omega}. \tag{2.10}$$

More precisely, every function $v_h \in S_h$ is identified with its coefficient vector relative to the nodal basis. This makes sense since $\|\cdot\|_{s,\Omega}$ and $\||\cdot\||_s$ are equivalent for $s = 0$ and $s = 1$ by (2.9) and (2.10), respectively. For $s > 1$, this is no longer the case (see Fig. 52).

As a consequence of the equivalence, we get estimates of the largest and smallest eigenvalues, as well as the condition number of the stiffness matrix.

2.6 Lemma. *Suppose that the hypotheses of Lemma 2.5 hold. Then there is a constant c independent of h such that*

$$\lambda_{\min}(A_h) \geq c^{-1}, \quad \lambda_{\max}(A_h) \leq ch^{-2}, \quad \kappa(A_h) \leq c^2h^{-2} \qquad (2.11)$$

where A_h is the system matrix corresponding to any H^1- or H_0^1-elliptic problem.

Proof. For positive definite matrices, the eigenvalues can be estimated in terms of the Rayleigh quotients. Using the inverse estimate II.6.8, we get $\|v_h\|_{1,\Omega} \leq ch^{-1}\|v_h\|_{0,\Omega}$, and

$$\lambda_{\max}(A_h) = \sup_x \frac{(x, A_h x)}{(x, x)} = \sup_{v_h \in S_h} \frac{|||v_h|||_1^2}{|||v_h|||_0^2} \leq c \sup_{v_h \in S_h} \frac{\|v_h\|_{1,\Omega}^2}{\|v_h\|_{0,\Omega}^2} \leq ch^{-2}.$$

Similarly, using $\|v_h\|_{1,\Omega} \geq \|v_h\|_{0,\Omega}$, we get

$$\lambda_{\min}(A_h) = \inf_x \frac{(x, A_h x)}{(x, x)} = \inf_{v_h \in S_h} \frac{|||v_h|||_1^2}{|||v_h|||_0^2} \geq c^{-1} \inf_{v_h \in S_h} \frac{\|v_h\|_{1,\Omega}^2}{\|v_h\|_{0,\Omega}^2} \geq c^{-1}.$$

Finally, the third assertion follows from $\kappa(A_h) = \lambda_{\max}(A_h)/\lambda_{\min}(A_h)$. $\qquad\square$

These estimates are sharp. The exponent of h^{-2} in (2.11) cannot be improved, since by Remark II.6.10, there exist functions in S_h for which $\|v_h\|_1 \approx ch^{-1}\|v_h\|_0$.

We also note that by the proof of the lemma, for every eigenfunction ϕ_h, the ratio $\|\phi_h\|_1^2/\|\phi_h\|_0^2$ gives the corresponding eigenvalue up to a constant. This shows that the oscillating eigenfunctions correspond to the large eigenvalues. In this connection, the situation is exactly the same as for the model Example 1.1.

Using Lemma 2.4 for $s = 0$, $t = 2$ and substituting the estimate (2.11) for $\lambda_{\max}$, we immediately get

2.7 Corollary. *(Smoothing Property) The iteration $x^{\nu+1} = (1 - \frac{1}{\omega}A_h)x^\nu$ with $\omega = \lambda_{\max}(A_h)$ satisfies*

$$|||x^\nu|||_2 \leq \frac{c}{\nu} h^{-2}\|x^0\|_0. \qquad (2.12)$$

Approximation Property

The quality of the coarse-grid correction in S_{2h} can be expressed in terms of the $||| \cdot |||_2$ norm, and the results resemble those in Ch. II, §7 if we replace the Sobolev norm $\| \cdot \|_{2,\Omega}$ by $||| \cdot |||_2$. The essential tool here is the duality argument of Aubin–Nitsche. By the approximation results of Ch. II, §6, in estimating the $\| \cdot \|_1$ norm of the error in terms of the $\| \cdot \|_2$ norm, we gain one power of h. In Corollary II.7.7, this gain was "propagated downwards" on the right-hand branch of the scale in Fig. 52. Specifically, it was shown that the same improvement occurs in estimating the $\| \cdot \|_0$ norm by the $\| \cdot \|_1$ norm. We now carry out the same process in the reverse direction, moving to the scale with the discrete norm, see Braess and Hackbusch [1983].

2.8 Lemma. *Given $v_h \in S_h$, let u_{2h} be the solution of the weak equation*

$$a(v_h - u_{2h}, w) = 0 \quad \text{for all } w \in S_{2h}.$$

In addition, let Ω be convex or have a smooth boundary. Then

$$\|v_h - u_{2h}\|_{1,\Omega} \le c\, 2h \||v_h\||_2, \tag{2.13}$$

$$\|v_h - u_{2h}\|_{0,\Omega} \le c\, 2h \|v_h - u_{2h}\|_{1,\Omega}. \tag{2.14}$$

Proof. By hypothesis, the problem is H^2-regular, and by Corollary II.7.7, (2.14) holds. In addition, recalling property 2.3(2), and a well-known argument from the proof of Céa's lemma, we have

$$\alpha \|v_h - u_{2h}\|_1^2 \le a(v_h - u_{2h}, v_h - u_{2h}) = a(v_h - u_{2h}, v_h) = (v_h - u_{2h}, A_h v_h)$$
$$\le \||v_h - u_{2h}\||_0 \||v_h\||_2 \le c_1 \|v_h - u_{2h}\|_0 \||v_h\||_2$$
$$\le c_1 c\, 2h \|v_h - u_{2h}\|_1 \||v_h\||_2 .$$

Dividing by $\|v_h - u_{2h}\|_1^2$, we get (2.13).

Second proof. Let $g \in S_h$ be defined by $(g, w)_{0,\Omega} = a(v_h, w)$ for all $w \in S_h$. Consider the auxiliary variational problem. Find $z \in H_0^1$ such that

$$a(z, w) = (g, w)_{0,\Omega} \quad \text{for all } w \in H_0^1(\Omega).$$

Obviously, v_h and u_{2h} are the finite element approximations of z in S_h an S_{2h}, respectively. From the H^2-regularity we conclude that

$$\|v_h - u_{2h}\|_{0,\Omega} \le \|v_h - z\|_{0,\Omega} + \|z - u_{2h}\|_{0,\Omega} \le ch^2 \|u\|_{2,\Omega} \le ch^2 \|g\|_{0,\Omega}.$$

On the other hand, we have $(A_h v_h, w) = a(v_h, w) = (g, w)_{0,\Omega}$ for $w \in S_h$, and the equivalence of the Euclidean norm and the L_2-norm yields

$$\|g\|_{0,\Omega}^2 = (A_h v_h, g) \le \|A_h v_h\| \|g\| \le \||v_h\||_2\, c \|g\|_{0,\Omega}.$$

After dividing by $\|g\|_{0,\Omega}$ and inserting the bound of $\|g\|_{0,\Omega}$ into the preceding estimate, we have

$$\|v_h - u_{2h}\|_{0,\Omega} \le ch^2 \||v_h\||_2$$

which is the desired estimate. □

We note that the second proof can be extended more easily to the nonconforming case, since v and u_{2h} are separated early in the proof by the application of the triangle inequality; cf. Braess and Verfürth [1992], Brenner [1991], and the axiomatic considerations by Braess, Dryja, and Hackbusch [1999].

Convergence Proof for the Two-Grid Method

The last two lemmas show that the smoothing property and the approximation property given in (2.1) and (2.2) in an abstract (general) form hold with the choices

$$\| \cdot \|_X = \||\cdot\||_2, \quad \| \cdot \|_Y = \||\cdot\||_0, \quad \beta = 2, \quad \text{and } \gamma = 1. \tag{2.15}$$

The following proof makes clear the fundamental importance of the properties (2.1) and (2.2). To keep the formalism from obscuring the key ideas, we give a concrete proof. – Note that for the two-grid method, the finer grid corresponds to the level $\ell = 1$, where $u_1 = u_{\ell_1}$ is the desired solution.

2.9 Convergence Theorem. *Under Hypotheses 2.1, the two-grid method using Jacobi relaxation (with $\lambda_{\max}(A_h) \leq \omega \leq c' \lambda_{\max}(A_h)$) satisfies*

$$\|u^{1,k+1} - u_1\|_{0,\Omega} \leq \frac{c}{\nu_1} \|u^{1,k} - u_1\|_{0,\Omega},$$

where c is a constant independent of h, and ν_1 is the number of pre-smoothings.

Proof. For smoothing with Richardson relaxation,

$$u^{1,k,1} - u_1 = (1 - \frac{1}{\omega} A_h)^{\nu_1} (u^{1,k} - u_1).$$

By Lemma 2.6,

$$\||u^{1,k,1} - u_1\||_2 \leq \frac{c}{\nu_1} h^{-2} \|u^{1,k} - u_1\|_{0,\Omega}. \tag{2.16}$$

By definition of the coarse-grid correction, $u^{1,k,2} = u^{1,k,1} + u_{2h}$ is characterized by

$$a(u^{1,k,1} + u_{2h}, w) = (f, w)_{0,\Omega} \quad \text{for all } w \in S_{2h}.$$

Moreover, the solution on level 1 satisfies the equation $a(u_1, w) = (f, w)_{0,\Omega}$ for $w \in S_h$. Since $S_{2h} \subset S_h$, subtracting the two equations gives

$$a(u^{1,k,1} - u_1 + u_{2h}, w) = 0 \quad \text{for all } w \in S_{2h}.$$

Now applying Lemma 2.8 to $v := u^{1,k,1} - u_1$, we get

$$\|u^{1,k,2} - u_1\|_{0,\Omega} \leq c h^2 \||u^{1,k,1} - u_1\||_2. \tag{2.17}$$

We can deal with the post-smoothing in a very rough way. Clearly, $\||(1 - \frac{1}{\omega} A_h)x\||_s \leq \||x\||_s$. This implies that $\||u^{1,k,3} - u_1\||_0 \leq \||u^{1,k,2} - u_1\||_0$, and because of the equivalence of the norms, we have

$$\|u^{1,k,3} - u_1\|_{0,\Omega} \leq c\|u^{1,k,2} - u_1\|_{0,\Omega}. \tag{2.18}$$

Now combining (2.16)–(2.18) and taking into consideration $u^{1,k+1} = u^{1,k,3}$, we get the assertion. $\square$

An Alternative Short Proof

If we are content with a convergence rate which is $\mathcal{O}(\nu^{-1/2})$, there is a much shorter proof. Then we do not need to introduce the scale of discrete norms (2.3), but the smoothing property of the matrix operations is less transparent without this scale. We refer to that proof for completeness; we need only rearrange some results known from the previous investigation.

The smoothing property and the approximation property also hold with

$$\| \cdot \|_X = \| \cdot \|_1, \quad \| \cdot \|_Y = \| \cdot \|_0, \quad \beta = 1, \text{ and } \gamma = \frac{1}{2}.$$

Indeed, the approximation property (2.13)

$$\|v - u_{2h}\|_0 \leq ch\|v - u_{2h}\|_1 \leq ch\|v\|_1$$

is immediate from the Aubin–Nitsche lemma. Moreover, when we apply Lemma 2.4 to the case $s = 0$, $s + t = 1$, we may prove the smoothing property

$$\|x^\nu\|_1 \leq c\,\nu^{-1/2}\|x^0\|_0$$

without reference to the discrete norms. The rest of proof proceeds as for Theorem 2.9.   □

Some Variants

It is easy to see the connection with the somewhat different terminology of Hackbusch [1985]. The matrix representation of the two-grid iteration is

$$u^{1,k+1} - u_1 = M(u^{1,k} - u_1) \tag{2.19}$$

with

$$
\begin{aligned}
M &= S^{\nu_2}(I - pA_{2h}^{-1}rA_h)S^{\nu_1} \\
&= S^{\nu_2}(A_h^{-1} - pA_{2h}^{-1}r)A_hS^{\nu_1}.
\end{aligned}
\tag{2.20}
$$

In particular, $pA_{2h}^{-1}rA_h$ describes the coarse-grid correction for a two-grid method. Writing the smoothness and approximation properties in the form

$$\|A_hS^\nu\| \leq \frac{c}{\nu}h^{-2}, \quad \|(A_h^{-1} - pA_{2h}^{-1}r)\| \leq ch^2 \tag{2.21}$$

and using $\|S\| \leq 1$, we get the contraction property $\|M\| \leq c/\nu < 1$ for sufficiently large ν. Since $\|\|Ax\|\|_0 = \|\|x\|\|_2$, the smoothing and approximation properties follow from (2.21) with the same norms and parameters as in (2.15).

The smoothing property is usually established as in Lemma 2.4 whenever the convergence of the multigrid method is carried out in Hilbert spaces. The proof of convergence w.r.t. the maximum norm by Reusken [1992] is different.

2.10 Reusken's Lemma. *Let* $\| \cdot \|$ *be a matrix norm which is associated with a vector norm. Moreover, assume that* $B = I - G^{-1}A$ *satisfies*

$$\|B\| \leq 1. \tag{2.22}$$

Then for $S = \frac{1}{2}(I + B)$,

$$\|AS^\nu\| \leq \sqrt{\frac{8}{\pi\nu}} \|G\|.$$

Thus, the smoothing property is obtained if we can find a matrix G such that (2.22) holds and $\|G\|$ can be estimated by the same power of h as $\|A\|$. It is interesting to note that in view of (2.22), B almost defines a convergent iterative process. However, for the purposes of smoothing, we take the average with the old vector. Note that $AS^\nu = 2^{-\nu}G(I - B)(I + B)^\nu$. The proof makes use of the formula $\|(I - B)(I+B)^\nu\| \leq 2^{\nu+1}\sqrt{\frac{2}{\pi\nu}}$, which in turn is verified via the Binomial equation. $\quad\square$

Finally, we note that Lemma 2.8 cannot be applied to nonconforming elements or to nonnested spaces (i.e. if $S_{2h} \not\subset S_h$). In order to establish the approximation property in these cases, the duality argument has to be modified; see Brenner [1989], Braess and Verfürth [1990]. Let $r_h \in S_h$ be a representation of the residual defined by $(r_h, w)_0 = \langle d, w \rangle := (f, w)_0 - a(u^{\ell,k,1}, w)$ for $w \in S_h$. Then $u_\ell - u^{\ell,k,1}$ and $v_{\ell-1}$ are the finite element approximations in S_ℓ and $S_{\ell-1}$, respectively, of

$$a(z, w) = (r_h, w)_0 \quad \text{for all } w \in H_0^1(\Omega). \tag{2.23}$$

Note that $\|u_\ell - v_{\ell-1} - u^{\ell,k,1}\| \leq \|u_\ell - u^{\ell,k,1} - z\| + \|v_{\ell-1} - z\|$, and the terms on the right-hand side are obtained from standard error estimates. The duality technique is needed to get sharp estimates. The representation of the residual in terms of $r_h \in S_\ell$ is chosen in order to use the discrete scale.

Problems

2.11 Show that for the scale of the Sobolev spaces, the analog

$$\|v\|_{s,\Omega}^2 \leq \|v\|_{s-1,\Omega} \|v\|_{s+1,\Omega}$$

of (2.5) holds for $s = 0$ and $s = 1$.

2.12 Hypothesis 2.1 requires H^2-regularity. For many problems with reentrant corners, we only have $H^{3/2}$-regularity, and instead of (2.13),

$$\|v - u_{2h}\|_{1,\Omega} \leq c\,h^{1/2} \||v\||_{3/2}.$$

Verify the smoothing and approximation properties for $\| \cdot \|_X = \|| \cdot \||_{3/2}$ and $\| \cdot \|_Y = \|| \cdot \||_{1/2}$.

2.13   The iteration in the two-grid algorithm 1.3 is a linear process with the iteration matrix M given in (2.20). Show that the spectral radius of M depends only on the sum $v_1 + v_2$, and not on how many a priori or a posteriori smoothings are carried out. – Does this also hold for several levels?

2.14 For every $\omega \le 1/\lambda_{max}$, the Jacobi iteration (1.4) leads to smoothing. Show that for $\omega = 1.9/\lambda_{max}$, it may converge faster as a stand-alone iteration [if $\kappa(A) > 20$], but is less effective as a smoother.

2.15 Let A be positive definite, and suppose $B = (I + A)^{-1}A$. Show that in the scale generated by A,

$$|||Bx|||_s \le |||x|||_s$$

for all $s \in \mathbb{R}$.

2.16   Suppose that for $s = 2$, two positive definite matrices A and B generate equivalent norms, i.e. norms which differ by at most a factor c. [For $s = 0$ they are trivially equivalent.] Show that the norms are equivalent for all s with $0 < s < 2$. – For $s > 2$ the assertion does not hold in general.

2.17 The smoothing property (2.12) for the Richardson iteration can be generalized to the solution of the saddle point problem

$$\begin{pmatrix} A & B^T \\ B & \end{pmatrix}\begin{pmatrix} u \\ p \end{pmatrix} = \begin{pmatrix} f \\ 0 \end{pmatrix}.$$

There is an iteration such that the successive approximants stay in $\ker B$. Let $\omega \ge \lambda_{max}(A)$ and consider the iteration

$$\begin{pmatrix} u^{v+1} \\ p^{v+1} \end{pmatrix} = \begin{pmatrix} u^v \\ p^v \end{pmatrix} + \begin{pmatrix} \omega I & B^T \\ B & \end{pmatrix}^{-1}\begin{pmatrix} f - Au^v - B^T p^v \\ -Bu^v \end{pmatrix} \qquad (2.24)$$

for $v = 0, 1, \ldots$ Let Q be the projector $Q := I - B^T(BB^T)^{-1}B$ and $M := Q(1 - \frac{1}{\omega}A)Q$. Show the following properties:

(1) $Bu^v = 0$ for $v > 0$.

(2) u^{v+1} and p^{v+1} depend only on u^v but do not depend on p^v.

(3) $u^{v+1} - u = Q(I - \frac{1}{\omega}A)(u^v - u)$.

(4) $A(u^v - u) + B^T(p^v - u) = \omega(I - \frac{1}{\omega}A)M(I - M)^{v-2}(I - \frac{1}{\omega}A)(u^0 - u)$.

Since $\|(I - \frac{1}{\omega}A)u\| \le \|u\|$ and the spectrum of M is contained in $[-1, +1]$, the last equation provides the smoothing property for the iteration (2.24).

The application of (2.24) to a multigrid algorithm for the Stokes problem was proposed by Braess and Sarazin [1996].

§ 3. Convergence for Several Levels

In the convergence theorem in the previous section we assumed that the coarse-grid correction was computed exactly. For algorithms using more than two levels, this is no longer the case. In this case we can think of the multigrid algorithm as a *perturbed two-grid algorithm*. It suffices to estimate the *size* of the perturbation; we don't need to know its details.

The goal of this section is to compute the convergence rate ρ_ℓ for the algorithm with ℓ changes of grid levels, i.e., with $\ell + 1$ levels:

$$\|u^{\ell,k+1} - u_\ell\| \leq \rho_\ell \|u^{\ell,k} - u_\ell\|. \tag{3.1}$$

Here u_ℓ is the solution in S_ℓ. We assume that we know the convergence rate ρ_1 for the two-grid method. We will find the rate ρ_ℓ from the rate ρ_1 by induction. First, we let $\| \cdot \|$ be an arbitrary norm.

Later, we sharpen the results by specializing to the energy norm, and in particular obtain convergence of the multigrid method already for a single smoothing step. Thus, we can dispense with the hypothesis that sufficiently many smoothing steps are carried out.

A Recurrence Formula for the W-Cycle

For smoothing with the Richardson method, clearly

$$\|u^{\ell,k,1} - u_\ell\| \leq \|u^{\ell,k} - u_\ell\|, \tag{3.2}$$

assuming an underlying discrete norm $\| \cdot \| := \|\| \cdot \|\|_s$. In the following, we will always assume that (3.2) is satisfied, since this property also holds in other important cases.

We compare the result $u^{\ell,k,2}$ of the actual coarse-grid correction with the exact coarse-grid correction $\hat{u}^{\ell,k,2}$. By (3.1), the two-grid rate is ρ_1, i.e.,

$$\|\hat{u}^{\ell,k,2} - u_\ell\| \leq \rho_1 \|u^{\ell,k} - u_\ell\|. \tag{3.3}$$

Together with (3.2), the triangle inequality yields

$$\|u^{\ell,k,1} - \hat{u}^{\ell,k,2}\| \leq (1 + \rho_1) \|u^{\ell,k} - u_\ell\|. \tag{3.4}$$

The left-hand side of (3.4) gives the size of the coarse-grid correction with exact computations. The real correction differs from the exact one by the error at level

$\ell - 1$. Thus, by the induction hypothesis, the relative error is at most $\rho_{\ell-1}^{\mu}$. Here $\mu = 1$ for the V-cycle, and $\mu = 2$ for the W-cycle, as usual. Hence,

$$\|u^{\ell,k,2} - \hat{u}^{\ell,k,2}\| \le \rho_{\ell-1}^{\mu} \|u^{\ell,k,1} - \hat{u}^{\ell,k,2}\|. \tag{3.5}$$

We now substitute (3.4) into (3.5). Using (3.3), we see that for the W-cycle,

$$\|u^{\ell,k,2} - u_\ell\| \le [\rho_1 + \rho_{\ell-1}^2(1 + \rho_1)] \|u^{\ell,k} - u_\ell\|.$$

Without post-smoothing, we have $u^{\ell,k+1} = u^{\ell,k,2}$, and (3.1) holds with rate ρ_ℓ, which can be estimated by (3.6).

3.1 Recurrence Formula. *For the multigrid method using the W-cycle, at level* $\ell \ge 2$ *we have*

$$\rho_\ell \le \rho_1 + \rho_{\ell-1}^2(1 + \rho_1). \tag{3.6}$$

Formula (3.6) leads to an estimate of the convergence rate independent of ℓ, provided ρ_1 is sufficiently small.

3.2 Theorem. *Suppose the two-grid rate is* $\rho_1 \le \frac{1}{3}$. *Then for the W-cycle,*

$$\rho_\ell \le \frac{5}{3}\rho_1 \le \frac{1}{3}, \quad for\ \ell = 2, 3, \ldots \tag{3.7}$$

Proof. For $\ell = 1$ the assertion is clear. It follows from the assertion for $\ell - 1$ and the recurrence formula (3.6) that

$$\rho_\ell \le \rho_1 + \frac{1}{3} \left(\frac{5}{3}\rho_1\right) \left(1 + \frac{1}{5}\right) = \frac{5}{3}\rho_1 \le \frac{1}{3}. \qquad \square$$

By Theorem 2.9, the convergence rate for the two-grid method is indeed smaller than 1/5 for sufficiently many smoothing steps. Now Theorem 3.2 implies the convergence of the multigrid method under the same hypothesis.

An Improvement for the Energy Norm

When referring to the energy norm, the recurrence formula (3.6) for the convergence rate can be replaced by a significantly better one. With respect to this norm, the exact coarse-grid correction yields the orthogonal projection of $u^{\ell,k,1} - u_\ell$ onto the subspace $S_{\ell-1}$. The error $\hat{u}^{\ell,k,2} - u_\ell$ after the exact coarse-grid correction is therefore orthogonal to $S_{\ell-1}$. In particular, it is then orthogonal to $u^{\ell,k,1} - \hat{u}^{\ell,k,2}$ and to $u^{\ell,k,2} - \hat{u}^{\ell,k,2}$ (see Fig. 53).

Thus, the estimate (3.4) can be replaced by

$$\|u^{\ell,k,1} - \hat{u}^{\ell,k,2}\|^2 = \|u^{\ell,k,1} - u_\ell\|^2 - \|\hat{u}^{\ell,k,2} - u_\ell\|^2.$$

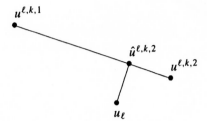

Fig. 53. The coarse-grid correction as an orthogonal projection

In addition, it follows from the orthogonality and (3.5) that

$$
\begin{aligned}
\|u^{\ell,k,2} - u_\ell\|^2 &= \|\hat{u}^{\ell,k,2} - u_\ell\|^2 + \|u^{\ell,k,2} - \hat{u}^{\ell,k,2}\|^2 \\
&\leq \|\hat{u}^{\ell,k,2} - u_\ell\|^2 + \rho_{\ell-1}^{2\mu}\|u^{\ell,k,1} - \hat{u}^{\ell,k,2}\|^2 \\
&= (1 - \rho_{\ell-1}^{2\mu})\|\hat{u}^{\ell,k,2} - u_\ell\|^2 + \rho_{\ell-1}^{2\mu}\|u^{\ell,k,1} - u_\ell\|^2. \quad (3.8)
\end{aligned}
$$

Now we make use of our knowledge of the two-grid rate. By (3.2),

$$
\|u^{\ell,k,2} - u_\ell\|^2 \leq [(1 - \rho_{\ell-1}^{2\mu})\rho_1^2 + \rho_{\ell-1}^{2\mu}]\|u^{\ell,k} - u_\ell\|^2.
$$

Thus, (3.1) holds with a rate which can be estimated by (3.9).

3.3 A Recurrence Formula. *The multigrid method with $\mu = 1$ for the V-cycle and $\mu = 2$ for the W-cycle satisfies*

$$
\rho_\ell^2 \leq \rho_1^2 + \rho_{\ell-1}^{2\mu}(1 - \rho_1^2) \quad (3.9)
$$

at level $\ell \geq 2$ with respect to the energy norm.

3.4 Theorem. *If the two-grid rate with respect to the energy norm satisfies $\rho_1 \leq \frac{1}{2}$, then*

$$
\rho_\ell \leq \frac{6}{5}\rho_1 \leq 0.6, \quad \text{for } \ell = 2, 3, \dots \quad (3.10)
$$

for the W-cycle.
Proof. For $\ell = 1$ there is nothing to prove. By the assertion for $\ell - 1$, it follows from the recurrence formula (3.9) that

$$
\begin{aligned}
\rho_\ell^2 &\leq \rho_1^2 + (\tfrac{6}{5}\rho_1)^4(1 - \rho_1^2) = \rho_1^2\{1 + (\tfrac{6}{5})^4[\rho_1^2(1 - \rho_1^2)]\} \\
&\leq \rho_1^2\{1 + (\tfrac{6}{5})^4 \tfrac{1}{4}\tfrac{3}{4}\} \leq \frac{36}{25}\rho_1^2 \leq 0.36.
\end{aligned}
$$

Taking the square root, we get the desired result.   □

The recurrence formula (3.9) gives unsatisfactory results for the V-cycle, as can be seen from the rapidly growing values of ρ_ℓ in Table 8.

Table 8. Convergence rates ρ_ℓ as in (3.9) for $\rho_1 = 1/5$.

$\ell =$	1	2	3	4	5	$\sup_\ell \rho_\ell$
W-cycle	0.2	0.2038	0.2041	0.2042	0.2042	0.2042
V-cycle	0.2	0.280	0.340	0.389	0.430	1.0

The Convergence Proof for the V-cycle

It is possible to establish a bound smaller than 1 for the convergence rate for the V-cycle, independent of the number of levels. To this end, we need a refinement of the method of proof which has an additional advantage: it shows that just one smoothing step suffices. Of course, the result also applies to the W-cycle.

The analysis of the two-grid method in §2 was based in essence on the scale of $\||\cdot\||_s$ norms for s in $[0, 2]$. Since we intend to make use of the energy norm, in the following we have to stay between $s = 1$ and $s = 2$. This halves the span between the maximal and minimal s. Here we present a simplified version of the original proof of Braess and Hackbusch [1983], although we obviously get weaker results with larger numbers of smoothing steps. (See, however, the remark at the end of this section.)

As before, we write $\|\cdot\|$ instead of $\||\cdot\||_1$ for the energy norm.

3.5 Convergence Theorem. *Under Hypotheses 2.1, the multigrid method with V-cycles or W-cycles satisfies*

$$\|u^{\ell,k+1} - u_\ell\| \le \rho_\ell \|u^{\ell,k} - u_\ell\|,$$
$$\rho_\ell \le \rho_\infty := \left(\frac{c}{c + 2\nu}\right)^{1/2}, \quad \ell = 0, 1, 2, \ldots, \tag{3.11}$$

w.r.t. the energy norm, assuming that Jacobi relaxation with $\lambda_{\max}(A) \le \omega \le c_0 \lambda_{\max}(A)$ is performed. Here c is a constant independent of ℓ and ν.

Before presenting the proof, we establish some lemmas. For abbreviation, let

$$w^m := u^{\ell,k,m} - u_\ell, \quad m = 0, 1, 2, \tag{3.12}$$

and define $\hat{w}^2$ in a similar way.

We now introduce a measure of the smoothness of the functions in the finite element space S_h:

$$\beta = \beta(v_h) := \begin{cases} 1 - \lambda_{\max}^{-1}(A) \dfrac{\||v_h\||_2^2}{\||v_h\||_1^2}, & \text{if } v_h \ne 0, \\ 0, & \text{if } v_h = 0. \end{cases} \tag{3.13}$$

Clearly, $0 \leq \beta < 1$. Smooth functions correspond to a number β near 1, and functions with a large oscillating part correspond to a small β. The factor β determines the amount of improvement for each smoothing step. Since the improvement always becomes successively smaller during the smoothing process, the size of the factor β *after* smoothing is decisive.

3.6 Lemma. *Smoothing using Jacobi relaxation satisfies*

$$\|\mathcal{S}^\nu v\| \leq [\beta(\mathcal{S}^\nu v)]^\nu \|v\| \quad \text{for all } v \in S_h.$$

Proof. Let $v = \sum_i c_i \phi_i$, where $\phi_1, \phi_2, \ldots$ are orthonormal eigenfunctions of A. In addition, let $\mu_i = 1 - \lambda_i / \lambda_{\max}$. By Hölder's inequality,

$$\sum_i \lambda_i \mu_i^{2\nu} |c_i|^2 \leq (\sum_i \lambda_i \mu_i^{2\nu+1} |c_i|^2)^{\frac{2\nu}{2\nu+1}} (\sum_i \lambda_i |c_i|^2)^{\frac{1}{2\nu+1}}.$$

In view of (2.6), this inequality is equivalent to

$$\|\mathcal{S}^\nu v\|^{2\nu+1} \leq \|\mathcal{S}^{\nu+\frac{1}{2}} v\|^{2\nu} \|v\|. \tag{3.14}$$

We abbreviate $w := \mathcal{S}^\nu v$, and divide (3.14) by $\|w\|^{2\nu}$. This gives

$$\|\mathcal{S}^\nu v\| \leq \left(\frac{\|\mathcal{S}^{1/2} w\|}{\|w\|}\right)^{2\nu} \|v\|. \tag{3.15}$$

Since $\mathcal{S}$ is self-adjoint and commutes with A, we also have

$$\|\mathcal{S}^{1/2} w\|^2 = (\mathcal{S}^{1/2} w, A\mathcal{S}^{1/2} w) = (w, A\mathcal{S} w)$$

$$= (w, Aw) - \frac{1}{\lambda_{\max}}(w, A^2 w) = \beta(w) \|w\|^2.$$

The assertion follows after substitution in (3.15). $\quad\square$

The quality of the coarse-grid correction can also be estimated in terms of the parameter β.

3.7 Lemma. *For the exact coarse-grid correction, we have*

$$\|\hat{w}^2\| \leq \min\{c\, \lambda_{\max}^{-1/2} \|\|w^1\|\|_2, \|w^1\|\}$$

$$= \min\{c\sqrt{1 - \beta(w^1)}, 1\} \|w^1\|. \tag{3.16}$$

Proof. The inequality (2.13) asserts that $\|\hat{w}^2\| \leq ch \|\|w^1\|\|_2$. Then using $\lambda_{\max} \leq ch^{-2}$, we can eliminate the factor h to get the bound $c\lambda_{\max}^{-1/2} \|\|w^1\|\|_2$. Moreover, the energy norm of the error is not increased by the coarse-grid correction, and the first assertion is proved. If we use (3.13) to eliminate $\|\|w^1\|\|_2$, we get the second assertion. $\quad\square$

The following formula is of central importance for the proof of the convergence theorem:

3.8 Recurrence Formula. *Suppose the hypotheses of Theorem 3.5 are satisfied. Then*

$$\rho_\ell^2 \le \max_{0 \le \beta \le 1} \beta^{2\nu} [\rho_{\ell-1}^{2\mu} + (1 - \rho_{\ell-1}^{2\mu}) \min\{1, c^2(1-\beta)\}]. \tag{3.17}$$

Here $\mu = 1$ for the V-cycle and $\mu = 2$ for the W-cycle, respectively, and c is the constant in Lemma 3.7.

Proof. By Lemma 3.6,

$$\|u^{\ell,k,1} - u_\ell\| \le \beta^\nu \|u^{\ell,k} - u_\ell\|,$$

where $\beta = \beta(u^{\ell,k,1} - u_\ell)$. By Lemma 3.7 with the same β, we have

$$\|\hat{u}^{\ell,k,2} - u_\ell\| \le \min\{c\sqrt{1-\beta}, 1\} \|u^{\ell,k,1} - u_\ell\|$$
$$\le \beta^\nu \min\{c\sqrt{1-\beta}, 1\} \|u^{\ell,k} - u_\ell\|.$$

We now insert this estimate in (3.8) to get

$$\|u^{\ell,k,2} - u_\ell\|^2 \le \beta^{2\nu}[(1 - \rho_{\ell-1}^{2\mu}) \min\{c^2(1-\beta), 1\} + \rho_{\ell-1}^{2\mu}] \|u^{\ell,k} - u_\ell\|^2.$$

Since $0 \le \beta < 1$, this proves the recurrence formula. □

Table 9. Convergence rate ρ_ℓ according to the recurrence formula (3.17) for $\nu = 2$

			V-cycle							W-cycle
c	$\ell =$ 1	2	3	4	5	6	7	8	∞	
0.5	.1432	.174	.189	.199	.205	.210	.214	.217	.243	.1437
1	.2862	.340	.366	.382	.392	.400	.406	.410	.448	.2904

Proof of Theorem 3.5. Since $\rho_0 = 0$, (3.11) holds for $\ell = 0$. To show how to get from $\ell - 1$ to ℓ, we insert $\rho_{\ell-1}^2 \le c^2/(c^2 + 2\nu)$ in the recurrence formula (3.17):

$$\rho_\ell^2 \le \max_{0 \le \beta \le 1} \{\beta^{2\nu} [\frac{c^2}{c^2 + 2\nu} + (1 - \frac{c^2}{c^2 + 2\nu}) c^2(1-\beta)]\}$$

$$= \frac{c^2}{c^2 + 2\nu} \max_{0 \le \beta \le 1} \{\beta^{2\nu} [1 + 2\nu(1-\beta)]\} \tag{3.18}$$

$$= \frac{c^2}{c^2 + 2\nu}.$$

Simple differentiation of the expression in the curly brackets shows that the maximum in (3.18) is attained for $\beta = 1$. □

In the multigrid method, the low frequency parts are handled more efficiently in the W-cycle than in the V-cycle. Hence, it is not surprising that in (3.18) the maximum is attained for $\beta = 1$. Thus, it makes sense to insert W-cycles once in a while if the number of levels is very large.

The proof shows that for large ν, the contraction number decreases only like

$$\nu^{-1/2}.$$

If both pre-smoothing and post-smoothing are used, the rate of decrease is ν^{-1}, as shown by the duality technique of Braess and Hackbusch [1983]. We remark that later convergence proofs for the V-cycle usually make use of an algebraic hypothesis instead of the H^2-regularity, see e.g. Bramble, Pasciak, Wang, and Xu [1991]. The question of whether appropriate algebraic hypotheses are really independent of H^2-regularity remains open, despite the paper of Parter [1987].

If the regularity hypothesis 2.1(2) is not satisfied, we have to expect a less favorable convergence rate. Then as suggested by Bank, Dupont, and Yserentant [1988], it makes more sense to use the multigrid method as a preconditioner for a CG method rather than as a stand-alone iteration. In fact these authors go one step further, and use the multigrid idea only for the construction of a so-called *hierarchical basis*. Then the convergence rate behaves like $1 - \mathcal{O}((\log \frac{1}{h})^{-p})$, which is still quite good for practical computations.

Problems

3.9 Show that for the W-cycle,

$$\sup_{\ell} \rho_\ell^2 \le \frac{\rho_1^2}{1 - \rho_1^2}, \quad \text{provided } \rho_1 < \sqrt{\frac{1}{2}}.$$

Hint: Use (3.9) to derive a recurrence formula for $1 - \rho_\ell^2$.

3.10 Show that for large c, the recurrence formula (3.17) gives the two-grid rate

$$\rho_1 \le (1 - \frac{1}{c^2})^\nu,$$

and compare with (3.11).

3.11 The amount of computation required for the W-cycle is approximately 50 % larger than for the V-cycle. Use the tables to compare the error reduction of three V-cycles and two W-cycles.

§ 4. Nested Iteration

So far we have treated the multigrid method as a pure iterative method. However, it turns out that the multigrid idea can also be used to find good starting values, so that one or two cycles of the multigrid iteration suffice. For this purpose there are two essential ideas.

1. The solution at level $\ell - 1$ is a good starting point for the iteration at level ℓ. This idea can be carried still further: we do not need to compute the function exactly since an approximation already provides a reasonable starting value.

2. The finite element approximation u_h is subject to the discretization error $\|u_h - u\|$. Thus, it makes little sense to carry out a lot of steps of the multigrid iteration to get an accuracy corresponding to the roundoff error. Instead, we should stop the iteration when

$$\|u^{h,k} - u_h\| \leq \frac{1}{2}\|u_h - u\|. \tag{4.1}$$

We will get only a marginal improvement of the total error $\|u^{h,k} - u\|$ by going any further.

Using the above ideas, we will create algorithms such that the amount of computation grows only linearly with the number of unknowns.

The starting value calculation is based on certain precursors of the multigrid method. The solution on the $2h$ grid was used as a starting value for classical iterative methods. Although the corresponding starting error has relatively small low frequency terms, classical relaxation methods still require too many steps; cf. Problem 4.6.

Computation of Starting Values

The following method for computing a starting value based on the multilevel concept is called *nested iteration*.

4.1 Algorithm NI$_\ell$ *for computing a starting value v^ℓ at level $\ell \geq 0$.*
If $\ell = 0$, find $v^0 = u_0 = A_0^{-1}b_0$, and exit the procedure.
Let $\ell > 0$.
Find an approximate solution $v^{\ell-1}$ of the equation $A_{\ell-1}u_{\ell-1} = b_{\ell-1}$ by applying NI$_{\ell-1}$.
Compute the prolongation of $v^{\ell-1}$, and set $v^{\ell,0} = p\,v^{\ell-1}$.

Using $v^{\ell,0}$ as a starting value, carry out one step (in general $q \geq 1$ steps) of the multigrid iteration $\mathbf{MGM}_\ell$ (see Fig. 54), and set

$$v^\ell = v^{\ell,1}.$$ ☐

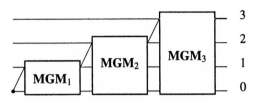

Fig. 54. Nested iteration $\mathbf{NI}_3$

For simplicity, in the following we assume that in Algorithm 4.1 only one cycle of the multigrid method is carried out. This is actually done for convergence rates $\rho < \frac{1}{4}$. Otherwise, we formally identify q cycles with one cycle having the convergence rate ρ^q, where q is such that $\rho^q < \frac{1}{4}$.

The accuracy of the resulting starting value can be computed easily. We restrict ourselves to the common case where the discretization error is of order $\mathcal{O}(h^2)$.

4.2 Theorem. *Assume that the finite element approximation $u_h \in S_h$ satisfies $\|u_h - u\| \leq c\,h^2$ for some constant $c > 0$. In addition, suppose the convergence rate ρ of the multigrid method w.r.t. the norm $\|\cdot\|$ is smaller than $1/4$. Then Algorithm 4.1 gives*

$$\|v^\ell - u_\ell\| \leq \frac{5\rho}{1 - 4\rho}\,c\,h_\ell^2. \tag{4.2}$$

Proof. Since $v^0 = u_0$, the formula (4.2) is immediate for $\ell = 0$.

Assuming (4.2) holds for $\ell - 1$, the fact that $h_{\ell-1} = 2h_\ell$ implies

$$\|v^{\ell-1} - u_{\ell-1}\| \leq \frac{5\rho}{1 - 4\rho}\,c(2h_\ell)^2.$$

Moreover, by the hypothesis on the discretization error,

$$\|u_{\ell-1} - u\| \leq c\,(2h_\ell)^2, \qquad \|u_\ell - u\| \leq c\,h_\ell^2.$$

Now the triangle inequality and $v^{\ell,0} = v^{\ell-1}$ give

$$\|v^{\ell,0} - u_\ell\| \leq \|v^{\ell-1} - u_{\ell-1}\| + \|u_{\ell-1} - u\| + \|u - u_\ell\|$$
$$\leq \frac{5\rho}{1 - 4\rho}4c\,h_\ell^2 + 5c\,h_\ell^2 = \frac{5}{1 - 4\rho}c\,h_\ell^2. \tag{4.3}$$

The multigrid cycle reduces the error by the factor ρ, i.e., $\|v^{\ell,1} - u_\ell\| \leq \rho\|v^{\ell,0} - u_\ell\|$. Using (4.3), we get the assertion (4.2). ☐

4.3 Remark. In many cases a convergence rate of $\rho \leq 1/6$ is realistic. Then Algorithm 4.1 produces an approximation with error $(5/2)c\,h_\ell^2$, and one additional cycle suffices to reduce the error to less than $\frac{1}{2}c\,h_\ell^2$.

Complexity

To estimate the computational complexity, we can assume that the amount of computation for

$$\left.\begin{array}{l} \text{smoothing in } S_\ell, \\ \text{prolongation of } S_{\ell-1} \text{ to } S_\ell, \text{ and} \\ \text{restriction of the residue } d_\ell \end{array}\right\} \qquad (4.4)$$

all involve cN_ℓ operations, where $N_\ell = \dim S_\ell$. The number of arithmetic operations for one smoothing step is proportional to the number of nonzero elements in the system matrix. For affine families of finite elements on uniform grids, this number is proportional to the number of unknowns. The prolongation and restriction matrices are even sparser. Thus, the operation count at level ℓ is

$$\leq (\nu + 1)c\,N_\ell, \qquad (4.5)$$

where ν is the number of smoothings.

For grids in $\mathbb{R}^2$, we collect the operations (4.4) from all levels. Each time we move to a coarser grid the number of unknowns decreases by approximately the factor 4. Summing the terms (4.5) for the multigrid iteration MGM_ℓ gives

$$(\nu+1)\,c\,(N_\ell + N_{\ell-1} + N_{\ell-2} + \cdots) \leq \frac{4}{3}(\nu+1)\,c\,N_\ell \quad \text{for the V-cycle,}$$

$$(\nu+1)\,c\,(N_\ell + 2N_{\ell-1} + 4N_{\ell-2} + \cdots) \leq 2(\nu+1)\,c\,N_\ell \quad \text{for the W-cycle.}$$
$$(4.6)$$

In addition, we must include the work needed to solve the system of equations on the coarsest grid. We ignore this additional computational effort for the moment. This is justified if the number of levels is large. The special case of a small number of levels will be treated separately later.

The computation of starting values can be analyzed in the same way. Because of the increase in dimension, each cycle of Algorithm 4.1 requires four times as much work as its predecessor. Since $\sum_k 4^{-k} = 4/3$, we have that

The operation count for computing the starting value is $\frac{4}{3}$ times
the operation count for one multigrid cycle on the finest grid.

In view of Remark 4.3, the operation count for the complete calculation is $\frac{7}{3}$ times the count for one multigrid cycle on the finest grid, provided the convergence rate is 1/6 or better. Thus, in particular, the complexity increases only linearly with the number of unknowns.

As mentioned before, in the V-cycle we distribute the smoothing steps equally between the phases before and after the coarse-grid correction. Thus, an analogous symmetric variant is recommended in calculating the starting values.

4.4 Algorithm NI$_\ell$ *to compute a starting value v^ℓ at the level $\ell \geq 0$*
(symmetric version).
If $\ell = 0$, find $v^0 = u_0 = A_0^{-1}b_0$, and exit the procedure.
Let $\ell > 0$.
With $v^{\ell,0} = 0$ carry out one step of the multigrid iteration **MGM$_\ell$**, and set $v^{\ell,1} = v^{\ell,0} + pv^{\ell-1}$.
Set $b_{\ell-1} = p(b_\ell - A_\ell v^{\ell,1})$.
Compute an approximate solution $v^{\ell-1}$ of the equation $A_{\ell-1}u_{\ell-1} = b_{\ell-1}$ using **NI$_{\ell-1}$**.
Compute the prolongation of $v^{\ell-1}$, and set $v^{\ell,2} = v^{\ell,1} + p\,v^{\ell-1}$.
With $v^{\ell,2}$ carry out one step (in general $p \geq 1$ steps) of the multigrid iteration **MGM$_\ell$**, and set

$$v^\ell = v^{\ell,3}. \qquad \qquad \Box$$

This variant compensates for the effect that in the V-cycle, the oscillating parts are handled better than the smoother ones. A problem whose smooth parts are relatively large is given to the block **NI$_{\ell-1}$**, so that the efficiency is increased.

Multigrid Methods with a Small Number of Levels

For many grids it is difficult to carry out more than two levels of coarsening. As we shall see, however, it still pays to use multigrid methods. In contrast to the case of a large number of levels, here the solution on the coarsest grid is a major part of the computational effort. To this effort we have to add the work required for smoothing, restrictions, and prolongations. This is, of course, less than with a large number of levels, and has already been estimated in (4.6) above.

For three levels, the number of unknowns on the coarsest grid is ca. 1/16 of that on the finest. In addition, the average bandwidth of the system matrix is reduced by a factor of about 4. Thus, the amount of work required for the Cholesky method is reduced by a factor $16 \cdot 4^2 = 256$. Now if we compute the starting value with Algorithm 4.1 and add a multigrid cycle, we have to solve

4 systems of equations for the V-cycle,

6 systems of equations for the W-cycle

on the coarsest grid (all with the same matrix). Even if the LU decomposition is recalculated each time, we still get a saving of a factor of 40 to 64, if we compare just the effort of the exact solver.[13]

Since the other transfer operations discussed earlier hardly affect the result, we clearly see the advantages of the multigrid method, even in those cases where at first glance it appears to be of marginal use.

The CASCADE Algorithm

We also mention the CASCADE algorithm of Bornemann and Deuflhard [1996]. It uses a different strategy. We begin with a CG method on the coarsest grid, and proceed from there successively from level to level to the finest, without ever going back to coarser grids. We choose a much larger number of iteration steps on the coarser grids (with the smaller dimensions) than on the finer grids.

The justification for this approach is the following result:

4.5 Recursion Relation. *Let u_ℓ denote the solution of the variational problem in S_ℓ and v_ℓ be the result of the CASCADE algorithm. If m_ℓ steps of the cg-algorithm are performed on the level ℓ, then*

$$\|v_\ell - u_\ell\|_1 \le \|v_{\ell-1} - u_{\ell-1}\|_1 + c\frac{h_\ell}{m_\ell}\|f\|_0. \tag{4.7}$$

From (4.7) we conclude that the error on the finest level is of the order of the discretization error provided that sufficiently many steps are performed on the coarse levels, say $m_\ell = 3^{\ell_{\max}-\ell} m_{\ell_{\max}}$. Nevertheless, the main part of the computing time is consumed on the finest level, and the larger number of steps for $\ell < \ell_{\max}$ does not spoil the efficiency.

The recursion relation can be established via the construction of polynomials with approximation properties that differ slightly from those in Ch. IV, §3. For details, see Bornemann and Deuflhard [1996] or Shaidurov [1996]. The investigations have been extended to saddle point problems by Braess and Dahmen [1999] and to nonconforming elements by Stevenson [1999].

For the computation of starting values the CASCADE algorithm is simpler than nested iteration because grid changes occur only in one direction. The algorithm can often replace even the complete multigrid procedure.

---

[13] Another advantage is that we can store the resulting small system of equations in fast memory, while larger systems have to be stored in external memory.

Problems

4.6 Suppose the finite element approximation $u_h \in S_h$ is such that $\|u_h - u\| \le c\,h^2$ for some constant $c > 0$. In addition, suppose the convergence rate for the two-grid method with $\nu = \nu_2$ post-smoothings satisfies $\rho < 1/10$.

Suppose a user (who perhaps is not familiar with the multigrid method) applies a classical relaxation method, starting with the solution on the $2h$ grid. Show that after ν steps,

$$\|u^{h,\nu} - u\| \le \frac{3}{2}c \cdot h^2.$$

Why doesn't a corresponding assertion hold for more than two grids?

4.7 Compare the operation counts of
(a) $\mathbf{NI}_\ell$ with the V-cycle,
(b) $\mathbf{NI}_\ell$ with the W-cycle,
(c) the symmetric version of $\mathbf{NI}_\ell$ with the V-cycle.

4.8 Suppose we want to insert a so-called F-cycle (see Fig. 55) between the V-cycle and the W-cycle as follows:

For $\ell = 2$, the F-cycle and W-cycle coincide.

For $\ell \ge 3$, perform both an F-cycle and a V-cycle at level $\ell - 1$.

Find the recurrence formula analogous to (3.17), and determine the rates numerically for $c = 1$, $c = \frac{1}{2}$, $\nu = 2$, and $\ell \le 8$.

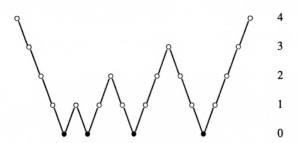

Fig. 55. F-cycle on five levels

4.9 Compare nested iteration $\mathbf{NI}_\ell$ using inner V-cycles with the F-cycle of $\mathbf{MGM}_\ell$. What is the difference?

§ 5. Multigrid Analysis via Space Decomposition

The multigrid analysis in §§2 and 3 is based on a smoothing property and an approximation property. The latter heavily depends on regularity assumptions. There are many variants and generalizations of Lemma 2.8, but one needs $H^{1+\alpha}$ regularity for some $\alpha > 0$.

This is different in the theory of Bramble, Pasciak, Wang, and Xu [1991]. For getting a connection with the previous theory we emphasize a decomposition property in the H^2 regular case although we did not state it explicitly. Let S_ℓ, $\ell = 0, 1, \ldots, L$, be a nested sequence of finite element spaces as in (1.5). Given $v \in S := S_L$, we may decompose it

$$v = \sum_{k=0}^{L} v_k, \quad v_k \in S_k \tag{5.1}$$

such that the partial sum $\sum_{k=0}^{\ell} v_k$ is the finite element solution to v in S_ℓ. It follows from Lemma 2.8 that

$$\|v_k\|_0 \leq ch_k\|v_k\|_1, \quad k = 1, 2, \ldots, L.$$

On the other hand, an inverse inequality yields

$$\|v_k\|_1 \leq ch_k^{-1}\|v_k\|_0, \quad k = 1, 2, \ldots, L.$$

From these inequalities and the orthogonality of the $v_k's$ we obtain the equivalence

$$\|v\|_1^2 = \sum_{k=0}^{L} \|v_k\|_1^2 \approx \|v_0\|_1^2 + \sum_{k=1}^{L} h_k^{-2}\|v_k\|_0^2. \tag{5.2}$$

A variational problem with a quadratic form that equals the right-hand side of (5.2) can be easily solved by successive solution in the subspaces because of the additive structure. Moreover, the smoothing procedures are efficient approximate solvers for the subspaces since the norms $\| \cdot \|_1$ and $h_k^{-1}\| \cdot \|_0$ are equivalent there. The multigrid method may be interpreted in this way.

Now Bramble, Pasciak, Wang, and Xu [1991] have observed that these and other similar relations may be derived from *properties of the function spaces* and that one does not need *regularity of the solution of the elliptic equation*. Oswald [1994] pointed out that Besov space properties are helpful to understand multilevel methods in this context; see below, In contrast to §3 the *multilevel iteration* is not treated as a perturbation of the 2-grid procedure.

We want to present the basic ideas of the theory of Bramble, Pasciak, Wang, and Xu. A complete theory without regularity is beyond the scope of this book; rather we demonstrate its advantage for another application which is not covered by the standard theory. The extension to locally refined meshes will be illustrated.

For convenience, we restrict ourselves to symmetric smoothing operators and to nested spaces. For more general results see Xu [1992], Wang [1994], and Yserentant [1993].

The norm without subscript refers to the energy norm $\| \cdot \| := (a(\cdot, \cdot))^{1/2}$ since the theory applies only to this norm. As mentioned before, the theory does not reflect the improvement of the convergence rate when the number of smoothing steps is increased.

Schwarz' Alternating Method

For a better understanding of space decomposition methods we first consider the *alternating method* which goes back to H.A. Schwarz [1869]. There is a simple geometrical interpretation, cf. Fig. 56, when the abstract formulation for the case of two subspaces is considered.

We are interested in the variational problem

$$a(u, v) = \langle f, v \rangle \quad \text{for } v \in H.$$

Here $a(., .)$ is the inner product of the Hilbert space H and $\| \cdot \|$ is the corresponding norm. Let H be the direct sum of two subspaces

$$H = V \oplus W,$$

and the determination of a solution in the subspaces V or W is assumed to be easy. Then an alternating iteration in the two subspaces is natural.

5.1 Schwarz Alternating Method. Let $u_0 \in H$.

When u_{2i} is already determined, find $v_{2i} \in V$ such that

$$a(u_{2i} + v_{2i}, v) = \langle f, v \rangle \quad \text{for } v \in V.$$

Set $u_{2i+1} = u_{2i} + v_{2i}$.
When u_{2i+1} is already determined, find $w_{2i+1} \in W$ such that

$$a(u_{2i+1} + w_{2i+1}, w) = \langle f, w \rangle \quad \text{for } w \in W.$$

Set $u_{2i+2} = u_{2i+1} + w_{2i+1}$. ◻

Obviously, projections onto the two subspaces alternate during the iteration. The *strengthened Cauchy inequality* (5.3) is crucial in the analysis.

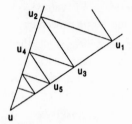

Fig. 56. Schwarz alternating iteration with one-dimensional subspaces V and W in Euclidean 2-space. The iterates $u_1, u_3, u_5, \ldots$ lie in $V^{\perp}$ and $u_2, u_4, \ldots$ in $W^{\perp}$. The angle between $V^{\perp}$ and $W^{\perp}$ is the same as between V and W.

5.2 Convergence Theorem. *Assume that there is a constant $\gamma < 1$ such that for the inner product in H*

$$|a(v, w)| \leq \gamma \|v\| \|w\| \quad for\ v \in V,\ w \in W. \tag{5.3}$$

Then we have for the iteration with the Schwarz alternating method the error reduction

$$\|u_{k+1} - u\| \leq \gamma \|u_k - u\| \quad for\ k \geq 1. \tag{5.4}$$

Proof. Because of the symmetry of the problem we may confine ourselves to even k. Since u_k is constructed by a minization in the subspace W to even k. By the construction of u_k we have

$$a(u_k - u, w) = 0 \quad \text{for } w \in W \tag{5.5}$$

We decompose $u_k - u = \hat{v} + \hat{w}$ with $\hat{v} \in V$, $\hat{w} \in W$. From (5.5) it follows with $w = \hat{w}$ that

$$a(\hat{v}, \hat{w}) = -\|\hat{w}\|^2. \tag{5.6}$$

By the strengthened Cauchy inequality (5.3) we have $a(\hat{v}, \hat{w}) = -\alpha_k \|\hat{v}\| \|\hat{w}\|$ with some $\alpha_k \leq \gamma$. Without loss of generality let $\alpha_k \neq 0$. It follows from (5.6) that $\|\hat{v}\| = \alpha_k^{-1} \|\hat{w}\|$ and $\|u_k - u\|^2 = \|\hat{v} + \hat{w}\|^2 = \|\hat{v}\|^2 - 2\|\hat{w}\|^2 + \|\hat{w}\|^2 = (\alpha_k^{-2} - 1)\|\hat{w}\|^2$.

Since u_{k+1} is the result of an optimization in V, we obtain an upper estimate from the simple test function $u_k + (\alpha_k - 1)\hat{v}$.

$$\begin{aligned}
\|u_{k+1} - u\|^2 &\leq \|u_k + (\alpha_k - 1)\hat{v}\|^2 \\
&= \|\alpha_k \hat{v} + \hat{w}\|^2 = (1 - \alpha_k^2)\|\hat{w}\|^2 = \alpha_k^2 \|u_k - u\|^2.
\end{aligned}$$

Noting that $\alpha_k \leq \gamma$, the proof is complete.   □

The bound in (5.4) is sharp. This becomes obvious from an example with one-dimensional spaces V and W depicted in Fig. 56 and also from Problem 5.8.

Algebraic Description of Space Decomposition Algorithms

The finite element spaces S_ℓ may be recursively constructed

$$S_0 = W_0,$$
$$S_\ell = S_{\ell-1} \oplus W_\ell, \quad \ell \geq 1, \tag{5.7}$$
$$S = S_L.$$

The finite element solution on the level ℓ is related to the operator $A_\ell : S_\ell \to S_\ell$ defined by

$$(A_\ell u, w) = a(u, w) \quad \text{for all } w \in S_\ell. \tag{5.8}$$

Moreover $A := A_L$. The corresponding Ritz projector $P_\ell : S \to S_\ell$ satisfies

$$a(P_\ell u, w) = a(u, w) \quad \text{for all } w \in S_\ell. \tag{5.9}$$

We note that the discussion below holds for any inner product $(\cdot, \cdot)$ in the Hilbert space S, but we will refer to the L_2 inner product or the ℓ_2 inner product when we deal with concrete examples. We recall that the L_2-norm is equivalent to the ℓ_2-norm of the associated vector representations, and the smoothing procedures refer to L_2-like operators. Therefore, we will use also the L_2-orthogonal projectors $Q_\ell : S \to S_\ell$,

$$(Q_\ell u, w) = (u, w) \quad \text{for all } w \in S_\ell. \tag{5.10}$$

It follows that

$$A_\ell P_\ell = Q_\ell A. \tag{5.11}$$

Indeed, for all $w \in S_\ell$ we obtain from (5.8)–(5.10) the equations $(A_\ell P_\ell u, w) = a(P_\ell u, w) = a(u, w) = a(u, Q_\ell w) = (Au, Q_\ell w) = (Q_\ell Au, w)$. Since $A_\ell P_\ell$ and $Q_\ell A$ are mappings into S_ℓ, this proves (5.11).

Assume that $\tilde{u}$ is an approximate solution of the variational problem in S, and let $A\tilde{u} - f$ be the residue. The solution of the variational problem in the subset $\tilde{u} + S_\ell$ is $\tilde{u} + A_\ell^{-1} Q_\ell(f - A\tilde{u})$. Therefore, the correction by the exact solution of the subproblem for the level ℓ is

$$A_\ell^{-1} Q_\ell(f - A\tilde{u}).$$

Since its computation is too expensive in general, the actual correction will be obtained from a computation with an approximate inverse B_ℓ^{-1}, i.e., the real correction will be

$$B_\ell^{-1} Q_\ell(f - A\tilde{u}). \tag{5.12}$$

The correction turns $\tilde{u}$ into $\tilde{u} + B_\ell^{-1} Q_\ell(f - A\tilde{u})$. For convenience, we will assume that

$$A_\ell \leq B_\ell, \tag{5.13}$$

i.e., $B_\ell - A_\ell$ is assumed to be positive semidefinite and tacitly B_ℓ is assumed to be symmetric. Often only the weaker condition $A_\ell \le \omega B_\ell$ with $\omega < 2$ is required, but we prefer to have the assumption without an extra factor in order to avoid some inconvenient factors in the estimates. Some standard techniques for dealing with the approximate solution above are found in Problems IV.4.14–17.

We recall (5.11) and following the standard notation, we define the linear mapping

$$T_\ell := B_\ell^{-1} Q_\ell A = B_\ell^{-1} A_\ell P_\ell. \tag{5.14}$$

From (5.12) we know that the correction of $\tilde{u}$ in the subspace S_ℓ yields the new iterate $\tilde{u} + T_\ell(u - \tilde{u})$, and its error is

$$(I - T_\ell)\,(u - \tilde{u}).$$

We consider the multigrid V-cycle with post-smoothing only. Consequently the error propagation operator for one complete cycle is

$$E := E_L$$

where

$$E_\ell := (I - T_\ell)\,(I - T_{\ell-1})\ldots(I - T_0), \quad \ell = 0, 1, \ldots, L, \tag{5.15}$$

and $E_{-1} := I$. This representation elucidates that the subspace corrections are applied in a multiplicative way.

Assumptions

The assumptions refer to the family of finite element spaces S_ℓ and the complementary spaces W_ℓ specified in (5.7).

Assumption A1. There exists a constant K_1 such that for all $v_\ell \in W_\ell$, $\ell = 0, 1, \ldots, L$,

$$\sum_{\ell=0}^{L} (B_\ell v_\ell, v_\ell) \le K_1 \Big\| \sum_{\ell=0}^{L} v_\ell \Big\|^2. \tag{5.16}$$

Assumption A2 *(Strengthened Cauchy–Schwarz Inequality).* There exist constants $\gamma_{k\ell} = \gamma_{\ell k}$ with

$$a(v_k, w_\ell) \le \gamma_{k\ell}\,(B_k v_k, v_k)^{1/2} (B_\ell w_\ell, w_\ell)^{1/2} \quad \text{for all } v_k \in S_k,\ w_\ell \in W_\ell \tag{5.17}$$

if $k \le \ell$. Moreover, there is a constant K_2 such that

$$\sum_{k,l=0}^{L} \gamma_{k\ell} x_k y_\ell \le K_2 \Big(\sum_{k=0}^{L} x_k^2\Big)^{1/2} \Big(\sum_{\ell=0}^{L} y_\ell^2\Big)^{1/2} \quad \text{for } x, y \in \mathbb{R}^{L+1}. \tag{5.18}$$

We postpone the verification of **A1**. – The verification of **A2** with a constant K_2 that is independent of the number of levels is not trivial. Therefore we provide a short proof of an estimate with a bound that increases only logarithmically. The standard Cauchy–Schwarz inequality and $A_\ell \leq B_\ell$ imply that we have $\gamma_{k\ell} \leq 1$ for all k, ℓ. Hence,

$$\sum_{k,l} \gamma_{k\ell} x_k y_\ell \leq \left(\sum_k |x_k|\right)\left(\sum_\ell |y_\ell|\right) \leq (L+1)\left(\sum_k x_k^2\right)^{1/2}\left(\sum_\ell y_\ell^2\right)^{1/2},$$

and (5.18) is obvious for

$$K_2 \leq L + 1 \leq c\,|\log h_L|. \tag{5.19}$$

Direct Consequences

From **A2** we conclude immediately that

$$\left\|\sum_{\ell=0}^{L} v_\ell\right\|^2 = \sum_{k,\ell} a(v_k, v_\ell)$$

$$\leq \sum_{k,\ell} \gamma_{k\ell}(B_k v_k, v_k)^{1/2}\,(B_\ell v_\ell, v_\ell)^{1/2} \leq K_2 \sum_{\ell=0}^{L}(B_\ell v_\ell, v_\ell). \tag{5.20}$$

Hence, the norms encountered in (5.16) and (5.20), are equivalent provided that **A1** and **A2** hold.

A direct consequence of **A1** is an analogue of an inequality which we considered in §2 in connection with logarithmic convexity. Note the asymmetry in the occurrence of the spaces in (5.21).

5.3 Lemma. *Let $w_\ell \in W_\ell$ and $u_\ell \in S = S_L$ for $\ell = 0, 1, \ldots, L$. Then we have*

$$\sum_{\ell=0}^{L} a(w_\ell, u_\ell) \leq \sqrt{K_1}\,\left\|\sum_{\ell=0}^{L} w_\ell\right\| \left(\sum_{\ell=0}^{L} a(T_\ell u_\ell, u_\ell)\right)^{1/2}. \tag{5.21}$$

Proof. Since $P_\ell w_\ell = w_\ell$, it follows from the Cauchy–Schwarz inequality in Euclidean space that

$$\sum_{\ell=0}^{L} a(w_\ell, u_\ell) = \sum_{\ell=0}^{L} a(w_\ell, P_\ell u_\ell)$$

$$= \sum_{\ell=0}^{L} (B_\ell^{1/2} w_\ell, B_\ell^{-1/2} A_\ell P_\ell u_\ell)$$

$$\leq \left(\sum_{\ell=0}^{L}(B_\ell w_\ell, w_\ell)\right)^{1/2} \left(\sum_{\ell=0}^{L}(A_\ell P_\ell u_\ell, B_\ell^{-1} A_\ell P_\ell u_\ell)\right)^{1/2}. \tag{5.22}$$

Next we derive an equality that is useful also in other contexts

$$\begin{aligned}
(B_\ell T_\ell w, T_\ell w) &= (T_\ell w, B_\ell B_\ell^{-1} A_\ell P_\ell w) \\
&= (T_\ell w, A_\ell P_\ell w) = a(T_\ell w, P_\ell w) = a(T_\ell w, w).
\end{aligned} \tag{5.23}$$

The first factor on the right-hand side of (5.22) can be estimated by **A1**. Since $T_\ell = B_\ell^{-1} A_\ell P_\ell$, we can insert (5.23) into the summands of the second factor, and the proof of the lemma is complete.   □

It is more than a coincidence that $a(T_\ell w, w)$ is a multiple of the discrete norm $\||P_\ell w\||_2$ that we encountered in §2 if B_ℓ is a multiple of the identity on the subspace S_ℓ.

Convergence of Multiplicative Methods

First we estimate the reduction of the error by the multigrid algorithm on the level ℓ from below.

5.4 Lemma. *Let $\ell \geq 1$. Then*

$$\|v\|^2 - \|(I - T_\ell)v\|^2 \geq a(T_\ell v, v). \tag{5.24}$$

Proof. From the binomial formula we obtain that the left-hand side of (5.24) equals

$$2a(T_\ell v, v) - a(T_\ell v, T_\ell v). \tag{5.25}$$

Next, we consider the second term using $A_\ell \leq B_\ell$ and (5.23)

$$a(T_\ell v, T_\ell v) \leq (B_\ell T_\ell v, T_\ell v) = a(T_\ell v, v).$$

Therefore the negative term in (5.25) can be absorbed by the term $a(T_\ell v, v)$ by subtracting 1 from the factor 2. and the proof is complete.   □

Now we turn to the central result of this §. It yields the convergence rate of the multigrid iteration in terms of the constants in the assumptions **A1** and **A2**.

5.5 Theorem. *Assume that A1 and A2 hold. Then the energy norm of the error propagation operator E of the multigrid iteration satisfies*

$$\|E\|^2 \leq 1 - \frac{1}{K_1(1 + K_2)^2}.$$

Proof. By applying Lemma 5.4 to $E_{\ell-1}v$ and noting that $E_\ell = (I - T_\ell)E_{\ell-1}$ we obtain

$$\|E_{\ell-1}v\|^2 - \|E_\ell v\|^2 \geq a(T_\ell E_{\ell-1}v, E_{\ell-1}v).$$

A summation over all levels is performed with telescoping

$$\|v\|^2 - \|Ev\|^2 \geq \sum_{\ell=0}^{L} a(T_\ell E_{\ell-1}v, E_{\ell-1}v). \tag{5.26}$$

Therefore the statement of the theorem will be clear if we verify

$$\|v\|^2 \leq K_1(1 + K_2)^2 \sum_{\ell=0}^{L} a(T_\ell E_{\ell-1}v, E_{\ell-1}v). \tag{5.27}$$

Indeed, (5.26) and (5.27) yield $\|v\|^2 \leq K_1(1 + K_2)^2(\|v\|^2 - \|Ev\|^2)$, and the rest of the proof is concerned with establishing this inequality.

To this end, let

$$v = \sum_{\ell=0}^{L} v_\ell, \quad v_\ell \in W_\ell,$$

be a (stable) decomposition. Obviously,

$$\|v\|^2 = \sum_{\ell=0}^{L} a(E_{\ell-1}v, v_\ell) + \sum_{\ell=1}^{L} a((I - E_{\ell-1})v, v_\ell). \tag{5.28}$$

Lemma 5.3 is used to deal with the first term

$$\sum_{\ell=0}^{L} a(E_{\ell-1}v, v_\ell) \leq \sqrt{K_1}\,\|v\| \left(\sum_{\ell=0}^{L} a(T_\ell E_{\ell-1}v, E_{\ell-1}v)\right)^{1/2}. \tag{5.29}$$

Next from $E_\ell - E_{\ell-1} = -T_\ell E_{\ell-1}$ it follows by induction that

$$I - E_{\ell-1} = \sum_{k=0}^{\ell-1} T_k E_{k-1}.$$

With the bound of the second term on the right-hand side of (5.28), we verify that the conditions for obtaining an improvement on the level ℓ are not affected much by the corrections in the previous steps. Here **A2** enters and

$$\sum_{\ell=1}^{L} a((I - E_{\ell-1})v, v_\ell)$$

$$= \sum_{\ell=1}^{L} \sum_{k=0}^{\ell-1} a(T_k E_{k-1}v, v_\ell)$$

$$\leq \sum_{\ell=1}^{L} \sum_{k=0}^{\ell-1} \gamma_{k\ell}(B_k T_k E_{k-1}v, T_k E_{k-1}v)^{1/2} (B_\ell v_\ell, v_\ell)^{1/2}$$

$$\leq K_2 \left(\sum_{k=0}^{L} (B_k T_k E_{k-1}v, T_k E_{k-1}v)\right)^{1/2} \left(\sum_{\ell=0}^{L} (B_\ell v_\ell, v_\ell)\right)^{1/2}.$$

From (5.23) and **A1** it follows that

$$\sum_{\ell=1}^{L} a((I - E_{\ell-1})v, v_\ell) \leq \sqrt{K_1} \, K_2 \, \|v\| \left(\sum_{k=0}^{L} a_k(T_k E_{k-1}v, E_{k-1}v) \right)^{1/2}. \quad (5.30)$$

Adding (5.29) and (5.30) and dividing by $\|v\|$ we obtain (5.27) completing the proof. $\qquad\qquad\qquad\qquad\qquad\qquad\qquad\qquad\qquad\qquad\qquad\qquad$ □

Verification of A1

In the case of full H^2 regularity and quasi-uniform triangulations optimal estimates are easily derived. We have an ideal case. Given $v \in S$, let u_ℓ be the finite element solution of v in S_ℓ, i.e., $u_\ell = P_\ell v$. Set

$$v = \sum_{\ell=0}^{L} v_\ell, \quad (5.31)$$

$$v_0 = P_0 v, \quad v_\ell = P_\ell v - P_{\ell-1} v = u_\ell - u_{\ell-1} \quad \text{for } \ell = 1, 2, \dots, L.$$

From the Galerkin orthogonality of finite element solutions we conclude that

$$\|v\|^2 = \sum_{\ell=0}^{L} \|v_\ell\|^2. \quad (5.32)$$

Since $u_{\ell-1}$ is also the finite element solution to u_ℓ in $S_{\ell-1}$ and $v_\ell = u_\ell - u_{\ell-1}$, it follows from the Aubin-Nitsche lemma that

$$\|v_\ell\|_0 \leq c \, h_{\ell-1} \|v_\ell\| \quad \text{for } \ell = 1, 2, \dots, L. \quad (5.33)$$

The approximate inverses for a multigrid algorithm with Richardson iteration as a smoother are given by

$$B_0 := A_0, \quad B_\ell := c \, \lambda_{\max}(A_\ell) I, \quad \ell = 1, 2, \dots, L. \quad (5.34)$$

The inverse estimates yield $\lambda_{\max}(A_\ell) \leq c h_\ell^{-2}$. Combining these facts and noting $h_{\ell-1} \leq c h_\ell$ we obtain

$$\sum_{\ell=0}^{L} (B_\ell v_\ell, v_\ell) \leq (A_0 v_0, v_0) + \sum_{\ell=1}^{L} c h_\ell^{-2}(v_\ell, v_\ell)$$

$$= \|v_0\|^2 + c \sum_{\ell=1}^{L} h_\ell^{-2} \|v_\ell\|_0^2 \quad (5.35)$$

$$\leq c \sum_{\ell=0}^{L} \|v_\ell\|^2 = c \, \|v\|^2.$$

This proves **A1** with a constant $K_1 = c$ that is independent of the number of levels.

In the cases with less regularity we perform the decomposition by applying the L_2-orthogonal projectors Q_ℓ instead of P_ℓ

$$v = \sum_{\ell=0}^{L} v_\ell, \tag{5.36}$$

$$v_0 = Q_0 v, \quad v_\ell = Q_\ell v - Q_{\ell-1} v \quad \text{for } \ell = 1, 2, \ldots, L.$$

From Lemma II.7.9 we have $\|Q_0 v\| \leq c\|v\|$. Next from (II.7.15) it follows that $\|v_\ell\|_0 \leq \|v - Q_\ell v\|_0 + \|v - Q_{\ell-1} v\|_0 \leq c h_\ell \|v\|$. Recalling the approximate solvers from (5.34) we proceed as in the derivation of (5.35)

$$\sum_{\ell=0}^{L} (B_\ell v_\ell, v_\ell) \leq \|v_0\|_1^2 + c \sum_{\ell=1}^{L} h_\ell^{-2} \|v_\ell\|_0^2 \tag{5.37}$$

$$\leq c\,(L+1)\|v\|^2.$$

This proves **A1** with a constant $K_1 \leq c(L+1)^{1/2} \leq c|\log h_L|^{1/2}$. Although this result is only suboptimal, it has the advantage that no regularity assumptions are required. As mentioned above, the logarithmic factor arises since we stay in the framework of Sobolev spaces. An analysis with the theory of Besov spaces shows that the factor can be dropped, see Oswald [1994].

Local Mesh Refinements

An inspection of the proof of Lemma II.7.9 shows that the estimate (5.37) remains true if the orthognal projector Q_ℓ is replaced by an operator of Clément type, e.g. we may choose $I_\ell = I_{h_\ell}$ from (II.6.19). That interpolation operator is nearly local.

This has a big advantage when we consider finite element spaces which arise from local mesh refinements. Assume that the refinement of the triangulation on the level ℓ is restricted to a subdomain $\Omega_\ell \subset \Omega$ and that

$$\Omega_L \subset \Omega_{L-1} \subset \ldots \subset \Omega_0 = \Omega. \tag{5.38}$$

Given $v \in S_L$, its restriction to $\Omega \setminus \Omega_\ell$ coincides there with some finite element function in S_ℓ. Now we modify $I_\ell v$ at the nodes outside Ω_ℓ and set

$$(I_\ell v)(x_j) := v(x_j) \quad \text{if } x_j \notin \Omega_\ell.$$

Specifically, when defining I_ℓ, the construction of the operator $\tilde{Q}_j$ in (II.6.17) is augmented by the rule (II.6.23). We have

$$(I_\ell v)(x) = v(x) \tag{5.39}$$

for x outside a neighborhood of Ω_ℓ, and from problem II.6.17 we know that the modification changes only the constants in the estimates of the L_2-error. The strip of $\Omega \setminus \Omega_\ell$, in which (5.39) does not hold, is small if rule II.8.1(1) is observed during the refinement process. Hence,

$$\|v - I_\ell v\|_0 \le ch_\ell \|v\|_1. \qquad (5.40)$$

We note that an estimate of this kind cannot be guaranted for the finite element solution in S_ℓ. So by using interpolation of Clément type we also obtain multigrid convergence in cases with local mesh refinements.

There is also a consequence for computational aspects of the multigrid method. Since

$$v_{\ell+1} = I_{\ell+1}v - I_\ell v = 0 \quad \text{outside a neighborhood of } \Omega_\ell,$$

the smoothing procedure on the levels $\ell+1$, $\ell+2$, ..., L may be restricted to the nodes in a neighborhood of Ω_ℓ. It is not necessary to perform the smoothing iteration at each level on the whole domain. For this reason the computing effort only increases linearly with the dimension of S_L. As was pointed out by Xu [1992] and Yserentant [1993], local refinements induce a faster increase of the computational complexity.

Problems

5.7 Let V, W be subspaces of a Hilbert space H. Denote the projectors onto V and W by P_V, P_W, respectively. Show that the following properties are equivalent:
(1) A strengthened Cauchy inequality (5.3) holds with $\gamma < 1$.
(2) $\|P_W v\| \le \gamma \|v\|$ holds for all $v \in V$.
(3) $\|P_V w\| \le \gamma \|w\|$ holds for all $w \in W$.
(4) $\|v + w\| \ge \sqrt{1 - \gamma^2}\, \|v\|$ holds for all $v \in V$, $w \in W$.
(5) $\|v + w\| \ge \sqrt{\frac{1}{2}(1 - \gamma)}\, (\|v\| + \|w\|)$ holds for all $v \in V$, $w \in W$.

5.8 Consider a sequence obtained by the Schwarz alternating method. Let α_k be the factor in the Cauchy inequality for the decomposition of the error in the iteration step k as in the proof of Theorem 5.2. Show that (α_k) is a nondecreasing sequence.

§ 6. Nonlinear Problems

Multigrid methods are also very useful for the numerical solution of nonlinear differential equations. We need only make some changes in the multigrid method for linear equations. These changes are typical for the efficient treatment of nonlinear problems. However, there is one essential idea involved which we might not otherwise encounter. We have to correct the right-hand side of the nonlinear equation on the coarse grid in order to compensate for the error which arises in moving between grids.

As an example of an important nonlinear diffferential equation, consider the Navier–Stokes equation

$$-\Delta u + \text{Re}\,(u\nabla)u - \text{grad } p = f \quad \text{in } \Omega,$$
$$\text{div}\, u = 0 \quad \text{in } \Omega, \qquad (6.1)$$
$$u = u_0 \quad \text{on } \partial\Omega.$$

If we drop the quadratic term in the first equation, we get the Stokes problem (III.6.1). Another typical nonlinear diffferential equation is

$$-\Delta u = e^{\lambda u} \quad \text{in } \Omega,$$
$$u = 0 \quad \text{on } \partial\Omega. \qquad (6.2)$$

It arises in the analysis of explosive processes. The parameter λ specifies the relation between the reaction heat and the diffusion constant. – Nonlinear boundary conditions are also of interest, in particular for problems in (nonlinear) elasticity.

We write a nonlinear boundary-value problem as an equation of the form $\mathcal{L}(u) = 0$. Suppose that for each $\ell = 0, 1, \ldots, \ell_{\max}$, the discretization at level ℓ leads to the nonlinear equation

$$\mathcal{L}_\ell(u_\ell) = 0 \qquad (6.3)$$

with $N_\ell := \dim S_\ell$ unknowns. In the sequel it is often more convenient to consider the formally more general equation

$$\mathcal{L}_\ell(u_\ell) = f_\ell \qquad (6.4)$$

with given $f_\ell \in \mathbb{R}^{N_\ell}$.

Within the framework of multigrid methods, there are two fundamentally different approaches:
1. The *multigrid Newton method* (MGNM), which solves the linearized equation using the multigrid method.
2. The *nonlinear multigrid method* (NMGM), which applies the multigrid method directly to the given nonlinear equation.

The Multigrid Newton Method

Newton iteration requires solving a linear system of equations for every step of the iteration. However, it suffices to compute an approximate solution in each step.

The following algorithm is a variant of the damped Newton method. We denote the derivative of the (nonlinear) mapping $\mathcal{L}$ by $D\mathcal{L}$.

6.1 Multigrid Newton Method.
Let $u^{\ell,0}$ be an approximation to the solution of the equation $\mathcal{L}_\ell(u_\ell) = f_\ell$.
For $k = 0, 1, \ldots$, carry out the following calculation:

1. *(Determine the direction)* Set $d^k = f_\ell - \mathcal{L}_\ell(u^{\ell,k})$. Perform one cycle of the algorithm $\mathbf{MGM}_\ell$ to solve

$$D\mathcal{L}_\ell(u^{\ell,k})\, v = d^k$$

   with the starting value $v^{\ell,0} = 0$. Call the result $v^{\ell,1}$.

2. *(Line search)* For $\lambda = 1, \frac{1}{2}, \frac{1}{4}, \ldots$, test if

$$\|\mathcal{L}_\ell(u^{\ell,k} + \lambda\, v^{\ell,1}) - f_\ell\| \le (1 - \frac{\lambda}{2})\, \|\mathcal{L}_\ell(u^{\ell,k}) - f_\ell\|. \qquad (6.5)$$

   As soon as (6.5) is satisfied, stop testing and set

$$u^{\ell,k+1} = u^{\ell,k} + \lambda\, v^{\ell,1}. \qquad\qquad \square$$

The direction to the next approximation is determined in the first step, and the distance to go in that direction is determined in the second step. If the approximations are sufficiently close to the solution, then we get $\lambda = 1$. In this case, step 2 can be replaced by the simpler classical method:

$2'$. Set $u^{\ell,k+1} = u^{\ell,k} + v^{\ell,1}$.

The introduction of the damping parameter λ and the associated test results in a stabilization; see Hackbusch and Reusken [1989]. Thus, the method is less sensitive to the choice of the starting value $u^{\ell,0}$.

It is known that the classical Newton method converges quadratically for sufficiently good starting values, provided the derivative $D\mathcal{L}_\ell$ is invertible at the solution. In Algorithm 6.1 we also have an extra linear error term, and the error $e^k := u^{\ell,k} - u_\ell$ satisfies the following recurrence formula:

$$\|e^{k+1}\| \le \rho\|e^k\| + c\|e^k\|^2.$$

Here ρ is the convergence rate of the multigrid algorithm.

This implies only linear convergence. At first glance this is a disadvantage, but quadratic convergence only happens in a neighborhood of the solution, and in particular only when the error $\|e^k\|$ is smaller than the discretization error. In view of the discussion in the previous section, this is no essential disadvantage.

The Nonlinear Multigrid Method

Methods based on applying the multigrid method directly to the nonlinear equation are often used instead of the multigrid Newton method. Here the calculation again involves smoothing steps and coarse-grid corrections. However, here the latter have a nonlinear character.

The simplest smoothing corresponds to the Jacobi method

$$v^\ell \longmapsto \mathcal{S}_\ell \, v^\ell := v^\ell + \omega[f_\ell - \mathcal{L}_\ell(v^\ell)]. \tag{6.6}$$

As in the linear case, the parameter ω is computed by estimating the largest eigenvalue of $D\mathcal{L}_\ell$.

The so-called *nonlinear Gauss–Seidel method* can be used to perform the smoothing. In order to reduce the amount of formalism, we restrict ourselves to an example, and consider the difference method for the equation (6.2) on a square grid. For the interior points, we have

$$u_i - h^2 e^{\lambda u_i} = G_i(u), \quad i = 1, 2, \dots N_\ell, \tag{6.7}$$

where $G_i(u)$ is $\frac{1}{4}$ the sum of the values at the neighboring nodes. For each $i = 1, 2, \dots$, we successively compute a refined value u_i by solving the i-th equation in (6.7) for u_i. This involves solving simple scalar nonlinear equations. More generally, we have

$$[\mathcal{L}_\ell(u_1^{k+1}, \dots, u_i^{k+1}, u_{i+1}^k, u_{i+2}^k, \dots)]_i = f_i, \quad i = 1, 2, \dots$$

The Gauss–Seidel method can also be used as a smoother in the nonlinear case.

The computation of the coarse-grid correction has to be done differently than in the linear case. We emphasize that in general, $u_{\ell-1} \neq r u_\ell$, where u_ℓ and $u_{\ell-1}$ are the finite element solutions in S_ℓ and $S_{\ell-1}$, respectively. This is why so far the coarse-grid correction has only been applied to the defect equation. Here we do something different. In passing between grids, *we compensate for the deviation of $u_{\ell-1}$ from $r u_\ell$ by including an additive term on the right-hand side*. This correction is the reason why we replaced the original equation (6.3) by the more general equation (6.4).

We also need the restriction of $u^{\ell,k,1}$ at the level $\ell - 1$. This could be done with a different operator from the one used to evaluate the restriction of the residue. Let r and $\tilde{r}$ be restriction operators for the residue and for the approximations, respectively, and let p be a prolongation. The operator $\mathcal{L}_{\ell-1}$ corresponds to the discretization of (6.3) at level $\ell - 1$.

6.2 Nonlinear Multigrid Iteration NMGM$_\ell$ *(k-th cycle at level $\ell \geq 1$):*

Let $u^{\ell,k}$ be a given approximation in S_ℓ.

  1. Pre-smoothing. Perform ν_1 smoothing steps:

$$u^{\ell,k,1} = S^{\nu_1} u^{\ell,k}.$$

  2. Coarse-grid correction. Set

$$d_\ell = f_\ell - \mathcal{L}_\ell(u^{\ell,k,1}),$$
$$u^{\ell-1,0} = \tilde{r}u^{\ell,k,1}, \tag{6.8}$$
$$f_{\ell-1} = \mathcal{L}_{\ell-1}(u^{\ell-1,0}) + rd_\ell,$$

and let $\hat{v}_{\ell-1}$ be the solution of

$$\mathcal{L}_{\ell-1}(v) = f_{\ell-1}. \tag{6.9}$$

If $\ell = 1$, find the solution, and set $v^{\ell-1} = \hat{v}_{\ell-1}$.
If $\ell > 1$, determine an approximation $v^{\ell-1}$ of $\hat{v}_{\ell-1}$ by carrying out μ steps of **NMGM$_{\ell-1}$** with the starting value $u^{\ell-1,0}$.
Set

$$u^{\ell,k,2} = u^{\ell,k,1} + p(v^{\ell-1} - u^{\ell-1,0}). \tag{6.10}$$

  3. Post-smoothing. Perform ν_2 smoothing steps using

$$u^{\ell,k,3} = S^{\nu_2} u^{\ell,k,2},$$

and set $u^{\ell,k+1} = u^{\ell,k,3}$.   □

The reader can verify that in the linear case we get Algorithm 1.7, independent of the choice of the restriction operator $\tilde{r}$.

    Note that the following diagram is *not commutative*:

$$
\begin{array}{ccc}
S_\ell & \xrightarrow{\ \mathcal{L}_\ell\ } & S_\ell \\
{\scriptstyle \tilde{r}}\downarrow & & \downarrow{\scriptstyle r} \\
S_{\ell-1} & \xrightarrow{\ \mathcal{L}_{\ell-1}\ } & S_{\ell-1}.
\end{array}
$$

More specifically, $f_{\ell-1} \neq rf_\ell$ in general. In fact,

$$f_{\ell-1} = rf_\ell + [\mathcal{L}_{\ell-1}(\tilde{r}u^{\ell,k,1}) - r\mathcal{L}_\ell(u^{\ell,k,1})]. \tag{6.11}$$

The *shift* by the extra term in the square brackets ensures that the solution u_ℓ of the equation (6.4) is a fixed point for the iteration. Assuming that $u^{\ell,k,1} = u_\ell$, it follows that $d_\ell = 0$ and $f_{\ell-1} = \mathcal{L}_{\ell-1}(u^{\ell-1,0})$. Thus, $v = u^{\ell-1,0}$ is a solution of (6.9), and $u^{\ell,k,2} = u^{\ell,k,1}$.

Starting Values

As explained in §4, for linear problems we can start on the coarsest grid and work toward the finest one. This is also possible for many nonlinear problems, but not for all. In particular, it can happen that the nonlinear problem only has the right number of solutions when the discretization is sufficiently fine.

For this reason, we now assume that we have a starting value which belongs to the domain of attraction of the desired solution. However, the error may still be much larger than the discretization error, and in fact by several orders of magnitude.

This is the usual case in practice, and we suggest proceeding as in Algorithm NI_ℓ. However, we have first to compute an appropriate right-hand side for the problems on the coarse grids.

In the following we use the notation of Algorithm 6.2.

6.3 Algorithm $NLNI_\ell$ $(\mathcal{L}_\ell, f_\ell, u^{\ell,0})$ *for improving a starting value $u^{\ell,0}$ for the equation $\mathcal{L}_\ell(u_\ell) = f_\ell$ at level $\ell \geq 0$, (such that the error of the result $\hat{u}^\ell$ is of the order of the discretization error).*

If $\ell = 0$, compute the solution $\hat{u}^\ell$ of the equation $\mathcal{L}_0(v) = f^0$, and exit the procedure.
Let $\ell > 0$.
Set $u^{\ell-1,0} = \tilde{r}u^{\ell,0}$ and

$$f_{\ell-1} = rf_\ell + [\mathcal{L}_{\ell-1}(u^{\ell-1,0}) - r\mathcal{L}_\ell(u^{\ell,0})]. \tag{6.12}$$

Find an approximate solution $\hat{u}_{\ell-1}$ of the equation $\mathcal{L}_{\ell-1}(v) = f_{\ell-1}$ by applying $NLNI_{\ell-1}$ $(\mathcal{L}_{\ell-1}, f_{\ell-1}, u^{\ell-1,0})$.
Determine the prolongation $u^{\ell,1} = p\hat{u}^{\ell-1}$.
Using $u^{\ell,1}$ as a starting value, carry out one step of the iteration $NLMG_\ell$. Denote the result as $u^{\ell,2}$. Set

$$\hat{u}^\ell = u^{\ell,2}. \qquad\qquad \square$$

Note that equation (6.12) has the same structure as (6.11).

Since we cannot proceed without reasonable starting values, for complicated problems, nonlinear multigrid methods are usually combined with continuation methods (also called incremental methods).

Problems

6.4 Verify that for linear problems, Algorithm 6.2 is equivalent to Algorithm 1.7.

6.5 The nonlinear equation (6.2) characterizes a solution of the nonquadratic variational problem

$$\int_\Omega [\frac{1}{2}(\nabla v)^2 - F(v)]dx \longrightarrow \min_{v \in H_0^1} !$$

(assuming a solution exists). Find a suitable function F on $\mathbb{R}$ by formally calculating the Euler equation corresponding to the variational problem.

Chapter VI

Finite Elements in Solid Mechanics

Finite element methods are the most widely used tools for computing the deformations and stresses of elastic and inelastic bodies subject to loads. These types of problems involve systems of differential equations with the following special feature: the equations are invariant under translations and orthogonal transformations since the elastic energy of a body does not change under so-called rigid body motions.

Practical problems in structural mechanics often involve small parameters which can appear in both obvious and more subtle ways. For example, for beams, membranes, plates, and shells, the thickness is very small in comparison with the other dimensions. On the other hand, for a cantilever beam, the part of the boundary on which Dirichlet boundary conditions are prescribed is very small. Finally, many materials allow only very small changes in density. These various cases require different variational formulations of the finite element computations. Using an incorrect formulation leads to so-called *locking*. Often, mixed formulations provide a suitable framework for both the computation and a rigorous mathematical analysis.

Most of the characteristic properties appear already in the so-called linear theory, i.e., for small deformations where no genuine nonlinear phenomenon occurs. However, strictly speaking, there is no complete linear elasticity theory, since the above-mentioned invariance under rigid body motions cannot be completely modeled in a linear theory. For this reason, we don't restrict ourselves to the linear theory until later.

§§1 and 2 contain a very compact introduction to elasticity theory. For more details, see Ciarlet [1988], Marsden and Hughes [1983], or Truesdell [1977]. Here we concentrate on those aspects of the theory which we need as background knowledge. In §3 we present several variational formulations for the linear theory, and also include an analysis of locking. Finally, we discuss membranes and plates. In particular, we explore the connection between two widely used plate models.

We limit ourselves to those elements whose construction or analysis is based on different approaches than the elements discussed in Chapters II and III. In particular, we will focus on the stability of the elements.

§ 1. Introduction to Elasticity Theory

Elasticity theory deals with the deformation of bodies under the influence of applied forces, and in particular, with the stresses and strains which result from deformations.

The three-dimensional case provides the foundation for the theory. The essential ingredients are the kinematics, the equilibrium equations, and the material laws.

Kinematics

We assume that we know a *reference configuration* $\bar{\Omega}$ for the body under consideration. Here $\bar{\Omega}$ is the closure of a bounded open set Ω. In general, $\bar{\Omega}$ is just the subset of $\mathbb{R}^3$ where the body is in an unstressed state (natural state). The current state is given by a mapping[14]

$$\phi : \bar{\Omega} \longrightarrow \mathbb{R}^3$$

where $\phi(x)$ represents the position of a point which was located at x in the reference configuration. We write

$$\phi = id + u, \tag{1.1}$$

and call u the *displacement*. Often we will assume that the displacements are small, and will neglect terms of higher order in u.

It is obvious that rigid body motions, i.e., translations and orthogonal transformations, do not alter the stresses in a body. This causes some difficulties since this invariance must be preserved in the finite element results – at least approximately.

In the following, we assume that the mapping ϕ is sufficiently smooth. ϕ represents a *deformation*, provided

$$\det(\nabla\phi) > 0.$$

Here $\nabla\phi$ is the *deformation gradient*, and its matrix representation is

$$\nabla\phi = \begin{bmatrix} \frac{\partial\phi_1}{\partial x_1} & \frac{\partial\phi_1}{\partial x_2} & \frac{\partial\phi_1}{\partial x_3} \\ \frac{\partial\phi_2}{\partial x_1} & \frac{\partial\phi_2}{\partial x_2} & \frac{\partial\phi_2}{\partial x_3} \\ \frac{\partial\phi_3}{\partial x_1} & \frac{\partial\phi_3}{\partial x_2} & \frac{\partial\phi_3}{\partial x_3} \end{bmatrix}. \tag{1.2}$$

---

[14] As before, we do not use any special notation to distinguish vectors, matrices, or tensors. In general, in this section we use lower case Latin letters for vectors, and capitals for tensors or matrices.

The word deformation suggests that subdomains with positive volume are mapped into subdomains with positive volume. Deformations are injective mappings locally.

The mapping ϕ induces

$$\phi(x+z) - \phi(x) = \nabla\phi(x) \cdot z + o(z).$$

In terms of the Euclidean distance,

$$\begin{aligned}\|\phi(x+z) - \phi(x)\|^2 &= \|\nabla\phi \cdot z\|^2 + o(\|z\|^2) \\ &= z' \nabla\phi^T \nabla\phi\, z + o(\|z\|^2).\end{aligned} \tag{1.3}$$

Thus, the matrix

$$C := \nabla\phi^T \nabla\phi \tag{1.4}$$

describes the transformation of the length element. It is called the *(right) Cauchy–Green strain tensor.* The deviation

$$E := \frac{1}{2}(C - I)$$

from the identity is called the *strain*, and is one of the most important concepts in the theory. Frequently, we will work with matrix representations of C and E. These matrices are obviously symmetric. Inserting (1.1) into (1.4) gives

$$E_{ij} = \frac{1}{2}\left(\frac{\partial u_i}{\partial x_j} + \frac{\partial u_j}{\partial x_i}\right) + \frac{1}{2}\sum_k \frac{\partial u_i}{\partial x_k}\frac{\partial u_j}{\partial x_k}. \tag{1.5}$$

In the linear theory we neglect the quadratic terms, leading to the following *symmetric gradient* as an approximation:

$$\varepsilon_{ij} := \frac{1}{2}\left(\frac{\partial u_i}{\partial x_j} + \frac{\partial u_j}{\partial x_i}\right). \tag{1.6}$$

1.1 Remark. Let Ω be connected. If the strain tensor associated with the deformation $\phi \in C^1(\Omega)$ satisfies the relation

$$C(x) = I \quad \text{for all } x \in \Omega,$$

then ϕ describes a rigid body motion, i.e., $\phi(x) = Qx + b$, where Q is an orthogonal matrix.

Sketch of a proof. Let Γ be a smooth curve in Ω. In view of (1.3) and $C(x) = I$, the rectifiable curves Γ and $\phi(\Gamma)$ always have the same length. This follows directly from the definition of the arc length via an integral. We now use it to establish the desired result.

Since ϕ is locally injective, if Ω is open, then $\phi(\Omega)$ is also open. For every $x_0 \in \Omega$, there exists a convex neighborhood U in Ω such that the convex hull of $\phi(U)$ is contained in $\phi(\Omega)$. The mapping $\phi|_U$ is globally distance preserving, i.e. for all pairs $x, y \in U$,

$$\|\phi(x) - \phi(y)\| = \|x - y\|. \tag{1.7}$$

To see this, let Γ be the line connecting the points x and y. Since $\phi(\Gamma)$ has the same length, $\|\phi(x) - \phi(y)\| \leq \|x - y\|$. The equality now follows by examining the preimage of the line connecting $\phi(x)$ and $\phi(y)$.

Because of (1.7), the auxiliary function

$$G(x, y) := \|\phi(y) - \phi(x)\|^2 - \|y - x\|^2$$

vanishes on $U \times U$. G is differentiable with respect to y, and $\frac{1}{2}\frac{\partial G}{\partial y_i}$ satisfies

$$\sum_k \frac{\partial \phi_k}{\partial y_i}(\phi_k(y) - \phi_k(x)) - (y_i - x_i) = 0.$$

This expression is differentiable with respect to x_j, and so

$$-\sum_k \frac{\partial \phi_k}{\partial y_i}\frac{\partial \phi_k}{\partial x_j} + \delta_{ij} = 0,$$

which is just the componentwise version of $\nabla\phi(y)^T \nabla\phi(x) = I$. Multiplying on the left by $\nabla\phi(y)$ and using $C = I$, we immediately get $\nabla\phi(x) = \nabla\phi(y)$. Thus, $\nabla\phi$ is constant on U, and ϕ is a linear transformation.

Now the result follows for the entire domain Ω by a covering argument.   □

The Equilibrium Equations

In mechanics we treat the influence of forces axiomatically. Euler and Cauchy both made essential contributions. For details, see Ciarlet [1988].

We assume that the interaction of the body with the outside world is described by two types of applied forces:

   (a) applied surface forces (forces distributed over the surface),
   (b) applied body forces (forces distributed over the volume).
A typical body force is the force of gravity, while the force caused by a load on a bridge (e.g. a vehicle) is a surface force.

The forces are distinguished by the work they do under deformations.

The body force $f : \Omega \longrightarrow \mathbb{R}^3$ results in a force $f \, dV$ acting on a volume element dV. Surface forces are specified by a function $t : \Omega \times S^2 \longrightarrow \mathbb{R}^3$ where S^2 denotes the unit sphere in $\mathbb{R}^3$: Let V be an arbitrary subdomain of Ω (with a sufficiently smooth boundary), and let dA be an area element on the surface with the unit outward-pointing normal vector n. Then the area element dA contributes $t(x, n) \, dA$ to the force, which also depends on the direction of n. The vector $t(x, n)$ is called the *Cauchy stress vector.*

The main axiom of mechanics asserts that in an equilibrium state, all forces and all moments add to zero. Here we must take into account both surface forces and body forces.

1.2 Axiom of Static Equilibrium. *(Stress principle of Euler and Cauchy)* Let B be a (deformed) body in equilibrium. Then there exists a vector field t such that in every subdomain V of B, the (volume) forces f and the stresses t satisfy

$$\int_V f(x) dx + \int_{\partial V} t(x, n) ds = 0, \tag{1.8}$$

$$\int_V x \wedge f(x) dx + \int_{\partial V} x \wedge t(x, n) ds = 0. \tag{1.9}$$

Here the symbol $\wedge$ stands for the vector product in $\mathbb{R}^3$.

Once the existence of the Cauchy stress vector is given, its exact dependence on the normal n can be determined. Here and in the sequel, we use the following sets of matrices:

$\mathbb{M}^3$, the set of 3×3 matrices,

$\mathbb{M}^3_+$, the set of matrices in $\mathbb{M}^3$ with positive determinants,

$\mathbb{O}^3$, the set of orthogonal 3×3 matrices,

$\mathbb{O}^3_+ := \mathbb{O}^3 \cap \mathbb{M}^3_+$,

$\mathbb{S}^3$, the set of symmetric 3×3 matrices,

$\mathbb{S}^3_>$, the set of positive definite matrices in $\mathbb{S}^3$.

1.3 Cauchy's Theorem. *Let* $t(\cdot, n) \in C^1(B, \mathbb{R}^3)$, $t(x, \cdot) \in C^0(S^2, \mathbb{R}^3)$, *and* $f \in C(B, \mathbb{R}^3)$ *be in equilibrium according to 1.2. Then there exists a symmetric tensor field* $T \in C^1(B, \mathbb{S}^3)$ *with the following properties:*

$$t(x, n) = T(x)n, \qquad x \in B, \; n \in S^2, \tag{1.10}$$

$$\operatorname{div} T(x) + f(x) = 0, \qquad x \in B, \tag{1.11}$$

$$T(x) = T^T(x), \qquad x \in B. \tag{1.12}$$

The tensor T is called the *Cauchy stress tensor.*

The key assertion of this famous theorem is the representability of the stress vector t in terms of the tensor T. Using the Gauss integral theorem, it follows from (1.8) that

$$\int_V f(x)dx + \int_{\partial V} T(x)nds = \int_V [f(x) + \operatorname{div} T(x)]dx = 0.$$

This relation also implies the differential equation (1.11). The equilibrium equations (1.9) for the moments imply the symmetry (1.12). ☐

The Piola Transform

We have formulated the equilibrium equations in terms of the coordinates of the deformed body B (as did Euler). Since these coordinates have to be computed in the first place, it is useful to transform the variables to the reference configuration. To distinguish the expressions, in the following we add a subscript R when referring to the reference configuration. In particular, $x = \phi(x_R)$.

The transformation of the body forces follows directly from the well-known transformation theorem for integrals, where the volume element is given by $dx = \det(\nabla\phi)dx_R$. The forces are proportional to density. Densities are transformed according to conservation of mass: $\rho(x)dx = \rho_R(x_R)dx_R$ which implies $\rho(\phi(x_R)) = \det(\nabla\phi^{-1})\rho_R(x_R)$. Consequently,

$$f(x) = \det(\nabla\phi^{-1})f_R(x_R). \tag{1.13}$$

The equation (1.13) makes implicit use of the assumption that under the deformation, point masses do not move to positions where we have a different force field. In this case we speak of a *dead load.*

The transformation of stress tensors is more complicated, but can be computed by elementary methods; see e.g. Ciarlet [1988]. In terms of the reference configuration, we have

$$\operatorname{div}_R T_R + f_R = 0 \tag{1.14}$$

with

$$T_R := \det(\nabla\phi)\, T\, (\nabla\phi)^{-T}. \tag{1.15}$$

Equation (1.14) is the analog of (1.11). However, in contrast to T, the so-called *first Piola–Kirchhoff stress tensor* T_R in (1.15) is not symmetric. To achieve symmetry, we introduce the *second Piola–Kirchhoff stress tensor*

$$\Sigma_R := \det(\nabla\phi)\, (\nabla\phi)^{-1}\, T\, (\nabla\phi)^{-T}. \tag{1.16}$$

Clearly, $\Sigma_R = (\nabla\phi)^{-1}T_R$.

The differences between the three stress tensors can be neglected for small deformation gradients.

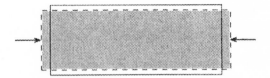

Fig. 57. A compression of a body in one direction leads to an expansion in the other directions. The relative size is given by the Poisson ratio ν.

Constitutive Equations

An important problem is to find the deformation of a body and the associated stresses corresponding to given external forces. The equilibrium equation (1.11) (respectively, (1.14)) gives only 3 equations. This does not determine the 6 components of the symmetric stress tensor. The missing equations arise from constitutive equations, which express how the deformations depend on properties of the material as well as the given forces.

1.4 Definition. A material is called *elastic* if there exists a mapping

$$\hat{T} : \mathbb{M}^3_+ \longrightarrow \mathbb{S}^3_+$$

such that for every deformed state,

$$T(x) = \hat{T}(\nabla\phi(x_R)). \tag{1.17}$$

The mapping $\hat{T}$ is called the *response function* for the Cauchy stress, and (1.17) is called the *constitutive equation.*

The constitutive equation implicitly contains the assumption that the stress depends on the displacement in a local way. In view of (1.16), we introduce the response function for the Piola–Kirchhoff stress,

$$\hat{\Sigma}(F) := \det(F)\, F^{-1}\hat{T}(F)\, F^{-T}. \tag{1.18}$$

(Formulae with the variables F will generally be applied with $F := \nabla\phi(x)$.)

For simplicity, we restrict ourselves to *homogeneous* materials, i.e., to materials for which $\hat{T}$ does not depend explicitly on x.

Response functions can be brought into a simpler form on the basis of physical laws. – First we make the simple observation that the components $\hat{T}_{ij}$ do not behave like scalar functions. Consider a rectangular parallelepiped whose faces are perpendicular to the coordinate axes. Suppose we press on the surfaces which are perpendicular to the x-axis as in Fig. 57. In addition to a compression in the x-direction, in general the material will react by stretching in the perpendicular directions in order to reduce the change in the volume or density, respectively.

1.5 Axiom of Material Frame-Indifference. The Cauchy stress vector $t(x, n) = T(x)n$ is independent of the choice of coordinates, i.e., $Qt(x, n) = t(Qx, Qn)$ for all $Q \in \mathbb{O}^3_+$.

A frame-indifferent material is also called *objective*.

1.6 Theorem. *Suppose the axiom of material frame-indifference holds. Then for every orthogonal transformation $Q \in \mathbb{O}^3_+$,*

$$\hat{T}(QF) = Q\,\hat{T}(F)\,Q^T. \tag{1.19}$$

Moreover, there exists a mapping $\tilde{\Sigma} : \mathbb{S}^3_> \longrightarrow \mathbb{S}^3$ such that

$$\hat{\Sigma}(F) = \tilde{\Sigma}(F^T F), \tag{1.20}$$

i.e., $\hat{\Sigma}$ depends only on $F^T F$.

Proof. Instead of rotating the coordinate system, we rotate the deformed body:

$$x \longmapsto Qx,$$
$$\phi \longmapsto Q\phi,$$
$$\nabla\phi \longmapsto Q\nabla\phi,$$
$$n \longmapsto Q^{-T}n = Qn,$$
$$t(x, n) \longmapsto Qt(x, n).$$

By Axiom 1.5, $t(Qx, Qn) = Qt(x, n)$, and thus $\hat{T}(QF)Q \cdot n = Q\hat{T}(F) \cdot n$. Replacing Qn by n and using $Q^T Q = I$, we get (1.19).

It follows from (1.18) and (1.19) after some elementary manipulations that

$$\hat{\Sigma}(QF) = \hat{\Sigma}(F) \quad \text{for } Q \in \mathbb{O}^3_+. \tag{1.21}$$

To prove (1.20), we consider the two nonsingular matrices F and G in $\mathbb{M}^3_+$ with $F^T F = G^T G$. Set $Q := FG^{-1}$. Then $Q^T Q = I$ and $\det(Q) > 0$. Now (1.21) implies $\hat{\Sigma}(F) = \hat{\Sigma}(G)$, and so in fact $\hat{\Sigma}$ depends only on the product $F^T F$. $\quad\square$

The axiom of frame-indifference holds for all materials. On the other hand, *isotropy* is purely a material property, which means that no direction in the material is preferred. Layered materials such as wood or crystal are not isotropic. Isotropy implies that the stress vectors do not change if we rotate the nondeformed body, i.e., before the deformation takes place.

1.7 Definition. A material is called *isotropic* provided

$$\hat{T}(F) = \hat{T}(FQ) \quad \text{for all } Q \in \mathbb{O}^3_+. \tag{1.22}$$

The different order of F and Q in (1.19) as compared to (1.22) is important. As in the proof of Theorem 1.6, it can be shown that (1.22) is equivalent to

$$\hat{T}(F) = \bar{T}(FF^T) \tag{1.23}$$

with a suitable function $\bar{T}$.

In view of the transformation properties, the response function depends in an essential way on the invariants of the matrix: every 3×3 matrix $A = (a_{ij})$ is associated with a triple of invariants $\iota_A = (\iota_1(A), \iota_2(A), \iota_3(A))$ defined by the corresponding characteristic polynomial

$$\det(\lambda I - A) = \lambda^3 - \iota_1(A)\lambda^2 + \iota_2(A)\lambda - \iota_3(A).$$

These principal invariants are closely related to the eigenvalues $\lambda_1, \lambda_2, \lambda_3$ of A:

$$\iota_1(A) := \sum_i a_{ii} = \text{trace}(A) = \lambda_1 + \lambda_2 + \lambda_3,$$

$$\iota_2(A) := \frac{1}{2}\sum_{ij}(a_{ii}a_{jj} - a_{ij}^2) = \frac{1}{2}[(\text{trace } A)^2 - \text{trace}(A^2)] \tag{1.24}$$

$$= \lambda_1\lambda_2 + \lambda_1\lambda_3 + \lambda_2\lambda_3,$$

$$\iota_3(A) := \det(A) = \lambda_1\lambda_2\lambda_3.$$

We can now formulate a famous theorem of elasticity theory. We employ the usual notation for diagonal matrices, $D = \text{diag}(d_{11}, d_{22}, \ldots, d_{nn})$.

1.8 Rivlin–Ericksen Theorem [1955]. *A response function $\hat{T} : \mathbb{M}^3_+ \longrightarrow \mathbb{S}^3$ is objective and isotropic if and only if it has the form $\hat{T}(F) = \bar{T}(FF^T)$, and*

$$\bar{T} : \mathbb{S}^3_> \longrightarrow \mathbb{S}^3$$
$$\bar{T}(B) = \beta_0(\iota_B)I + \beta_1(\iota_B)B + \beta_2(\iota_B)B^2. \tag{1.25}$$

Here β_0, β_1, and β_2 are functions of the invariants of B.

Proof. By (1.23), $\hat{T}(F) = \bar{T}(FF^T)$ with a suitable function $\bar{T}$. It remains to give the proof for the special form (1.25).

(1) First, let $B = \text{diag}(\lambda_1, \lambda_2, \lambda_3)$ be a diagonal matrix, and let $FF^T = B$, e.g. $F = B^{1/2}$. In addition, let $T = (T_{ij}) = \hat{T}(F)$. The matrix $Q := \text{diag}(1, -1, -1)$ is orthogonal, and by Theorem 1.6,

$$\hat{T}(QF) = QTQ^T = \begin{pmatrix} T_{11} & -T_{12} & -T_{13} \\ -T_{21} & T_{22} & T_{23} \\ -T_{31} & T_{32} & T_{33} \end{pmatrix}. \tag{1.26}$$

On the other hand, $QF(QF)^T = QBQ^T = B$, and thus by hypothesis, $\hat{T}(QF) = \hat{T}(F) = T$. By (1.26), this can happen only if $T_{12} = T_{13} = 0$. A similar argument with $Q = \text{diag}(-1, -1, +1)$ shows that $T_{23} = 0$. Thus $T(B)$ is diagonal if B is a diagonal matrix.

(2) Suppose again that B is a diagonal matrix. If $B_{ii} = B_{jj}$, then $T_{ii} = T_{jj}$. To verify this we consider the case $B_{11} = B_{22}$, and choose

$$Q = \begin{pmatrix} 0 & 1 & \\ 1 & 0 & \\ & & -1 \end{pmatrix}.$$

Then $QBQ^T = B$, and analogously to part (1), we deduce that $T_{11} = (QTQ^T)_{11} = T_{22}$.

Thus, we can represent T in the form

$$T = \beta_0 I + \beta_1 B + \beta_2 B^2 \tag{1.27}$$

with suitable coefficients $\beta_0, \beta_1, \beta_2$. Now if we permute the diagonal elements of B, then as we have seen, the elements of T have to be permuted in the same way. This gives the representation (1.27) for the new matrix with the same coefficients β_0, β_1 and β_2 as before. Thus, β_0, β_1 and β_2 are *symmetric* functions of λ_i, and the theorem is proved in the case of a diagonal matrix B.

(3) Suppose $F \in \mathbb{M}^3_+$ and $B = FF^T$ is not diagonal. There exists an orthogonal matrix Q such that $QBQ^{-1} = D$ is a diagonal matrix. Replacing Q by $-Q$ if necessary, we can assume $\det Q > 0$. Note that $\iota_B = \iota_D$. By the above considerations and the material frame-indifference, we deduce that

$$\begin{aligned}
\hat{T}(F) &= Q^{-1}\hat{T}(QF)Q^{-T} \\
&= Q^{-1}\bar{T}(D)Q \\
&= Q^{-1}[\beta_0 I + \beta_1 D + \beta_2 D^2]Q \\
&= \beta_0 I + \beta_1 B + \beta_2 B^2,
\end{aligned}$$

and the proof is complete. $\square$

1.9 Remarks. In the special case where FF^T is a multiple of the unit matrix, $\hat{T}(F)$ is also a multiple of the unit matrix. Then the stress has the character of a pure pressure.

For the transfer of the result on the Cauchy tensor to a corresponding formula for the second Piola–Kirchhoff tensor, we make use of the formula of Cayley–Hamilton: $B^3 - \iota_1(B)B^2 + \iota_2(B)B - \iota_3(B)I = 0$. Eliminating I from (1.25), we get

$$\bar{T}(B) = \tilde{\beta}_1 B + \tilde{\beta}_2 B^2 + \tilde{\beta}_3 B^3$$

with different coefficients. Multiplying on the left by F^{-1} and on the right by F^{-T}, with the notation of Theorem 1.6 we get a reformulation in terms of the Cauchy–Green stress tensor C.

1.10 Corollary. $\Sigma(\nabla\phi) = \tilde{\Sigma}(\nabla\phi^T \nabla\phi)$ *for an isotropic and objective material,* *where*

$$\tilde{\Sigma}(C) = \gamma_0 I + \gamma_1 C + \gamma_2 C^2, \tag{1.28}$$

and where $\gamma_0, \gamma_1, \gamma_2$ *are functions of the invariants* ι_C.

Linear Material Laws

The stress-strain relationship can be described in terms of two parameters in the neighborhood of a strain-free reference configuration. Setting $C = I + 2E$ in (1.28), $\tilde{\Sigma}(I + 2E) = \gamma_0(E) I + \gamma_1(E) E + \gamma_2(E) E^2$, where we have not changed the notation for the functions.

1.11 Theorem. *Suppose that in addition to the hypotheses of Corollary 1.10, γ_0, γ_1 and γ_2 are differentiable functions of $\iota_1(E), \iota_2(E)$ and $\iota_3(E)$. Then there exist numbers π, λ, μ with*

$$\tilde{\Sigma}(I + 2E) = -\pi I + \lambda \operatorname{trace}(E) I + 2\mu E + o(E) \quad \text{as } E \to 0.$$

Sketch of a proof. First note that $\tilde{\Sigma}(1+2E) = \gamma_0(E) I + \gamma_1 E + o(E)$. In particular, only the constant term in γ_1 is used. By Remarks 1.9, we know that $\tilde{\Sigma}(I) = -\pi I$ with a suitable $\pi \geq 0$. By (1.24), we deduce that $\iota_2 = \mathcal{O}(E^2)$ and $\iota_3 = \mathcal{O}(E^3)$, and only the constants and the trace remain in the terms of first order in $\gamma_0(E)$. □

Normally, the situation $C = I$ corresponds to an unstressed condition, and $\pi = 0$. The other two constants are called *Lamé constants*. If we ignore the terms of higher order, we are led to the *linear material law of Hooke*:

$$\tilde{\Sigma}(I + 2E) = \lambda \operatorname{trace}(E) I + 2\mu E. \tag{1.29}$$

A material which satisfies (1.29) in general and not just for small strains is called a *St. Venant–Kirchhoff material*. Note that in the approximation (1.6),

$$\operatorname{trace}(\varepsilon) = \operatorname{div} u, \tag{1.30}$$

and thus the Lamé constant λ describes the stresses due to change in density. The other Lamé constant μ is sometimes called the *shear modulus of the material*.

If we use a different set of frequently used parameters, namely the *modulus of elasticity* E and the *Poisson ratio* v, we have the following connection:

$$v = \frac{\lambda}{2(\lambda + \mu)}, \qquad E = \frac{\mu(3\lambda + 2\mu)}{\lambda + \mu},$$
$$\lambda = \frac{Ev}{(1 + v)(1 - 2v)}, \qquad \mu = \frac{E}{2(1 + v)}. \tag{1.31}$$

It follows from physical considerations that $\lambda > 0$, $\mu > 0$, and $E > 0$, $0 < \nu < \frac{1}{2}$.

The Poisson ratio ν describes the influence of stresses on displacements in the orthogonal directions shown in Fig. 57. For many materials, $\nu \approx 1/3$. On the other hand, for *nearly incompressible materials*, $\lambda \gg \mu$, i.e., ν is very close to $1/2$.

The deformations, stresses, and strains are defined by the kinematics, the equilibrium equations, and the constitutive equations. In principle, only the equilibrium equations (for the Cauchy stress tensor) are linear.

If we assume small deformations, and replace the strain E by the linearization ε, it suffices to work with the so-called *geometrically linear theory*. However, for practical everyday calculation, the complete linear theory where we also assume that the constitutive equations are linear and work with isotropic media is of the greatest importance.

Problem

1.12 Often a *polar factorization* of the deformation gradient

$$F = RU \quad \text{or} \quad F = VR$$

with positive definite Hermitean matrices U, V, and an orthogonal Matrix R is considered. In this way the invariance properties are accentuated. Show that

$$U := (F^T F)^{1/2}, \quad R := FU^{-1}, \quad \text{and} \quad V := RUR^T$$

yield the desired factorization and that it is unique. Here F is assumed to be nonsingular.

§ 2. Hyperelastic Materials

By Cauchy's theorem, the equilibrium state of an elastic body is characterized by

$$- \operatorname{div} T(x) = f(x), \quad x \in \Omega, \tag{2.1}$$

and the boundary conditions

$$\begin{aligned} \phi(x) &= \phi_0(x), \quad x \in \Gamma_0, \\ T(x) \cdot n &= g(x), \quad x \in \Gamma_1. \end{aligned} \tag{2.2}$$

Here f is the applied body force and g is the surface traction on the part Γ_1 of the boundary. Γ_0 denotes the part of the boundary on which the displacement is given.

We regard these equations as a boundary-value problem for the deformation ϕ, and write

$$\begin{aligned} - \operatorname{div} \hat{T}(x, \nabla \phi(x)) &= f(x), \quad x \in \Omega, \\ \hat{T}(x, \nabla \phi(x)) n &= g(x), \quad x \in \Gamma_1, \\ \phi(x) &= \phi(x_0), \quad x \in \Gamma_0. \end{aligned} \tag{2.3}$$

For simplicity, we neglect the dependence of the forces f and g on ϕ, i.e., we consider them to be dead loads; cf. Ciarlet [1988, §2.7].

To be more precise, Ω is the domain occupied by the deformed body, and is also unknown. For simplicity, we identify Ω with the reference configuration, and restrict ourselves to an approximation which makes sense for small deformations.

2.1 Definition. An elastic material is called *hyperelastic* if there exists an energy functional $\hat{W} : \Omega \times M_+^3 \longrightarrow \mathbb{R}$ such that

$$\hat{T}(x, F) = \frac{\partial \hat{W}}{\partial F}(x, F) \quad \text{for } x \in \Omega, \ F \in M_+^3 .$$

There is a variational formulation corresponding to the boundary-value problem (2.3) for hyperelastic materials, provided that the vector fields f and g can be written as gradient fields: $f = \operatorname{grad} \mathcal{F}$ and $g = \operatorname{grad} \mathcal{G}$. In this case the solutions of (2.3) are stationary points of the total energy

$$I(\psi) = \int_\Omega [\hat{W}(x, \nabla \psi(x)) - \mathcal{F}(\psi(x))] dx + \int_{\Gamma_1} \mathcal{G}(\psi(x)) dx. \tag{2.4}$$

As deformations we admit functions ψ which satisfy Dirichlet boundary conditions on Γ_0 along with the local injectivity condition $\det(\nabla\psi(x)) > 0$. – We introduce appropriate function spaces later.

The expression (2.4) refers to the variational formulation for the displacements. We note that frequently the stresses are also included as variables in the variational problem. Because of the coupling of the kinematics with the constitutive equations, we get a saddle point problem, and thus mixed methods need to be applied.

2.2 Remark. The properties of the material laws discussed in §1 may be rediscovered in analogous properties of the energy functionals. To save space, we present them without proof.

For an objective material, $\hat{W}(x, \cdot)$ is a function of only $C = F^T F$:

$$\hat{W}(x, F) = \tilde{W}(x, F^T F)$$

and

$$\tilde{\Sigma}(x, C) = 2\frac{\partial \tilde{W}(x, C)}{\partial C} \quad \text{for all } C \in \mathbb{S}_{>}^3.$$

The dependence of C can be made more precise. $\tilde{W}$ depends only on the principal invariants of C, i.e., $\tilde{W}(x, C) = \dot{W}(x, \iota_C)$ for $C \in \mathbb{S}_{>}^3$. Analogously, for isotropic materials, we have

$$\hat{W}(x, F) = \hat{W}(x, FQ) \quad \text{for all } F \in \mathbb{M}_{+}^3, \ Q \in \mathbb{O}_{+}^3.$$

In particular, for small deformations,

$$\tilde{W}(x, C) = \frac{\lambda}{2}(\text{trace } E)^2 + \mu\, E : E + o(E^2) \tag{2.5}$$

with $C = I + 2E$. Here, as usual,

$$A : B := \sum_{ij} A_{ij} B_{ij} = \text{trace}(A^T B),$$

for any two matrices A and B.

2.3 Examples. (1) For St. Venant–Kirchhoff materials,

$$\begin{aligned}
\hat{W}(x, F) &= \frac{\lambda}{2}(\text{trace } F - 3)^2 + \mu\, F : F \\
&= \frac{\lambda}{2}(\text{trace } E)^2 + \mu\, \text{trace } C.
\end{aligned} \tag{2.6}$$

(2) For so-called *neo-Hookean materials*,

$$\tilde{W}(x, C) = \frac{1}{2}\mu[\text{trace}(C - I) + \frac{2}{\beta}\{(\det C)^{-\beta/2} - 1\}], \qquad (2.7)$$

where $\beta = \frac{2\nu}{1-2\nu}$.

We note that (2.6) is restricted to strains which are not too large. Indeed, we expect that

$$\hat{W}(x, F) \longrightarrow \infty \quad \text{as } \det F \to 0, \qquad (2.8)$$

since $\det F \to 0$ means that the density of the deformed material becomes very large. The condition (2.8) implies that $\hat{W}$ is not a convex function of F. Indeed, the set of matrices

$$B = \{F \in \mathbb{M}^3; \ \det F > 0\} \qquad (2.9)$$

is not a convex set; see Problem 2.4. There are many matrices F_0 with $\det F_0 = 0$ which are the convex combination of two matrices F_1 and F_2 with positive determinants. By the continuity of $\hat{W}$ at F_1 and F_2, we would get the boundedness in a neighborhood of F_0 whenever $\hat{W}$ is assumed to be convex.

Problems

2.4 Show that (2.9) does not define a convex set by considering the convex combinations of the matrices

$$\begin{pmatrix} 2 & & \\ & 2 & \\ & & 2 \end{pmatrix} \quad \text{and} \quad \begin{pmatrix} -1 & & \\ & -4 & \\ & & 1 \end{pmatrix}.$$

2.5 Consider a St. Venant–Kirchhoff material with the energy function (2.6), and show that there would exist negative energy states if $\mu < 0$ were to hold.

2.6 Consider the neo-Hookean material for small strains, and establish Hooke's law with the same parameters μ and ν.

2.7 Often the energy functional depends on $J := \det F$. Show that the derivate is given by

$$D_F J : \delta F = J \ \text{trace}(F^{-1}\delta F), \qquad D_C J : \delta C = \frac{1}{2}J \ \text{trace}(C^{-1}\delta C).$$

Hint. For $F = I$ we have obviously $D_I J \ \delta F = \text{trace}(\delta F)$ and $\det(F + \delta F) = \det F \det(I + F^{-1}\delta F)$.

§ 3. Linear Elasticity Theory

In the linearized equations of elasticity theory we take account only of terms of first order in the displacement u while terms of higher order are neglected. This affects the kinematics in terms of the approximation (1.6), and the constitutive equations in terms of (1.29) or (2.6). Here we restrict ourselves to the isotropic case for two reasons: to keep the discussion more accessible, and because this case is more important in practice. In this framework, we do not have to distinguish between different stress tensors. In order to make this clear, we write

$$\sigma \text{ instead of } \Sigma \quad \text{and} \quad \varepsilon \text{ instead of } E.$$

We begin with a short overview, and then in the framework of three-dimensional elasticity theory consider various formulations of the variational problems, including mixed methods in particular.

In order to make this discussion as independent of the previous sections as possible, we first recall the necessary equations.

The Variational Problem

In the framework of the linear theory, the variational problem is to minimize the energy

$$\Pi := \int_\Omega [\frac{1}{2}\varepsilon : \sigma - f \cdot u]dx + \int_{\Gamma_1} g \cdot u \, dx. \tag{3.1}$$

Here $\varepsilon : \sigma := \sum_{ik} \varepsilon_{ik}\sigma_{ik}$. The variables σ, ε and u in (3.1) are not independent, but instead are coupled by the kinematic equations

$$\varepsilon_{ij} = \frac{1}{2}\left(\frac{\partial u_i}{\partial x_j} + \frac{\partial u_j}{\partial x_i}\right) \tag{3.2}$$
$$\text{or} \quad \varepsilon = \varepsilon(u) =: \nabla^{(s)}u,$$

where $\nabla^{(s)}$ is the symmetric gradient, and the linear constitutive equations

$$\varepsilon = \frac{1+\nu}{E}\sigma - \frac{\nu}{E}\text{trace}\,\sigma\,I. \tag{3.3}$$

In order to establish the connection between (3.1) and (2.4), we first invert (3.3). Since trace $I = 3$, it follows from (3.3) that trace $\varepsilon = (1 - 2\nu)/E$ trace σ, and solving for σ gives

$$\sigma = \frac{E}{1+\nu}\left(\varepsilon + \frac{\nu}{1-2\nu}\text{trace}\,\varepsilon\,I\right). \tag{3.4}$$

In contrast to (1.29), the constants here are expressed in terms of the modulus of elasticity and the Poisson ratio. Moreover, $\varepsilon : I = \text{trace } \varepsilon$, and hence

$$\frac{1}{2} \sigma : \varepsilon = \frac{1}{2} (\lambda \text{ trace } \varepsilon \, I + 2\mu\varepsilon) : \varepsilon = \frac{\lambda}{2} (\text{trace } \varepsilon)^2 + \mu \, \varepsilon : \varepsilon \qquad (3.5)$$

coincides with the energy functional in (2.6).

We note that the equation (3.4) is often written componentwise:

$$\begin{bmatrix} \sigma_{11} \\ \sigma_{22} \\ \sigma_{33} \\ \sigma_{12} \\ \sigma_{13} \\ \sigma_{23} \end{bmatrix} = \frac{E}{(1+\nu)(1-2\nu)} \begin{bmatrix} 1-\nu & \nu & \nu & & & \\ \nu & 1-\nu & \nu & & 0 & \\ \nu & \nu & 1-\nu & & & \\ & & & 1-2\nu & & \\ & 0 & & & 1-2\nu & \\ & & & & & 1-2\nu \end{bmatrix} \begin{bmatrix} \varepsilon_{11} \\ \varepsilon_{22} \\ \varepsilon_{33} \\ \varepsilon_{12} \\ \varepsilon_{13} \\ \varepsilon_{23} \end{bmatrix}$$

$$\text{or} \quad \sigma = C\varepsilon, \qquad (3.6)$$

see Problem 4.7.[15] The fact that the matrix C is positive definite for $0 \le \nu < \frac{1}{2}$ can be seen by applying the Gerschgorin theorem to the inverse:

$$C^{-1} = \frac{1}{E} \begin{bmatrix} 1 & -\nu & -\nu & & & \\ -\nu & 1 & -\nu & & 0 & \\ -\nu & -\nu & 1 & & & \\ & & & 1+\nu & & \\ & 0 & & & 1+\nu & \\ & & & & & 1+\nu \end{bmatrix}. \qquad (3.7)$$

Clearly, (3.1), (3.2), and (3.3) lead to a mixed variational formulation. We can now eliminate one or two variables. Thus, there are three distinct formulations in the engineering literature; see Stein and Wunderlich [1973]. Before treating them in detail, we give a short overview.

(1) The displacement formulation.
We eliminate σ with the help of (3.6), and then ε using (3.2):

$$\Pi(v) = \int_\Omega [\frac{1}{2} \nabla^{(s)} v : C \, \nabla^{(s)} v - f \cdot v] dx + \int_{\Gamma_1} g \cdot v \, dx$$

$$= \int_\Omega [\mu \, \varepsilon(v) : \varepsilon(v) + \frac{\lambda}{2} (\text{div } v)^2 - f \cdot v] dx + \int_{\Gamma_1} g \cdot v \, dx \longrightarrow \text{min!}$$

$$(3.8)$$

---

[15] In engineering references the nondiagonal components of ε are usually normalized so that they differ from (3.2) by a factor of 2. In this case some of our constants will be changed by a factor 2.

Here $\partial\Omega$ is divided into Γ_0 and Γ_1 depending on the boundary conditions as in (2.2). Assuming for simplicity that zero boundary conditions are specified on Γ_0, we have to find the minimum over

$$H_\Gamma^1 := \{v \in H^1(\Omega)^3;\ v(x) = 0 \text{ for } x \in \Gamma_0\}.$$

The associated weak formulation is the following: Find $u \in H_\Gamma^1$ with

$$\int_\Omega \nabla^{(s)}u : C\,\nabla^{(s)}v\,dx = (f, v)_0 - \int_{\Gamma_1} g \cdot v\,dx \quad \text{for all } v \in H_\Gamma^1.$$

In terms of the L_2-scalar product for matrix-valued functions, we can write these equations in the short form

$$(\nabla^{(s)}u, C\,\nabla^{(s)}v)_0 = (f, v)_0 - \int_{\Gamma_1} g \cdot v\,dx \quad \text{for all } v \in H_\Gamma^1(\Omega), \qquad (3.9)$$

and in particular, for St. Venant–Kirchhoff materials as

$$2\mu(\nabla^{(s)}u, \nabla^{(s)}v)_0 + \lambda(\text{div}\,u, \text{div}\,v)_0 = (f, v)_0 - \int_{\Gamma_1} g \cdot v\,dx. \qquad (3.10)$$

The associated classical elliptic differential equation is the Lamé differential equation

$$\begin{aligned}
-2\mu \,\text{div}\,\varepsilon(u) - \lambda \,\text{grad div}\,u &= f \quad \text{in } \Omega, \\
u &= 0 \quad \text{on } \Gamma_0, \\
\sigma(u) \cdot n &= g \quad \text{on } \Gamma_1.
\end{aligned} \qquad (3.11)$$

(2) The mixed method of Hellinger and Reissner

In this method, also called the Hellinger–Reissner principle, the displacement and stresses remain as unknowns, while the strains are eliminated:

$$\begin{aligned}
(C^{-1}\sigma - \nabla^{(s)}u, \tau)_0 &= 0 &&\text{for all } \tau \in L_2(\Omega), \\
-(\sigma, \nabla^{(s)}v)_0 &= -(f, v)_0 + \int_{\Gamma_1} g \cdot v\,dx &&\text{for all } v \in H_\Gamma^1(\Omega).
\end{aligned} \qquad (3.12)$$

The equivalence of (3.9) and (3.12) can be seen as follows: Let u be a solution of (3.9). Since $u \in H^1$,

$$\sigma := C\,\nabla^{(s)}u \in L_2. \qquad (3.13)$$

Because of the symmetry of C, (3.9) implies the second equation of (3.12). The first equation of (3.12) is just the weak formulation of (3.13). As soon as we establish that the two variational problems are uniquely solvable, we have the equivalence.

We can write (3.12) as a classical differential equation in the form

$$\begin{aligned} \operatorname{div}\sigma &= -f && \text{in } \Omega, \\ \sigma &= C\nabla^{(s)}u && \text{in } \Omega, \\ u &= 0 && \text{on } \Gamma_0, \\ \sigma \cdot n &= g && \text{on } \Gamma_1. \end{aligned}$$

In particular, in view of the second equation, σ is a symmetric tensor.[16]

The equations fit (at least formally) in the general framework of Ch. III in the following canonical form:[17]

$$X = L_2(\Omega), \quad M = H_\Gamma^1(\Omega),$$
$$a(\sigma, \tau) = (C^{-1}\sigma, \tau)_0, \quad b(\tau, v) = -(\tau, \nabla^{(s)}v)_0.$$

As for the mixed formulation of the Poisson equation (see Ch. III, §5), there is an alternative: Fix

$$X := H(\operatorname{div}, \Omega), \quad M = L_2(\Omega),$$
$$a(\sigma, \tau) = (C^{-1}\sigma, \tau)_0, \quad b(\tau, v) = (\operatorname{div}\sigma, v)_0, \tag{3.14}$$

where $H(\operatorname{div}, \Omega)$ is once again the closure of $C^\infty(\Omega, \mathbb{S}^3)$ w.r.t. the norm (III.5.4),

$$\|\tau\|_{H(\operatorname{div}, \Omega)} := (\|\tau\|_0^2 + \|\operatorname{div}\tau\|_0^2)^{1/2}.$$

Integrating by parts, we get $b(\tau, v) = (\operatorname{div}\tau, v)_0$. Which formulation makes the most sense depends among other things on the boundary conditions (see below). The connection with the Cauchy equilibrium equations (1.11) is clear from the second version.

---

[16] If we give up the linearization in the kinematics, we get the nonlinear system

$$\begin{aligned} \partial_j(\sigma_{ij} + \sigma_{kj}\partial_k u_i) &= -f_i && \text{in } \Omega, \\ \sigma &= CE(u) && \text{in } \Omega, \\ u &= 0 && \text{on } \Gamma_0, \\ (\sigma_{ij} + \sigma_{kj}\partial_k u_i)n_j &= g_i && \text{on } \Gamma_1. \end{aligned}$$

Here the sums are to be taken over the double indices using the so-called Einstein convention.

[17] For simplicity, we do not write the more precise formulation of $\varepsilon, \sigma \in L_2(\Omega, \mathbb{S}^3)$.

(3) The mixed method of Hu and Washizu (Hu–Washizu principle).
Here all three variables remain in the equations:

$$
\begin{aligned}
(C\varepsilon - \sigma, \eta)_0 &= 0 &&\text{for all } \eta \in L_2(\Omega), \\
(\varepsilon - \nabla^{(s)}u, \tau)_0 &= 0 &&\text{for all } \tau \in L_2(\Omega), \\
-(\sigma, \nabla^{(s)}v)_0 &= -(f, v)_0 + \int_{\Gamma_1} g \cdot v\, dx &&\text{for all } v \in H_\Gamma^1(\Omega).
\end{aligned}
\tag{3.15}
$$

In comparison with (3.12), we have now added the relation $\varepsilon := C^{-1}\sigma \in L_2(\Omega)$, so that (3.12) and (3.15) are equivalent. To fit this in the general framework, we set

$$
\begin{aligned}
X &:= L_2(\Omega) \times L_2(\Omega), \quad M := H_\Gamma^1(\Omega), \\
a(\varepsilon, \sigma, \eta, \tau) &= (C\varepsilon, \eta)_0, \quad b(\eta, \tau, v) = (\tau, \nabla^{(s)}v - \varepsilon)_0.
\end{aligned}
$$

We consider all three approaches in more detail below.

The simplest of the three is the displacement formulation. Establishing the validity of the Babuška–Brezzi condition for the mixed methods is considerably more difficult than for the Stokes problem; see below. On the other hand, for applications we are mostly interested in computing the stresses with more accuracy than the displacements. Thus, we look for approaches where the stresses are computed directly rather than via subsequent evaluation of derivatives. Hence, we turn to mixed methods despite their complexity, and in fact prefer the Hellinger–Reissner rather than the Hu–Washizu principle.

The Hellinger–Reissner principle arose from leaving all components of the stress tensor in the equations. Versions where only some special terms with strains and stresses remain in the equations are also important in practice. This typically leads to mixed methods with penalty terms. As an example, we later discuss a method for nearly incompressible material.

The Displacement Formulation

It follows from (3.8) that the energy for the displacement method is H^1-elliptic, provided that the quadratic form $\int \varepsilon(v) : \varepsilon(v)\, dx$ has this property. This is the content of a famous inequality. Here we do not restrict the dimension d to be 3.

3.1 Korn's Inequality *(Korn's first inequality). Let Ω be an open bounded set in $\mathbb{R}^d$ with piecewise smooth boundary. Then there exists a number $c = c(\Omega) > 0$ such that*

$$
\int_\Omega \varepsilon(v) : \varepsilon(v)\, dx + \|v\|_0^2 \geq c\|v\|_1^2 \quad \text{for all } v \in H^1(\Omega)^d.
\tag{3.16}
$$

For a proof, see Duvaut and Lions [1976] and Nitsche [1981]. Its structure is similar to the proof that the divergence satisfies an inf-sup condition as a mapping of $H^1(\Omega)^d$ into $L_2(\Omega)$; cf. Ch. III, §6. A special case in which the inequality can be easily verified is dealt with in Remark 3.5 below.

3.2 Remark. If the strain tensor E of a deformation is trivial, then by Remark 1.1 the deformation is an affine distance-preserving transformation. An analogous assertion holds for the linearized strain tensor ε: *Let $\Omega \subset \mathbb{R}^3$ be open and connected. Then for $v \in H^1(\Omega)$,*

$$\varepsilon(v) = 0,$$

if and only if

$$v(x) = a \wedge x + b \quad \text{with } a, b \in \mathbb{R}^3. \tag{3.17}$$

For the proof, we note that

$$\frac{\partial^2}{\partial x_i \partial x_j} v_k = \frac{\partial}{\partial x_i} \varepsilon_{jk} + \frac{\partial}{\partial x_j} \varepsilon_{ik} - \frac{\partial}{\partial x_k} \varepsilon_{ij} = 0$$

in $H^{-1}(\Omega)$ if $\varepsilon(v) = 0$. From this we conclude that every component v_k must be a linear function. But then a simple computation shows that a displacement of the form $v(x) = Ax + b$ can only be compatible with $\varepsilon(v) = 0$ if A is skew-symmetric. This leads to (3.17). – On the other hand, it is easy to verify that the linear strains for the displacements of the form (3.17) vanish. $\square$

Korn's inequality is simplified for functions which satisfy a zero boundary condition. In the sense of Remark II.1.6, it is only necessary that v vanishes on a part Γ_0 of the boundary, and that Γ_0 possesses a positive $(n - 1)$-dimensional measure.

3.3 Korn's Inequality *(Korn's 2nd inequality). Let $\Omega \subset \mathbb{R}^3$ be an open bounded set with piecewise smooth boundary. In addition, suppose $\Gamma_0 \subset \partial\Omega$ has positive two-dimensional measure. Then there exists a positive number $c' = c'(\Omega, \Gamma_0)$ such that*

$$\int_\Omega \varepsilon(v) : \varepsilon(v) dx \geq c' \|v\|_1^2 \quad \text{for all } v \in H_\Gamma^1(\Omega). \tag{3.18}$$

Here $H_\Gamma^1(\Omega)$ is the closure of $\{v \in C^\infty(\Omega)^3; \; v(x) = 0 \text{ for } x \in \Gamma_0\}$ w.r.t. the $\| \cdot \|_1$-norm.

Proof. Suppose that the inequality is false. Then there exists a sequence $(v_n) \in H_\Gamma^1(\Omega)$ with

$$\|\varepsilon(v_n)\|_0^2 := \int \varepsilon(v_n) : \varepsilon(v_n) dx \leq \frac{1}{n} \quad \text{and} \quad |v_n|_1 = 1.$$

Because of the hypothesis on Γ_0, Friedrichs' inequality implies $\|v_n\|_1 \leq c_1$ for all n and some suitable $c_1 > 0$. Since $H^1(\Omega)$ is compact in $H^0(\Omega)$, there is a subsequence of (v_n) which converges w.r.t. the $\|\cdot\|_0$-norm. With the constant c from Theorem 3.1, we have $c\|v_n - v_m\|_1^2 \leq \|\varepsilon(v_n - v_m)\|_0^2 + \|v_n - v_m\|_0^2 \leq 2\|\varepsilon(v_n)\|_0^2 + 2\|\varepsilon(v_m)\|^2 + \|v_n - v_m\|_0^2 \leq \frac{2}{n} + \frac{2}{m} + \|v_n - v_m\|_0^2$.

The L_2-convergent subsequence is thus a Cauchy sequence in $H^1(\Omega)$, and so converges in the sense of H^1 to some u_0. Hence, $\|\varepsilon(u_0)\| = \lim_{n\to\infty} \|\varepsilon(v_n)\| = 0$, and $|u_0|_1 = \lim_{n\to\infty} |v_n|_1 = 1$. By Remark 3.2, we deduce from $\varepsilon(u_0) = 0$ that u_0 has the form (3.17). In view of the zero boundary condition on Γ_0, it follows that $u_0 = 0$. This is a contradiction to $|u_0|_1 = 1$. $\square$

Korn's inequality asserts that the variational problem (3.8) is elliptic. Thus, the general theory immediately leads to

3.4 Existence Theorem. *Let $\Omega \subset \mathbb{R}^3$ be a domain with piecewise smooth boundary, and suppose Γ_0 has positive two-dimensional measure. Then the variational problem (3.8) of linear elasticity theory has exactly one solution.*

3.5 Remark. In the special case where Dirichlet boundary conditions are prescribed (i.e., $\Gamma_0 = \Gamma$ and $H^1_\Gamma = H^1_0$), the proof of Korn's first inequality is simpler. In this case

$$|v|_{1,\Omega} \leq \sqrt{2}\, \|\varepsilon(v)\|_{0,\Omega} \quad \text{for all } v \in H^1_0(\Omega)^3. \tag{3.19}$$

It suffices to show the formula for smooth vector fields. In this case we have

$$2\nabla^{(s)}v : \nabla^{(s)}v - \nabla v : \nabla v = \mathrm{div}[(v\nabla)v - (\mathrm{div}\,v)v] + (\mathrm{div}\,v)^2. \tag{3.20}$$

Here $(v\nabla)$ is to be interpreted as $\sum_i v_i \frac{\partial}{\partial x_i}$. The formula (3.20) can be verified, for example, by solving for all terms in the double sum. Since $v = 0$ on $\partial\Omega$, it follows from the Gauss integral theorem that

$$\int_\Omega \mathrm{div}[(v\nabla)v - (\mathrm{div}\,v)v]dx = \int_{\partial\Omega} [(v\nabla)v - (\mathrm{div}\,v)v]n\,ds = 0.$$

Integrating (3.20) over Ω, we have

$$2\|\nabla^{(s)}v\|_0^2 - |v|_1^2 = \int_\Omega (\mathrm{div}\,v)^2 dx \geq 0,$$

and (3.19) is proved. $\square$

Note that the constant in (3.19) is independent of the domain. If we are given Neumann boundary conditions on a part of the boundary, the constant can easily depend on Ω. We will see the consequences in connection with the locking effect for the cantilever beam shown in Fig. 58. — On the other hand, for the pure traction problem, i.e., for $\Gamma_0 = \emptyset$, there is again a compatibility condition; see Problem 3.17.

The Mixed Method of Hellinger and Reissner

The mixed method of Hellinger and Reissner (see Reissner [1950]) has many similarities to the mixed formulation of the Poisson equation in Ch. III, §5. The variational formulation according to (3.12) is

$$(\mathcal{C}^{-1}\sigma, \tau)_0 - (\tau, \nabla^{(s)}u)_0 = 0 \qquad \text{for all } \tau \in L_2(\Omega),$$

$$-(\sigma, \nabla^{(s)}v)_0 \qquad = -(f, v)_0 + \int_{\Gamma_1} g \cdot v \, dx \quad \text{for all } v \in H^1_\Gamma(\Omega),$$

$$(3.21)$$

which corresponds to the standard displacement formulation. Since $v < \frac{1}{2}$, $\mathcal{C}$ is positive definite and the bilinear form $(\mathcal{C}^{-1}\sigma, \tau)_0$ is L_2-elliptic. The following lemma shows that the inf-sup condition follows from Korn's inequality.

3.6 Lemma. *Suppose the hypotheses for Korn's second inequality are satisfied. Then for all $v \in H^1_\Gamma(\Omega)$,*

$$\sup_{\tau \in L_2(\Omega, \mathbb{S}^3)} \frac{(\tau, \nabla^{(s)}v)_0}{\|\tau\|_0} \geq c'\|v\|_1,$$

where c' is the constant in (3.18).

Proof. Given $v \in H^1_\Gamma(\Omega)$, $\tau := \nabla^{(s)}v$ is a symmetric L_2-tensor. Moreover, by (3.18), $\|\tau\|_0 = \|\nabla^{(s)}v\|_0 \geq c'\|v\|_1$. It suffices to consider the case $v \neq 0$:

$$\frac{(\tau, \nabla^{(s)}v)_0}{\|\tau\|_0} = \frac{\|\nabla^{(s)}v\|_0^2}{\|\nabla^{(s)}v\|_0} \geq c'\|v\|_1,$$

which establishes the inf-sup condition. □

The formulation with the spaces as in (3.21) is almost equivalent to the displacement formulation. Specifically, it can be understood as a displacement formulation combined with a softening of the energy. It is suitable for the method of *enhaced assumed strains* by Simo and Rifai [1990]; see Ch. III, §5. As in the discretization of the Poisson equation using the Raviart–Thomas element, generally the pairing (3.14) is more appropriate. We seek $\sigma \in H(\text{div}, \Omega)$ and $u \in L_2(\Omega)^3$ with

$$(\mathcal{C}^{-1}\sigma, \tau)_0 + (\text{div}\,\tau, u)_0 = 0 \qquad \text{for all } \tau \in H(\text{div}, \Omega), \ \tau n = 0 \text{ on } \Gamma_1,$$

$$(\text{div}\,\sigma, v)_0 \qquad = -(f, v)_0 \text{ for all } v \in L_2(\Omega)^3,$$

$$\sigma n = g \qquad \text{on } \Gamma_1.$$

$$(3.22)$$

We assume that an inhomogeneous boundary condition has been reduced to a homogeneous one in the sense of Ch. II, §2. The equations (3.22) are the Euler equations for the saddle point problem

$$(\mathcal{C}^{-1}\sigma, \sigma)_0 \longrightarrow \min_{\sigma \in H(\text{div}, \Omega)} !$$

with the restriction

$$\text{div } \sigma = f$$

and the boundary condition $\sigma n = g$ on Γ_1. This is often called the *dual mixed method*.

Just as in $H_0^1(\Omega)$ where boundary values for the function are prescribed, in the (less regular) space $H(\text{div}, \Omega)$ we can specify the normal components on the boundary. This becomes clear from the jump conditions in Problem II.5.14. Here we assume that the boundary is piecewise smooth.

Although in (3.22) formally we required only that $u \in L_2(\Omega)^3$, in fact the solution satisfies $u \in H_\Gamma^1(\Omega)$. It follows from (3.22) that $\varepsilon(u) = C^{-1}\sigma \in L_2(\Omega)$. Indeed, suppose $i, j \in \{1, 2, 3\}$ and that only $\tau_{ij} = \tau_{ji}$ are nonzero. In addition, let $\tau_{ij} \in C_0^\infty(\Omega)$. Then writing w instead of τ_{ij}, it follows from (3.22) that

$$\frac{1}{2}\int_\Omega (u_i \frac{\partial w}{\partial x_j} + u_j \frac{\partial w}{\partial x_i})dx = -\int_\Omega (C^{-1}\sigma)_{ij}w\,dx.$$

Recalling Definition II.1.1, we see that the symmetric gradient $(\nabla^{(s)}u)_{ij}$ exists in the weak sense, and coincides with $(C^{-1}\sigma)_{ij} \in L_2(\Omega)$. Now Korn's first inequality implies $u \in H^1(\Omega)^3$. Finally, we apply Green's formula. Because of the symmetry, it follows that for all test functions τ as in (3.22),

$$\int_{\partial\Omega} u \cdot \tau n\, ds = \int_\Omega \nabla u : \tau\, dx + \int_\Omega u \cdot \text{div } \tau\, dx$$

$$= \int_\Omega \nabla^{(s)}u : \tau\, dx + \int_\Omega u \cdot \text{div } \tau\, dx$$

$$= \int_\Omega C^{-1}\sigma : \tau\, dx + \int_\Omega u \cdot \text{div } \tau\, dx = 0.$$

Since this holds for all test functions, it follows that $u = 0$ on $\Gamma_0 = \partial\Omega\backslash\Gamma_1$. $\qquad\square$

The inf-sup condition and the V-ellipticity follow exactly as in Ch. III, §5. However, we emphasize that they are by no means trivial for the finite element spaces, and it is not easy to find stable pairings of finite element spaces. We explore the consequences for the much simpler two-dimensional case in §4.

We remark that we obtain different natural boundary conditions for the two formulations: with (3.21) they are $\sigma n = g$ on Γ_1, while with (3.22) we have $u = 0$ on Γ_0.

3.7 Remark. When $\Gamma_0 = \Gamma$, $\Gamma_1 = \emptyset$, i.e., for pure displacement boundary conditions, we need an extra argument for the Hellinger–Reissner principle. In this case the stresses lie in the subspace

$$\hat{H}(\text{div}, \Omega) := \{\tau \in H(\text{div}, \Omega); \int_\Omega \text{trace } \tau\, dx = 0\}. \tag{3.23}$$

Indeed, combining (1.30), (3.3), the Gauss integral theorem, and the fact that $u = 0$ on the boundary, we have

$$\int_\Omega \text{trace } \sigma \, dx = \frac{E}{1 - 2\nu} \int_\Omega \text{trace } \varepsilon \, dx = \frac{E}{1 - 2\nu} \int_\Omega \text{div } u \, dx$$

$$= \frac{E}{1 - 2\nu} \int_{\partial\Omega} u \cdot n \, ds = 0.$$

The Mixed Method of Hu and Washizu

In the Hu–Washizu principle, the stresses take the role of the Lagrange multipliers [Washizu 1968]. Let

$$X := L_2(\Omega) \times H^1_\Gamma(\Omega), \quad M := L_2(\Omega),$$

$$a(\varepsilon, u; \eta, v) = (\varepsilon, C\eta)_0, \quad b(\varepsilon, u; \tau) = -(\varepsilon, \tau)_0 + (\nabla^{(s)}u, \tau)_0. \tag{3.24}$$

We seek $(\varepsilon, u) \in X$ and $\sigma \in M$ with

$$
\begin{aligned}
(\varepsilon, C\eta)_0 \quad\quad - (\eta, \sigma)_0 &= 0 && \text{for all } \eta \in L_2(\Omega), \\
(\nabla^{(s)}v, \sigma)_0 &= (f, v)_0 - \int_{\Gamma_1} g \cdot v \, dx && \text{for all } v \in H^1_\Gamma(\Omega), \\
-(\varepsilon, \tau)_0 + (\nabla^{(s)}u, \tau)_0 \quad\quad &= 0 && \text{for all } \tau \in L_2(\Omega).
\end{aligned}
$$

By the definiteness of C, there exists $\beta > 0$ such that

$$
\begin{aligned}
a(\eta, v; \eta, v) = (\eta, C\eta)_0 &= \frac{1}{2}(\eta, C\eta)_0 + \frac{1}{2}(\nabla^{(s)}v, C\nabla^{(s)}v)_0 \\
&\geq \frac{\beta}{2}(\|\eta\|_0^2 + \|\nabla^{(s)}v\|_0^2) \\
&\geq \beta'(\|\eta\|_0^2 + \|v\|_1^2)
\end{aligned}
$$

on the subspace

$$V = \{(\eta, v) \in X; \ -(\eta, \tau)_0 + (\nabla^{(s)}v, \tau)_0 = 0 \text{ for } \tau \in M\},$$

and thus a is V-elliptic. Here we have again used Korn's inequality.

The inf-sup condition is very easy to verify. We need only evaluate b with $\eta = \tau$ and $v = 0$.

As a second possibility, using the same bilinear form a, we can work with the pairing

$$X := L_2(\Omega) \times L_2(\Omega)^3, \quad M := \{\tau \in H(\text{div}, \Omega); \ \tau n = 0 \text{ on } \Gamma_1\},$$

$$b(\varepsilon, u; \tau) = -(\varepsilon, \tau)_0 - (u, \text{div } \tau)_0. \tag{3.25}$$

The argument is the same as in the second formulation of the Hellinger–Reissner principle.

In regard to the finite element approximation, we should mention one difference as compared with the Stokes problem. The bilinear form a is elliptic on the entire space X only for the first version of the Hellinger–Reissner principle, while in the other cases it is only V-elliptic. The ellipticity on V_h can only be obtained if the space X_h is not too large in comparison with M_h; see Problem III.4.18. On the other hand, since the inf-sup condition requires that X_h be sufficiently large, the finite element spaces X_h and M_h have to fit together.

There is one more reason why it is not easy to provide stable, genuine elements for the Hu-Washizu principle. Here elements are said to be genuine if they are not equivalent to some elements for the Hellinger-Reissner theory or for the displacement formulation.

3.8 First Limit Principle of Stolarski and Belytschko [1966]. *Assume that $u_h \in V_h, \varepsilon_h \in E_h$, and $\sigma_h \in S_h$ constitute the finite element solution of a problem by the Hu-Washizu method. If the finite element spaces satisfy the relation*

$$S_h \subset C E_h, \tag{3.26}$$

then (σ_h, u_h) is the finite element solution of the Hellinger-Reissner formulation with the (same) spaces S_h and V_h.

Proof. The arguments in the proof are purely algebraic and apply to the pairings (3.24) and (3.14), or (3.25) and (3.15), respectively. In order to be specific we restrict ourselves to the first case and assume that

$$
\begin{aligned}
(\varepsilon_h, C\eta)_0 \qquad\qquad - (\eta, \sigma_h)_0 &= 0 &&\text{for all } \eta \in E_h, \\
(\nabla^{(s)}v, \sigma_h)_0 &= (f, v)_0 - \int_{\Gamma_1} g \cdot v \, dx &&\text{for all } v \in V_h, \\
- (\varepsilon_h, \tau)_0 + (\nabla^{(s)}u_h, \tau)_0 &= 0 &&\text{for all } \tau \in S_h.
\end{aligned}
\tag{3.27}
$$

From the first equation and the symmetry of the bilinear forms we conclude that

$$(\varepsilon_h - C^{-1}\sigma_h, C\eta)_0 = 0 \quad \text{for all } \eta \in E_h.$$

By the assumption (3.26), we may set $\eta := \varepsilon_h - C^{-1}\sigma_h$ and obtain

$$(\varepsilon_h - C^{-1}\sigma_h, C(\varepsilon_h - C^{-1}\sigma_h))_0 = 0.$$

Since C is positive definite, it follows that $\varepsilon_h = C^{-1}\sigma_h$. When inserting this into the other equations of (3.27), we see that σ_h and u_h are finite element solutions of (3.21). $\qquad\qquad\qquad\qquad\qquad\qquad\qquad\qquad\qquad\qquad \square$

Since the same problems occur for the membrane, we delay a discussion of suitable finite elements until later.

Nearly Incompressible Material

The mixed methods discussed in this section thus far refer to standard saddle point formulations. There are situations in which saddle point problems with penalty terms are the appropriate tool. We start with a typical example that can serve as a model problem.

Some materials such as rubber are nearly incompressible. It requires a great deal of energy to produce a small change in density. This results in a large difference in the magnitude of the Lamé constants:

$$\lambda \gg \mu.$$

The bilinear form in the displacement formulation (3.10),

$$a(u, v) := \lambda(\operatorname{div} u, \operatorname{div} v)_0 + 2\mu(\varepsilon(u), \varepsilon(v))_0,$$

is indeed H^1-elliptic, since in principle,

$$\alpha \|v\|_1^2 \leq a(v, v) \leq C\|v\|_1^2 \quad \text{for all } v \in H_\Gamma^1(\Omega), \tag{3.28}$$

where $\alpha \leq \mu$ and $C \geq \lambda + \mu$. Therefore, C/α is very large. Since by Céa's lemma the ratio C/α enters in the error estimate, we can expect errors which are significantly larger than the approximation error. This phenomenon is frequently observed in finite element computations, and is called *Poisson locking* or *volume locking*. This is a special case of a *locking effect*, and we now examine it in a preparation for a more general discussion of the effect.

One way to overcome locking is to use a formulation involving a *mixed problem with a penalty term*. We start with (3.10), and write the linear functional in more abstract form as

$$\lambda(\operatorname{div} u, \operatorname{div} v)_0 + 2\mu(\varepsilon(u), \varepsilon(v))_0 = \langle \ell, v \rangle \quad \text{for all } v \in H_\Gamma^1. \tag{3.29}$$

Substituting

$$\lambda \operatorname{div} u = p, \tag{3.30}$$

and using the weak version of (3.30), we are led to the following problem: *Find* $(u, p) \in H_\Gamma^1(\Omega) \times L_2(\Omega)$ *such that*

$$
\begin{aligned}
2\mu(\varepsilon(u), \varepsilon(v))_0 + (\operatorname{div} v, p)_0 &= \langle \ell, v \rangle & \text{for all } v \in H_\Gamma^1(\Omega), \\
(\operatorname{div} u, q)_0 - \frac{1}{\lambda}(p, q)_0 &= 0 & \text{for all } q \in L_2(\Omega).
\end{aligned}
\tag{3.31}
$$

Since the bilinear form $(\varepsilon(u), \varepsilon(v))_0$ is elliptic on H_Γ^1, (3.31) is very similar to the Stokes problem; see Ch. III, §6. As we observed there, in the case where $\Gamma_0 = \partial\Omega$

(more precisely if the two-dimensional measure of Γ_1 vanishes), $\int_\Omega p\,dx = 0$, and $L_2(\Omega)$ should be replaced by $L_2(\Omega)/\mathbb{R}$.

We know from the theory of mixed problems with penalty terms that the stability is the same as for the problem

$$2\mu(\varepsilon(u), \varepsilon(v))_0 + (\operatorname{div} v, p)_0 = \langle \ell, v \rangle,$$
$$(\operatorname{div} u, q)_0 \qquad\qquad\quad = 0,$$

as $\lambda \to \infty$. The situation here is simple, since the quadratic form $(\varepsilon(v), \varepsilon(v))_0$ is *coercive on the entire space* and not just for divergence-free functions. Moreover, the penalty term is a regular perturbation. Therefore, we can solve (3.31) using the same elements as for the Stokes problem. Since the inverse of the associated operator

$$L : H^1_\Gamma \times L_2 \to (H^1_\Gamma \times L_2)'$$

is bounded independently of the parameter λ, the finite element solution converges *uniformly in λ*.

To be more specific, consider the discretization

$$2\mu(\varepsilon(u_h), \varepsilon(v))_0 + (\operatorname{div} v, p_h)_0 = \langle \ell, v \rangle \quad \text{for all } v \in X_h,$$
$$(\operatorname{div} u_h, q)_0 - \lambda^{-1}(p, q_h)_0 = 0 \qquad \text{for all } q \in M_h. \tag{3.31$_h$}$$

with $X_h \subset H^1_\Gamma(\Omega)$, $M_h \subset L_2(\Omega)$. The commonly used Stokes elements have the following approximation property. Given $v \in H^1_\Gamma(\Omega)$ and $q \in L_2(\Omega)$, there exist $P_h v \in X_h$ and $Q_h q \in M_h$ such that

$$\|v - P_h v\|_1 \le ch\|v\|_2,$$
$$\|q - Q_h q\|_0 \le ch\|q\|_1. \tag{3.32}$$

For convenience, we restrict ourselves to pure displacement boundary conditions. Good approximation is guaranteed by the following regularity result. *If Ω is a convex polygonal domain, or if Ω has a smooth boundary, then*

$$\|u\|_2 + \lambda\|\operatorname{div} u\|_1 \le ch\|f\|_0; \tag{3.33}$$

see Theorem A.1 in Vogelius [1983]. Following the usual procedure (see e.g. Theorem III.4.5) we have

$$2\mu(\varepsilon(u_h - P_h u), \varepsilon(v))_0 + (\operatorname{div} v, p_h - Q_h p)_0 = \langle \ell_u, v \rangle \quad \text{for all } v \in X_h,$$
$$(\operatorname{div}(u_h - P_h u), q)_0 - \lambda^{-1}(p_h - Q_h p, q)_0 = \langle \ell_p, q \rangle \quad \text{for all } q \in M_h.$$

The functionals ℓ_u and ℓ_p can be expressed in terms of $u - P_h u$ and $p - Q_h p$. From the estimates (3.32) and (3.33) we immediately obtain $\|\ell_u\|_{-1} + \|\ell_p\|_0 \le ch\|f\|_0$. Hence,

$$\|u - u_h\|_1 + \|\lambda \operatorname{div} u - p_h\|_0 \le ch\|f\|_0, \tag{3.34}$$

where c is a constant independent of λ. The finite element method is robust.

Nonconforming methods are also very popular for treating nearly incompressible materials. The finite element discretization $(3.31)_h$ of the saddle point problem can also be interpreted as a nonconforming method, and this discretization will serve as a model for analyzing nonconforming methods for the problem with a small parameter.

3.9 Remark. *Let $u_h \in X_h \subset H^1_\Gamma(\Omega)$ and $p_h \in M_h \subset L_2(\Omega)$ be the solution of $(3.31)_h$. Define a discrete divergence operator by*

$$\begin{aligned} &\mathrm{div}_h : H^1(\Omega) \to M_h \\ &(\mathrm{div}_h v, q)_0 = (\mathrm{div}\, v, q)_0 \quad \textit{for all } q \in M_h. \end{aligned} \tag{3.35}$$

Then u_h is also the solution of the variational problem

$$2\mu \|\varepsilon(v)\|^2_0 + \lambda \|\mathrm{div}_h v\|^2_0 - \langle \ell, v \rangle \longrightarrow \min_{v \in X_h}! \tag{3.36}$$

Indeed, the solution u_h of (3.36) is characterized by

$$2\mu(\varepsilon(u_h), \varepsilon(v))_0 + \lambda(\mathrm{div}_h u_h, \mathrm{div}_h v)_0 = \langle \ell, v \rangle \qquad \text{for all } v \in X_h.$$

Setting $p_h := \lambda\, \mathrm{div}_h u_h$ by analogy to (3.30), we have

$$\begin{aligned} 2\mu(\varepsilon(u_h), \varepsilon(v))_0 + (\mathrm{div}_h v, p_h)_0 &= \langle \ell, v \rangle \quad \text{for all } v \in X_h, \\ (\mathrm{div}_h u_h, q)_0 - \lambda^{-1}(p_h, q)_0 &= 0 \qquad \text{for all } q \in M_h. \end{aligned}$$

Here the operator div_h is met only in inner products with the other factor in M_h. Now, from the definition (3.35) we know that the operator div_h may be replaced here by div. Therefore, u_h together with $p_h := \lambda\, \mathrm{div}_h u_h$ satisfy $(3.31)_h$. □

The inequality $(3.32)_2$ implies the estimate $\| \mathrm{div}\, v - \mathrm{div}_h v\|_0 \le ch\| \mathrm{div}\, v\|_1$, but the analogy does not extend to most of the nonconforming methods. Moreover, the discrete divergence does not result from an orthogonal projection in many cases. A basis for an alternative is offered by the following regularity result in Theorem 3.1 of Arnold, Scott, and Vogelius [1989]. *Given u with $\mathrm{div}\, u \in H^1(\Omega)$, there exists $w \in H^2(\Omega) \cap H^1_0(\Omega)$ such that*

$$\mathrm{div}\, w = \mathrm{div}\, u \quad \text{and} \quad \|w\|_2 \le c\| \mathrm{div}\, u\|_1. \tag{3.37}$$

3.10 Lemma. *Assume that (3.34) and (3.35) hold. Let the mapping $\mathrm{div}_h : X_k \to L_2(\Omega)$ satisfy*

$$\| \mathrm{div}\, v - \mathrm{div}_h v\|_0 \le ch\|v\|_2 \tag{3.38}$$

and

$$\mathrm{div}_h v = 0 \quad \textit{if } \mathrm{div}\, v = 0. \tag{3.39}$$

Then we have

$$\lambda \| \operatorname{div} u - \operatorname{div}_h u \|_0 \leq c'h \| f \|_0 \qquad (3.40)$$

where u denotes the solution of the variational problem (3.31).

Proof. By (3.34) and (3.35) there exists $w \in H^2(\Omega) \cap H_0^1(\Omega)$ such that div $w =$ div u and

$$\| w \|_2 \leq c \| \operatorname{div} u \|_1 \leq c\lambda^{-1} \| f \|_0.$$

From (3.38) we conclude that

$$\| \operatorname{div} w - \operatorname{div}_h w \|_0 \leq c'h \| w \|_2 \leq c'h\lambda^{-1} \| f \|_0. \qquad (3.41)$$

Since div$(w - u) = 0$, it follows from (3.37) that div$_h(w - u) = 0$ and div $u -$ div$_h u =$ div $w -$ div$_h w$. Combining this with (3.39) we obtain

$$\| \operatorname{div} u - \operatorname{div}_h u \|_0 = \| \operatorname{div} w - \operatorname{div}_h w \|_0 \leq c'h\lambda^{-1} \| f \|_0,$$

and the proof is complete. □

Now the discretization error for nearly incompressible material is estimated by the lemma of Berger, Scott, and Strang [1972]. The term for the approximation error is the crucial one. Let $P_h : H^2(\Omega) \cap H_0^1(\Omega) \to X_h$ be an interpolation operator. It is sufficient to make provision for

$$\| v - P_h v \|_1 + \| \operatorname{div} v - \operatorname{div}_h v \|_0 \leq ch \| v \|_2,$$
$$\| \operatorname{div}_h v_h \|_0 \leq c \| v_h \|_1 \quad \text{for all } v_h \in X_h.$$

These inequalities are clear for the model problem in (3.35). Moreover, let the mesh-dependent bilinear form a_h be defined by polarization of the quadratic form in (3.36). It follows from Lemma 3.10 that we have for $w_h \in X_h$

$$\begin{aligned} a_h(u - P_h u, w_h) &= \mu(\varepsilon(u - P_h u), \varepsilon(w_h))_0 + \lambda(\operatorname{div} u - \operatorname{div}_h u, \operatorname{div}_h w_h)_0, \\ &\leq ch \| u \|_2 \| w_h \|_1 + ch \| u \|_2 \| \operatorname{div}_h w_h \|_0, \\ &\leq ch \| u \|_2 \| w_h \|_1. \end{aligned}$$

Finally, the consistency error is given by the term $\lambda(\operatorname{div} u - \operatorname{div}_h u, \operatorname{div}_h w_h)_0$. An estimate of this expression is already included in the formula above, and we have a robust estimate with a constant c that does not depend on λ, namely

$$\| u - u_h \|_1 \leq ch \| f \|_0.$$

The following theory describes certain discretizations for which uniform convergence cannot be expected. The equation (3.29) is included as a special case if we set $X := H_\Gamma^1(\Omega)$, $a_0(u, v) := 2\mu(\varepsilon(u), \varepsilon(v))_0$, $Bv := \operatorname{div} v$, and $t^2 = 1/\lambda$.

Locking

The concept of *locking effect* is used frequently by engineers to describe the case where a finite element computation produces significantly smaller displacements than it should. In addition to volume locking, we also have *shear locking, membrane locking*, and *thickness locking* as well as others with no special names. The essential point is that because of a small parameter t, as in (3.28) the quotient C/α grows, and the convergence of the finite element solution to the true solution is *not uniform in t* as $h \to 0$. The papers of Arnold [1981], Babuška and Suri [1992] and Suri, Babuška, and Schwab [1995] have made fundamental contributions to the understanding of locking effects.

Let X be a Hilbert space, $a_0 : X \times X \to \mathbb{R}$ a continuous, symmetric, coercive bilinear form with $a_0(v, v) \geq \alpha_0 \|v\|^2$, and $B : X \to L_2(\Omega)$ a continuous linear mapping. Generally, B has a nontrivial kernel and dim ker $B = \infty$. In addition, let t be a parameter $0 < t \leq 1$. Given $\ell \in X'$, we seek a solution $u := u_t \in X$ of the equation

$$a_0(u_t, v) + \frac{1}{t^2}(Bu_t, Bv)_{0,\Omega} = \langle \ell, v \rangle \quad \text{for all } v \in X. \tag{3.42}$$

The existence and uniqueness are guaranteed by the coercivity of $a(u, v) := a_0(u, v) + t^{-2}(Bu, Bv)_{0,\Omega}$.

Suppose there exists $u_0 \in X$ with

$$Bu_0 = 0, \quad d := \langle \ell, u_0 \rangle > 0. \tag{3.43}$$

After multiplying u_0 by a suitable factor, we can assume that $a_0(u_0, u_0) \leq \langle \ell, u_0 \rangle$. In particular, the energy of the minimal solution satisfies

$$\Pi(u_t) \leq \Pi(u_0) = \frac{1}{2}[a_0(u_0, u_0) + \frac{1}{t^2}\|Bu_0\|^2_{0,\Omega}] - \langle \ell, u_0 \rangle \leq -\frac{1}{2}d$$

with a bound that is independent of t. Thus, $\langle \ell, u_t \rangle \geq -\Pi(u_t) \geq \frac{1}{2}d$, and so

$$\|u_t\| \geq \|\ell\|^{-1} \frac{1}{2}d \quad \text{for all } t > 0 \tag{3.44}$$

is bounded below, where $\|\ell\| := \|\ell\|_{X'}$.

We now consider the solution of the variational problem in the finite element space $X_h \subset X$. Looking at the results of Babuška and Suri [1992] or Braess [1998], we recognize that the locking effect occurs when $X_h \cap \ker B = \emptyset$ and[18]

$$\|Bv_h\|_{0,\Omega} \geq C(h)\|v_h\|_X \quad \text{for all } v_h \in X_h. \tag{3.45}$$

---

[18] To be precise, we have to exclude functions from X_h which are polynomials in Ω. This is why on the right-hand side of (3.45) we should replace $\|v_h\|_X$ by $\inf\{\|v_h - q\|_X; q \in \mathcal{P}_k\}$ with a suitable degree k.

The coercivity of a on X_h follows from (3.45) with the ellipticity constant $\alpha = \alpha_0 + t^{-2}C(h)^2$. By the stability result II.4.1,

$$\|u_h\| \leq \alpha^{-1}\|\ell\| \leq t^2 C(h)^{-2} \|\ell\|. \tag{3.46}$$

For a small parameter t this gives a solution which is too small in contrast to (3.44) – and this is what engineers recognize as locking. The convergence cannot be uniform in t as $h \to 0$. On the other hand a finite element method is called *robust* for a problem with a small parameter t provided that the convergence is uniform in t.

A poor approximation of the kernel as specified by (3.45) will be verified for the Timoshenko beam and linear elements in (3.55) below. Similarly, for bilinear elements on rectangular grids one has

$$\| \operatorname{div} v_h\|_0 \geq \inf_{\substack{p\in\Pi_1^2 \\ \operatorname{div} p=0}} \frac{h}{12 \operatorname{diam}(\Omega)}|v_h - p|_{1,\Omega} \quad \text{for all } v_h \in (Q_1)^2; \tag{3.47}$$

cf. Braess [1996]. This shows Poisson locking of bilinear elements and elucidates that special means as described above are indeed required.

3.11 Remark. From a mathematical standpoint, it would be better to say we have a *poorly conditioned* problem than to call it locking. The key point is the appearance of the large ratio C/α in the constant in Céa's lemma (for example, as in (3.28)). In this case the condition number $\|L\| \cdot \|L^{-1}\|$ of the corresponding isomorphism $L : X \to X'$ is large.

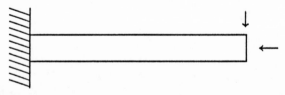

Fig. 58. The poor conditioning is clear for a cantilever beam. A vertical load leads to a significantly larger deformation than an equally large horizontal one. The constant in Korn's inequality is very small since Dirichlet boundary conditions are prescribed on only a small part of the boundary.

There is a qualitative difference between a poor approximation due to using too coarse a grid in the discretization and a poor approximation due to locking. It can be understood in terms of the associated eigenvalue problem; cf. Braess [1988]. As illustrated in Problem I.4.8, the lower eigenvalues are normally approximated well. This observation was also an important argument in the multigrid theory. On the other hand, in the discretization even the lower eigenvalues will clearly be shifted once we have a tendency toward locking.

3.12 Remark. There are several approaches to reducing the effects of locking:

1. Convert the variational problem (3.42) to a saddle point problem with a penalty term (as explained above for nearly incompressible materials). With $p = t^{-2} Bu$, we get

$$
\begin{aligned}
a_0(u, v) + (Bv, p)_{0,\Omega} &= \langle \ell, v \rangle && \text{for all } v \in X, \\
(Bu, q)_{0,\Omega} - t^2(p, q)_{0,\Omega} &= 0 && \text{for all } q \in L_2(\Omega).
\end{aligned}
\tag{3.48}
$$

2. Use *selective reduced integration*. Then in setting up the system matrix, the term

$$
t^{-2}(Bu_h, Bv_h)_{0,\Omega}
$$

will be relaxed as in (3.36) so that the constant in (3.45) is also reduced. This makes sense if the process can be understood as one where the approximating functions v_h are replaced in a neighborhood by others for which Bv is small. Strict mathematical proofs usually involve working with equivalent mixed formulations; see §6.

3. Simo and Rifai [1990] expand the space in the spirit of Remark II.5.7 by including nonconforming approximating functions so that sufficiently many functions are available in order to make $\|Bv\|_{0,h}$ small (if an appropriate discrete norm is chosen).

The first two methods listed in 3.12 are partially equivalent; cf. Remark 3.9. The finite element approximation of (3.48),

$$
\begin{aligned}
a_0(u_h, v) + (Bv, p_h)_{0,\Omega} &= \langle \ell, v \rangle && \text{for all } v \in X_h, \\
(Bu_h, q)_{0,\Omega} - t^2(p, q)_{0,\Omega} &= 0 && \text{for all } q \in M_h,
\end{aligned}
\tag{3.49}
$$

is equivalent to the minimization problem

$$
\frac{1}{2} a_0(u_h, u_h) + \frac{1}{2} t^{-2} \|R_h Bu_h\|_0 - \langle \ell, u_h \rangle \longrightarrow \min_{u_h \in X_h} !
\tag{3.50}
$$

Here $R_h : L_2(\Omega) \to M_h$ is the orthogonal L_2-projector. In practice other projectors are used along with so-called *selected reduced integration*.

Here we have concentrated on linear problems. For a discussion of nonlinear problems, see Stein and Wriggers [1997].

Locking of the Timoshenko Beam

Shear locking has been observed when computations for the Timoshenko beam are performed with finite elements which are piecewise polynomials of low degree. The analysis is easily done on the basis of (3.45) when P_1 elements are used.

We will see in §5 that the stored energy of a beam is given by

$$\Pi(\theta, w) := \frac{1}{2} \int_0^b (\theta')^2 dx + \frac{t^{-2}}{2} \int_0^b (w' - \theta)^2 dx, \qquad (3.51)$$

if b is the length of the beam and t is the thickness (multiplied by a correction factor). Here, the rotation θ and the deflection w are in $H_0^1(0, b)$ and in the above setting,

$$B(\theta, w) := w' - \theta. \qquad (3.52)$$

Given $g \in H_0^1(0, b)$, we obtain a pair of functions with $B(\theta, w) = 0$ by defining

$$\begin{aligned} \theta(x) &:= g(x) - \frac{6}{b^3} x(b - x) \int_0^b g(\xi) d\xi, \\ w(x) &:= \int_0^x \theta(\xi) d\xi. \end{aligned} \qquad (3.53)$$

Hence, the kernel of B is infinite dimensional.

Now assume that the interval $[0, b]$ is divided into subintervals of length h and that $\theta_h, w_h \in \mathcal{M}_{0,0}^1$. Note that

$$\int_\xi^{\xi+h} (\alpha x + \beta)^2 dx \geq \frac{h^3}{6} \alpha^3 = \frac{h^2}{6} \int_\xi^{\xi+h} \alpha^2 dx,$$

whenever $\alpha, \beta \in \mathbb{R}$. From this inequality it follows that on each subinterval of the partition

$$\int_\xi^{\xi+h} (w_h' - \theta_h)^2 dx \geq \frac{h^2}{6} \int_\xi^{\xi+h} (\theta_h')^2 dx.$$

After summing over all subintervals we have

$$\|w_h' - \theta_h\|_0 \geq \frac{h}{3} |\theta_h|_1. \qquad (3.54)$$

From Friedrichs' inequality and $|w_h|_1 \leq \|w_h' - \theta_h\|_0 + \|\theta_h\|_0$ we conclude that

$$\|w_h' - \theta_h\|_0 \geq ch(\|\theta_h\|_1 + \|w_h\|_1). \qquad (3.55)$$

This proves (3.45) with $C(h) = ch$, and it follows from the preceding investigations that the P_1 elements lock.

On the other hand, it is clear that the construction (3.53) of pairs (θ, w) with $B(\theta, w) = 0$ can be approximated by piecewise polynomials of degree 2 and then the locking is avoided.

Problems

3.13 Verify that in the case of pure Dirichlet boundary conditions, i.e., $H_\Gamma^1(\Omega)$ $= H_0^1(\Omega)$, the classical solution of (3.10) is in fact a solution of the differential equation (3.11).

3.14 Does $\operatorname{div} u = 0$ imply the relation $\operatorname{div} \sigma = 0$, or does the converse hold? Hint: The connection between $\operatorname{div} u$ and trace ε is useful.

3.15 Verify Green's formula

$$\int_\Omega \tau : \varepsilon(v)\, dx = -\int_\Omega v \cdot \operatorname{div} \tau\, dx + \int_{\partial\Omega} v \cdot \tau n\, ds$$

for symmetric tensors $\tau \in H(\operatorname{div}, \Omega)$ and $v \in H^1(\Omega)$. Why don't the boundary terms play a role in passing from (3.21) to (3.22)?

3.16 Verify by explicit computation that the expressions $(\operatorname{div} v)^2$ and $\varepsilon(v) : \varepsilon(v)$ are invariant under orthogonal transformations.

3.17 Consider the Hellinger–Reissner principle for a nearly incompressible material. Is the introduction of the variables $p = \lambda \operatorname{div} u$ simpler for the formulation (3.21) or for (3.22)?

3.18 Consider the pure traction problem, i.e., the problem with pure Neumann boundary conditions. Show that the displacement problem (3.9) has only a solution if the compatibility condition

$$-\int_\Omega f \cdot v\, dx + \int_{\partial\Omega} g \cdot v\, dx = 0 \quad \text{for all } v \in \mathrm{RM}$$

holds. Here RM denotes the space of rigid body motions, i.e. the set of all functions v of the form (3.17). — How many compatibility conditions do we encounter for $d = 2$ and $d = 3$, respectively?

304

§ 4. Membranes

In solving three-dimensional problems, it is often possible to work in two (or even one) dimensions since the length of the domain in one of more space directions is very small. In such cases it is useful to consider the problem for the lower-dimensional continuum, and then discretize afterwards. Some typical examples are bars, beams, membranes, plates, and shells. The simplest two-dimensional example is the membrane. However, this example already shows that the reduction in dimension cannot be accomplished by simply eliminating one coordinate. Moreover, the boundary conditions have some influence on the reduction process. There are two cases depending on the boundary condition.

Plane Stress States

Let $\omega \subset \mathbb{R}^2$ be a domain, and suppose $t > 0$ is a number which is significantly smaller than the diameter of ω. We suppose that there are only external forces operating on the body $\Omega = \omega \times (-\frac{t}{2}, +\frac{t}{2})$, and that their z-components vanish so that they depend only on x and y.[19] If the membrane is thin and a deformation in the z-direction is possible, we have the so-called *plane stress state*, i.e.,

$$\sigma_{ij}(x, y, z) = \sigma_{ij}(x, y), \quad i, j = 1, 2,$$
$$\sigma_{i3} = \sigma_{3i} = 0, \qquad i = 1, 2, 3. \tag{4.1}$$

Then in particular $\varepsilon_{i3} = \varepsilon_{3i} = 0$ for $i = 1, 2$. In order for $\sigma_{33} = 0$, by the constitutive equations (3.6) we have

$$\varepsilon_{33} = -\frac{\nu}{1-\nu}(\varepsilon_{11} + \varepsilon_{22}). \tag{4.2}$$

If we now eliminate the strain ε_{33}, we get the constitutive equations for the plane stress state:

$$\begin{bmatrix} \sigma_{11} \\ \sigma_{22} \\ \sigma_{12} \end{bmatrix} = \frac{E}{1-\nu^2} \begin{bmatrix} 1 & \nu & 0 \\ \nu & 1 & 0 \\ 0 & 0 & 1-\nu \end{bmatrix} \begin{bmatrix} \varepsilon_{11} \\ \varepsilon_{22} \\ \varepsilon_{12} \end{bmatrix}$$

or

$$\sigma = \frac{E}{1+\nu}[\varepsilon + \frac{\nu}{1-\nu}(\varepsilon_{11} + \varepsilon_{22})I]. \tag{4.3}$$

---

[19] It is easier to visualize the situation if we exchange the y- and z-coordinates. Suppose the middle surface of a thin wall lies in the (x, y)-plane, and that the displacement in the y-direction is very small. Now suppose the wall is subject to a load in the vertical direction so that the external forces operate in a direction parallel to the (middle surface of the) wall.

Finally, the kinematics are compatible with (4.1) and (4.2) provided

$$u_i(x, y, z) = u_i(x, y), \qquad i = 1, 2,$$
$$u_3(x, y, z) = z \cdot \varepsilon_{33}(x, y)$$

and terms of order $\mathcal{O}(z)$ are neglected in the strain term.

Plane Strain States

If boundary conditions are enforced at $z = \pm t/2$ which ensure that the z-component of the displacement vanishes, then we have the *plane strain state*:

$$\varepsilon_{ij}(x, y, z) = \varepsilon_{ij}(x, y), \quad i, j = 1, 2,$$
$$\varepsilon_{i3} = \varepsilon_{3i} = 0, \qquad i = 1, 2, 3. \tag{4.4}$$

The associated displacements satisfy $u_i(x, y, z) = u_i(x, y)$ for $i = 1, 2$, and $u_3 = 0$. It follows from $\varepsilon_{33} = 0$ along with (3.7) that

$$\sigma_{33} = \nu(\sigma_{11} + \sigma_{22}), \tag{4.5}$$

and σ_{33} can be eliminated. We obtain the constitutive equations for the plane strain state if we restrict (3.6) to the remaining components:

$$\begin{bmatrix} \sigma_{11} \\ \sigma_{22} \\ \sigma_{12} \end{bmatrix} = \frac{E}{(1+\nu)(1-2\nu)} \begin{bmatrix} 1-\nu & \nu & 0 \\ \nu & 1-\nu & 0 \\ 0 & 0 & 1-2\nu \end{bmatrix} \begin{bmatrix} \varepsilon_{11} \\ \varepsilon_{22} \\ \varepsilon_{12} \end{bmatrix}. \tag{4.6}$$

Membrane Elements

Both plane elasticity problems lead to a two-dimensional problem with the same structure as the full three-dimensional elasticity problem.

The displacement model thus involves the (two-dimensional and isoparametric versions of the) conforming elements which also play a role for scalar elliptic problems of second order:

(a) bilinear quadrilateral elements,
(b) quadratic triangular elements,
(c) biquadratic quadrilateral elements,
(d) eight-node quadrilateral elements in the serendipity class.

On the other hand, the simplest linear triangular elements are frequently unsatisfactory. For practical problems, there are often preferred directions because of certain geometric relationships. In this case, higher order elements or quadrilateral elements prove to be more flexible.

The PEERS Element

Mixed methods have not been heavily used for problems in plane elasticity, since in the finite element approximation of the Hellinger–Reissner principle (3.22), two stability problems occur simultaneously. The bilinear form $a(\sigma, \tau) := (C^{-1}\sigma, \tau)_0$ is not elliptic on the entire space $X := H(\text{div}, \Omega)$, but only on the kernel V. Here we are using the notation of the general theory in Ch. III, §4.

In order to ensure the ellipticity on V_h, currently the best possibility is to choose $V_h \subset V$ which assures that the condition in III.4.7 is satisfied. As Brezzi and Fortin [1991, p. 284] have shown via a dimensional argument, the *symmetry of the stress tensor* σ is a major difficulty. For this reason, Arnold, Brezzi and Douglas [1984] have developed the PEERS element (plane elasticity element with reduced symmetry). It has been studied further by Stenberg [1988], among others. The so-called BDM elements of Brezzi, Douglas, and Marini [1985] are constructed in a similar way. All of these elements are also useful for nearly incompressible materials.

There is no satisfactory mathematical theory for the other mixed methods. Often there exist nontrivial functions with zero energy (so-called *zero energy modes*), which must be filtered out, since otherwise we get the (hour-glass) instabilities discussed in Ch. III, §7, due to the violation of the inf-sup condition.

In discussing the PEERS elements, for simplicity we restrict ourselves to pure displacement boundary conditions. In this case (3.22) simplifies to

$$
\begin{aligned}
(C^{-1}\sigma, \tau)_0 + (\text{div}\,\tau, u)_0 &= 0 & &\text{for all } \tau \in H(\text{div}, \Omega), \\
(\text{div}\,\sigma, v)_0 &= -(f, v)_0 & &\text{for all } v \in L_2(\Omega)^2.
\end{aligned}
\tag{4.7}
$$

Since we allow unsymmetric tensors in the following, the *antisymmetric part*

$$
as(\tau) := \tau - \tau^T \in L_2(\Omega)^{2\times 2},
$$

i.e. $as(\tau)_{ij} = \tau_{ij} - \tau_{ji}$, plays a role. Since $as(\tau)$ is already completely determined by its (2,1)-component, we will refer to this component.

We consider the following saddle point problem:
Find $\sigma \in X := H(\text{div}, \Omega)^{2\times 2}$ and $(u, \gamma) \in M := L_2(\Omega)^2 \times L_2(\Omega)$ such that

$$
\begin{aligned}
(C^{-1}\sigma, \tau)_0 + (\text{div}\,\tau, u)_0 + (as(\tau), \gamma) &= 0, & &\tau \in H(\text{div}, \Omega)^{2\times 2}, \\
(\text{div}\,\sigma, v)_0 &= -(f, v)_0, & &v \in L_2(\Omega)^2, \\
(as(\sigma), \eta) &= 0, & &\eta \in L_2(\Omega).
\end{aligned}
\tag{4.8}
$$

Here $(as(\tau), \eta) := \int_\Omega (\tau_{12} - \tau_{21})\eta\, dx$.

Note that the rotations of scalar- and vector-valued functions in $\mathbb{R}^2$ are defined differently:[20]

$$\operatorname{curl} p := \begin{pmatrix} \frac{\partial p}{\partial x_2} \\ -\frac{\partial p}{\partial x_1} \end{pmatrix}, \qquad \operatorname{rot}\begin{pmatrix} u_1 \\ u_2 \end{pmatrix} := \frac{\partial u_2}{\partial x_1} - \frac{\partial u_1}{\partial x_2}. \tag{4.9}$$

4.1 Lemma. *The saddle point problems (4.7) and (4.8) are equivalent. If (σ, u, γ) is a solution of (4.8), then (σ, u) is a solution of (4.7). Conversely, if (σ, u) is a solution of (4.7), then $(\sigma, u, \gamma = \frac{1}{2}\operatorname{rot} u)$ is a solution of (4.8).*

Proof. (1) Let (σ, u, γ) be a solution of (4.8). The third equation asserts that σ is symmetric. For symmetric τ we have $(as(\tau), \gamma) = 0$, and the first equation in (4.8) reduces to the first one in (4.7). Since the second relation in (4.7) can be read off directly from (4.8), we have shown that (4.7) holds.

(2) Let (σ, u) be a solution of (4.7). By the discussion following (3.21), it follows that $u \in H_0^1(\Omega)^2$, and in the same way as in the derivation of (3.20), we deduce that

$$(C^{-1}\sigma, \tau)_0 - (\tau, \nabla^{(s)}u)_0 = 0 \tag{4.10}$$

for all *symmetric* fields τ. On the other hand, the symmetry of the expressions $C^{-1}\sigma$ and $\nabla^{(s)}u$ implies that (4.10) also holds for all skew-symmetric fields. Now the decomposition of ∇u gives

$$\begin{aligned}
\nabla^{(s)}u &= \nabla u - \frac{1}{2}as(\nabla u) \\
&= \nabla u - \frac{1}{2}\left(\frac{\partial u_2}{\partial x_1} - \frac{\partial u_1}{\partial x_2}\right)\begin{pmatrix} 0 & -1 \\ +1 & 0 \end{pmatrix} \\
&= \nabla u - \frac{1}{2}\operatorname{rot} u \begin{pmatrix} 0 & -1 \\ +1 & 0 \end{pmatrix}.
\end{aligned}$$

$\varepsilon(u)$ is the symmetric gradient. The skew-symmetric part is coupled with the rotation. Finally, another application of Green's formula gives

$$\begin{aligned}
\int_\Omega \tau : \nabla^{(s)}u\, dx &= \int_\Omega \tau : \nabla u\, dx + \frac{1}{2}\int_\Omega (\tau_{21} - \tau_{12})\operatorname{rot} u\, dx \\
&= -\int_\Omega \operatorname{div} \tau\, u\, dx + \frac{1}{2}\int_\Omega (\tau_{21} - \tau_{12})\operatorname{rot} u\, dx.
\end{aligned}$$

Combining this with (4.10) leads to the first relation in (4.8). Since the second equation can be taken from (4.7) and the third is obvious because of the symmetry of σ, this establishes (4.8). □

---

[20] We follow the convention of writing curl p rather than rot p if p is a scalar function. The reader should also be aware that the operators in (4.9) sometimes appear in the literature with different signs.

The equivalence does not immediately imply that the saddle point problem satisfies the hypotheses of Theorem III.4.5. Clearly, the kernel is

$$V = \{\tau \in H(\text{div}, \Omega)^{2\times 2}; \ (\text{div}\,\tau, v)_0 = 0 \text{ for } v \in L_2(\Omega)^2,$$
$$(as(\tau), \eta) = 0 \text{ for } \eta \in L_2(\Omega)\}$$

just as for the saddle point problem (4.7). Now the ellipticity of the quadratic form $(\mathcal{C}^{-1}\sigma, \sigma)_0$ can be carried over.

To establish the inf-sup condition, suppose we are given a pair $(v, \eta) \in L_2(\Omega)^2 \times L_2(\Omega)$. Then we construct $\tau \in H(\text{div}, \Omega)^{2\times 2}$ with

$$\text{div}\,\tau = v,$$
$$as(\tau) = \eta, \tag{4.11}$$
$$\|\tau\|_{H(\text{div},\Omega)} \le c(\|v\|_0 + \|\eta\|_0),$$

where we write $as(\tau)$ instead of $as(\tau)_{21}$. In view of Remark 3.7, we want

$$\int_\Omega \text{trace}\,\tau\,dx = 0. \tag{4.12}$$

First we determine $\tau_0 \in H(\text{div}, \Omega)^{2\times 2}$ which satisfies just the equation $\text{div}\,\tau_0 = v$. For example, let $\psi \in H_0^1(\Omega)$ be a solution of the Poisson equation $\Delta\psi = v$ and $\tau_0 := \nabla\psi$. Then $\|\psi\|_1 \le c\|v\|_0$ immediately implies $\|\tau_0\|_{H(\text{div},\Omega)} \le \|\tau_0\|_0 + \|\text{div}\,\tau_0\|_0 = \|\nabla\psi\|_0 + \|v\|_0 \le c\|v\|_0$.

In order to satisfy the second equation in (4.11), we set

$$s := \int_\Omega [\eta - as(\tau_0)]dx \ / \int_\Omega dx,$$
$$\beta := \eta - as(\tau_0) - s,$$

and construct $q \in H^1(\Omega)^2$ with

$$\text{div}\,q = \beta$$

via a Neumann problem in the same way as we constructed τ_0. In particular, $\|q\|_1 \le c\|\beta\|_0 \le c(\|\eta\|_0 + \|v\|_0)$. Finally, let

$$\tau := \tau_0 + \begin{pmatrix} \text{curl}\,q_1 \\ \text{curl}\,q_2 \end{pmatrix} + \frac{s}{2}\begin{pmatrix} 0 & -1 \\ 1 & 0 \end{pmatrix}.$$

To show that τ is a solution of (4.11), we first recall that the divergence of a rotation vanishes. Thus, $\text{div}\,\tau = \text{div}\,\tau_0 = v$. Moreover, by construction of q,

$$as(\tau) = as(\tau_0) + as\begin{pmatrix} \frac{\partial q_1}{\partial y} & -\frac{\partial q_1}{\partial x} \\ \frac{\partial q_2}{\partial y} & -\frac{\partial q_2}{\partial x} \end{pmatrix} + s$$
$$= as(\tau_0) + \text{div}\,q + s = \eta.$$

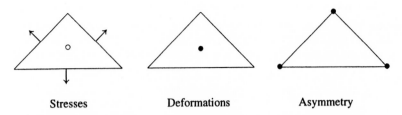

| Stresses | Deformations | Asymmetry |

Fig. 59. PEERS element (o stands for the rotation of the bubble function)

Since q is a gradient field, $\frac{\partial q_1}{\partial y} - \frac{\partial q_2}{\partial x} = 0$, the constraint (4.12) is easy to verify. Now (4.11) implies that

$$(\operatorname{div} \tau, v)_0 + (as(\tau), \eta) = \|v\|_0^2 + \|\eta\|_0^2$$
$$\geq c\|\tau\|_{H(\operatorname{div}, \Omega)} (\|v\|_0 + \|\eta\|_0),$$

and thus the inf-sup condition holds.   □

4.2 Remark. If the traction is prescribed on a part of the boundary, i.e., Γ_1 is nonempty, then the proof of the inf-sup condition is more complicated. More precisely, the solution of $\operatorname{div} \tau_0 = v$ and $\operatorname{div} q = \beta$ requires more care. We recall that the Neumann boundary condition is not a natural boundary condition for the mixed formulation (4.7) and we have

$$\sigma \in \{\tau \in H(\operatorname{div}, \Omega); \ \tau n = 0 \ \text{ on } \Gamma_1\}.$$

We cannot proceed as we have done after (4.12). We first construct τ with $\tau_0 \cdot n = 0$ on Γ_1 by prescribing a Neumann boundary condition there, i.e., $\nabla \psi \cdot n = 0$. Next observe that $\operatorname{rot} q \cdot n = 0$ implies

$$\nabla q \cdot t = 0; \quad \text{thus } q = const \ \text{ on } \Gamma_1.$$

We can find q by solving the Stokes problem $\int (\nabla q)^2 dx \longrightarrow$ min! under the constraints $\operatorname{div} q = \beta$ and $q = 0$ on Γ_1. The inf-sup condition for the elasticity problem (4.8) then follows from the inf-sup condition for the Stokes problem in view of the fact that the auxiliary problem is well defined.

The analysis of (4.8) was performed independently of Korn's inequality. On the other hand, the inf-sup condition for the divergence operator was used. Now it follows from the equivalence described in Lemma 4.1 that the inf-sup condition for the Stokes problem implies Korn's inequality in $\mathbb{R}^2$. — A different proof of this implication was presented by Falk [1991].

To define a simple member of the family of PEERS elements, we employ the usual notation. (See also Fig. 59.) Let $\mathcal{T}$ be a triangulation of Ω, and let

$$\mathcal{M}^k := \{v \in L_2(\Omega);\ v|_T \in \mathcal{P}_k \quad \text{for every } T \in \mathcal{T}\},$$
$$\mathcal{M}_0^k := \mathcal{M}^k \cap H^1(\Omega), \qquad \mathcal{M}_{0,0}^k := \mathcal{M}^k \cap H_0^1(\Omega),$$
$$RT_k := \{v \in (\mathcal{M}^{k+1})^2 \cap H(\text{div}, \Omega); \tag{4.13}$$
$$v|_T = \begin{pmatrix} p_1 \\ p_2 \end{pmatrix} + p_3 \begin{pmatrix} x \\ y \end{pmatrix},\ p_1, p_2, p_3 \in \mathcal{P}_k\},$$
$$B_3 := \{v \in \mathcal{M}_0^3;\ v(x) = 0 \text{ on every edge of the triangulation}\}.$$

The PEERS element is the one with the smallest number of local degrees of freedom:

$$\sigma_h \in X_h := (RT_0)^2 \oplus \text{curl}(B_3)^2,$$
$$v_h \in W_h := (\mathcal{M}^0)^2,$$
$$\gamma_h \in \Gamma_h := \mathcal{M}_0^1.$$

Note that the divergence of the functions in $\text{curl}(B_3)^2$ vanishes, and that the divergence of a piecewise differentiable function is an L_2 function if and only if the normal components are continuous on the inter-element boundaries.

By construction, $\text{div}\,\tau_h \in W_h$. Thus, $\text{div}\,\tau_h = 0$ follows immediately from $(\text{div}\,\tau_h, v_h) = 0$ for all $v_h \in W_h$. Thus, the condition in III.4.7 is satisfied, and the form $(C^{-1}\sigma_h, \tau_h)$ is elliptic on the kernel. The inf-sup condition can be established in a similar way as for the continuous problem. Since (4.11) must be replaced by finite element approximations, however, here the details are more involved, see Arnold, Brezzi, and Douglas [1984].

The implementation and postprocessing described by Arnold and Brezzi [1985], see Ch. III, §5 is also advantegeous for computations with the PEERS element. The postprocessing was also used to estimate the local error by adaptive grid refinement, see Braess, Klaas, Niekamp, Stein, and Wobschal [1995].

Problems

4.3 Show that we get the constitutive equations (4.3) for plane stress states by restricting the relationship $\varepsilon = C^{-1}\sigma$ to the components with $i = 1, 2$. – What comparable assertion holds for plane strain states?

4.4 For a nearly incompressible material, ν is close to $\frac{1}{2}$. This causes difficulties in the denominators $(1 - 2\nu)$ of (1.31) and (3.6), respectively. On the other hand, in view of (4.3), for the plane stress states, $\nu = \frac{1}{2}$ does not cause any problem. Give an explanation.

4.5   Let Ω be a domain in $\mathbb{R}^2$ with piecewise smooth boundary, and suppose $\psi \in H^1(\Omega)$ satisfies the Neumann boundary condition $\frac{\partial \psi}{\partial n} = 0$ on $\partial\Omega$. What can you say about rot ψ on the boundary?

Hint: First suppose $\psi \in C^1(\bar{\Omega})$, and consider the tangential component of the rotation on the boundary.

4.6   Does the relation

$$\Delta u = -\,\text{rot rot}\, u + \text{grad div}\, u$$

between the derivatives hold for vector fields in $\mathbb{R}^2$ and $\mathbb{R}^3$?

4.7   It would be mathematically cleaner if (3.6) were written in the form

$$\sigma_{ij} = \sum_{k\ell} C_{ijk\ell}\, \varepsilon_{k\ell}. \tag{4.14}$$

Give a formula for the elasticity tensor – or more precisely for its components $C_{ijk\ell}$ – which involves only the Lamé constants and the Kronecker symbol so that (4.14) describes the material law $\sigma = 2\mu\varepsilon + \lambda\,\text{trace}\,\varepsilon\,I$ for a St. Venant–Kirchhoff material. – Do this also for the plane strain state and the plane stress state.

4.8   Let $\Omega \subset \mathbb{R}^2$, and let $H(\text{rot}, \Omega)$ be the completion of $C^\infty(\Omega)^n$ w.r.t. the norm

$$\|v\|^2_{H(\text{rot})} := \|v\|^2_{0,\Omega} + \|\,\text{rot}\,v\|^2_{0,\Omega}.$$

Show that a set $S_h \subset L_2(\Omega)^2$ of piecewise polynomials lies in $H(\text{rot})$ if and only if the component $v \cdot \tau$ in the direction of the tangent is continuous on the edges of the element.

Hint: See Problem II.5.14 for $H(\text{div}, \Omega)$.

312

§ 5. Beams and Plates: The Kirchhoff Plate

A plate is a thin continuum subject to applied forces which – in constrast to the case a membrane – are orthogonal to the middle surface. We distinguish between two cases. The *Kirchhoff plate* leads to a fourth order elliptic problem. Usually it is solved using nonconforming or mixed methods.

The *Mindlin–Reissner plate* involves somewhat weaker hypotheses. It is described by a differential equation of second order, and so at first glance its numerical treatment appears to be simpler. However, it turns out that the calculations for the Mindlin plate are actually more difficult, and that the problems plaguing the Kirchhoff plate are still present, although concealed. In particular, the Mindlin plate tends to shear locking, and using standard elements leads to poor numerical results.

The analogous reduction of thin membranes, i.e., of membranes with one very small dimension, leads to beams.

After introducing both plate models, we turn our attention first to a discussion of the Kirchhoff plate, and in particular to the *clamped plate*.

The Hypotheses

We consider a thin plate of constant thickness t whose middle surface coincides with the (x, y)-plane. Thus, $\Omega = \omega \times (-\frac{t}{2}, +\frac{t}{2})$ with $\omega \subset \mathbb{R}^2$. We suppose that the plate is subject to external forces which are orthogonal to the middle surface.

5.1 Hypotheses of Mindlin and Reissner.
H1. *Linearity hypothesis.* Segments lying on normals to the middle surface are linearly deformed and their images are segments on straight lines again.
H2. The displacement in the z-direction does not depend on the z-coordinate.
H3. The points on the middle surface are deformed only in the z-direction.
H4. The normal stress σ_{33} vanishes.

Under hypotheses H1–H3 the displacements have the form

$$u_i(x, y, z) = -z\theta_i(x, y), \quad \text{for } i = 1, 2,$$
$$u_3(x, y, z) = w(x, y). \tag{5.1}$$

We call w the *transversal displacement* or *(normal) deflection*, and $\theta = (\theta_1, \theta_2)$ the *rotation*.

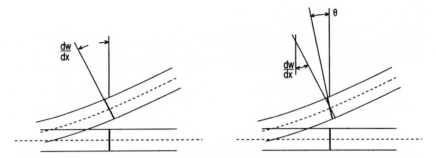

Fig. 60. Beam or section through a plate with and without the normal hypothesis

5.2 The Hypotheses of Kirchhoff–Love. Suppose that in addition to hypotheses H1–H4, we have:

H5. *Normal hypothesis.* The deformations of normal vectors to the middle surface are again orthogonal to the (deformed) middle surface.

This hypothesis is often found under the shorter names *Love's hypothesis* or *Kirchhoff's hypothesis*.

The normal hypothesis implies that the rotations are no longer independent of the deflections (see Fig. 60).

$$\left.\begin{aligned} \theta_i(x, y) &= \frac{\partial}{\partial x_i} w(x, y), \\ u_i(x, y, z) &= -z\frac{\partial w}{\partial x_i}(x, y), \end{aligned}\right\} \quad i = 1, 2. \tag{5.2}$$

We restrict ourselves to body forces which we assume to be independent of z. The associated strains are then

$$(\nabla^{(s)}u)_{ij} = -z(\nabla^{(s)}\theta)_{ij}, \quad (\nabla^{(s)}u)_{i3} = \frac{1}{2}(\frac{\partial}{\partial x_i} w - \theta_i), \quad i, j = 1, 2. \tag{5.3}$$

In view of the hypothesis $\sigma_{33} = 0$, we can make use of the formulae (4.2) and (4.3) for the plane stresses, when we evaluate the bilinear form from the energy functional (3.1):

$$\begin{aligned} \varepsilon : \sigma &= \sum_{i,j=1}^{2} \varepsilon_{ij}\sigma_{ij} + 2\sum_{j=1}^{2} \varepsilon_{3j}\sigma_{3j} \\ &= \frac{E}{1+\nu}\Big[\sum_{i,j=1}^{2} \varepsilon_{ij}^2 + \frac{\nu}{1-\nu}(\varepsilon_{11} + \varepsilon_{22})^2 + 2\sum_{j=1}^{2} \varepsilon_{3j}^2\Big] \\ &= 2\mu \sum_{\substack{i,j=1 \\ (i,j)\neq(3,3)}}^{3} \varepsilon_{ij}^2 + \lambda\frac{2\mu}{\lambda+2\mu}(\varepsilon_{11} + \varepsilon_{22})^2. \end{aligned}$$

With the model (5.1) and the derivatives (5.3), the integration in (3.1) over the z variable is easily evaluated:[21]

$$\Pi(u) := \Pi(\theta, w) = \frac{t^3}{12}a(\theta, \theta) + \frac{\mu t}{2}\int_\omega |\nabla w - \theta|^2 dx_1 dx_2 - t\int_\omega f w\, dx_1 dx_2 \tag{5.4}$$

with

$$a(\theta, \psi) := \int_\omega [2\mu\, \varepsilon(\theta) : \varepsilon(\psi) + \frac{\lambda}{2}\frac{2\mu}{\lambda + 2\mu}\, \text{div}\,\theta\,\text{div}\,\psi]\, dx_1 dx_2. \tag{5.5}$$

The symmetric gradient

$$\varepsilon_{ij}(\theta) := \frac{1}{2}\Big(\frac{\partial\theta_i}{\partial x_j} + \frac{\partial\theta_j}{\partial x_i}\Big), \quad i, j = 1, 2 \tag{5.6}$$

and the divergence are now based on functions of two variables. The first term in (5.4) contains the *bending part* of the energy, and the second term contains the *shear term*. Clearly the latter vanishes in the Kirchhoff model.

The solution of the variational problem does not change if the energy functional is multiplied by a constant. Without altering the notation, we multiply by t^{-3}, replace $t^2 f$ by f in the load, and normalize μ, leading to the (dimensionless) expression

$$\Pi(u) = \frac{1}{2}a(\theta, \theta) + \frac{t^{-2}}{2}\int_\Omega |\nabla w - \theta|^2 dx - \int_\Omega f w\, dx. \tag{5.7}$$

To stay with the usual notation, we have written Ω instead of ω.

Now in the framework of the Kirchhoff model, (5.2) and (5.6) imply

$$\varepsilon_{ij}(\theta) = \partial_{ij}w,$$

and thus (5.5) gives the bilinear form

$$a(\nabla w, \nabla v) = \int_\Omega [\mu \sum_{i,j} \partial_{ij} w\, \partial_{ij}v + \lambda'\Delta w\Delta v]dx, \tag{5.8}$$

with a suitable constant λ'. This is a variational problem of fourth order of the same structure as the variational formulation of the biharmonic equation.

---

[21] Since $D_{33}u = 0$, the model (5.1) is consistent with hypothesis H4 and with (4.2), only if $\text{div}\,\theta = 0$. The two hypotheses lead to slightly different factors in (5.5). Therefore in computations of plates some so-called *shear-correction factors* may be found, but this has no effect on our analysis.

By Korn's inequality, the bilinear form $a(\nabla w, \nabla v)$ is H^2-elliptic on $H_0^2(\Omega)$. For a conforming treatment, the variational problem

$$\frac{1}{2}a(\nabla v, \nabla v) - (f, v) \longrightarrow \min_{v \in H_0^2(\Omega)} ! \tag{5.9}$$

for the clamped Kirchhoff plate requires C^1 elements, which is computationally expensive. In this respect, the numerical treatment of the Mindlin plate appears at first glance to be simpler since the problem (5.7) is obviously H^1-elliptic for $(w, \theta) \in H_0^1(\Omega) \times H_0^1(\Omega)^2$. However, as we shall see, the Mindlin plate contains a small parameter.

Finally, we would like to mention the so-called *Babuška paradox*, see e.g. Babuška and Pitkäranta [1990]. If we approximate a domain with a smooth boundary by polygonal domains, then the solutions for Kirchhoff plates on these domains usually do not converge to the solution for the original domain. This holds for the clamped plates as well as for some other boundary conditions.

Note on Beam Models

While plate models refer to elliptic problems in 2-space, the beam models lead to boundary value problems with ordinary differential equations. The beam with the Kirchhoff hypothesis is called the *Bernoulli beam*, and the beam which corresponds to the Mindlin plate is the *Timoshenko beam*. If we eliminate the Lamé constants, we obtain the energy of the Timoshenko beam by a reduction of (5.7) to the one-dimensional case:

$$\Pi(\theta, w) := \frac{1}{2}\int_0^b (\theta')^2 dx + \frac{t^{-2}}{2}\int_0^b (w' - \theta)^2 dx - \int fw\, dx.$$

Here $\theta, w \in H_0^1(0, b)$ where b denotes the length of the beam.

We have already considered this model in §3 when illustrating shear locking. It should be emphasized that the computation and the analysis of plates is much more involved and cannot be understood as a simple generalization; cf. Problem 5.13.

Mixed Methods for the Kirchhoff Plate

Nonconforming and mixed methods play an important role in the theory of Kirchhoff plates. We begin with mixed methods since we will also use them for the analysis of nonconforming elements. In the following, $a(\cdot, \cdot)$ always denotes the H^1-elliptic bilinear form on $H_0^1(\Omega)^2$ defined in (5.5).

The minimization of

$$\frac{1}{2}a(\theta, \theta) - (f, w) \tag{5.10}$$

subject to the constraint

$$\nabla w = \theta \quad \text{in } \Omega$$

leads to the following saddle point problem: Find $(w, \theta) \in X$ and $\gamma \in M$ such that

$$
\begin{aligned}
a(\theta, \psi) + (\nabla v - \psi, \gamma)_0 &= (f, v) \quad \text{for all } (v, \psi) \in X, \\
(\nabla w - \theta, \eta)_0 &\hphantom{==} = 0 \quad\;\; \text{for all } \eta \in M.
\end{aligned}
\tag{5.11}
$$

In choosing the spaces X and M for the plate, our first choice would be

$$X := H_0^2(\Omega) \times H_0^1(\Omega)^2, \quad M := H^{-1}(\Omega)^2. \tag{5.12}$$

Clearly, the bilinear forms in (5.11) are continuous. In view of Korn's inequality and the constraint $\nabla v = \psi$, we have

$$
\begin{aligned}
a(\psi, \psi) &\geq c\|\psi\|_1^2 \\
&= \frac{c}{2}\|\psi\|_1^2 + \frac{c}{2}\|\nabla v\|_1^2 \\
&\geq c'(\|\psi\|_1^2 + \|v\|_2^2),
\end{aligned}
\tag{5.13}
$$

and a satisfies the ellipticity condition for a saddle point problem. Moreover,

$$\sup_{v, \psi} \frac{(\nabla v - \psi, \eta)_0}{\|v\|_2 + \|\psi\|_1} \geq \sup_{\psi \in H_0^1} \frac{-(\psi, \eta)_0}{\|\psi\|_1} = \|\eta\|_{-1},$$

and the inf-sup condition is also satisfied. The solution of (5.11) can now be estimated from the general theory using Theorem III.4.3:

$$\|w\|_2 + \|\theta\|_1 + \|\gamma\|_{-1} \leq c\|f\|_{-2}. \tag{5.14}$$

More importantly, the regularity theory for convex domains Ω yields the sharper result

$$\|w\|_3 + \|\theta\|_2 + \|\gamma\|_0 \leq c\|f\|_{-1}, \tag{5.15}$$

see Blum and Rannacher [1980]. Since $H_0^2(\Omega)$ is dense in $H_0^1(\Omega)$, (5.11) even holds for all $v \in H_0^1(\Omega)$ provided the domain is convex and $f \in H^{-1}(\Omega)$.

The following alternative pairing is also important since we do not need C^1 elements for a conforming treatment of the normal deflections.

$$X := H_0^1(\Omega) \times H_0^1(\Omega)^2, \quad M := H^{-1}(\text{div}, \Omega). \tag{5.16}$$

Here $H^{-1}(\text{div}, \Omega)$ is the completion of $C^\infty(\Omega)^2$ with respect to the norm

$$\|\eta\|_{H^{-1}(\text{div},\Omega)} := (\|\eta\|_{-1}^2 + \|\text{div}\,\eta\|_{-1}^2)^{\frac{1}{2}}. \tag{5.17}$$

If the domain Ω satisfies the hypotheses of Theorem II.1.3, then $H^{-1}(\text{div}, \Omega)$ can be viewed as the set of H^{-1} functions whose divergences lie in $H^{-1}(\Omega)$.

If the constraint $\nabla v = \psi$ is satisfied, then by (5.13) we have $a(\psi, \psi) \geq c'(\|\psi\|_1^2 + \|v\|_1^2)$, and the ellipticity of a is clear. Moreover,

$$\sup_{v,\psi} \frac{(\nabla v - \psi, \eta)_0}{\|v\|_1 + \|\psi\|_1} = \frac{1}{2}\sup_{v,\psi} \frac{(\nabla v - \psi, \eta)_0}{\|v\|_1 + \|\psi\|_1} + \frac{1}{2}\sup_{v,\psi} \frac{(\nabla v - \psi, \eta)_0}{\|v\|_1 + \|\psi\|_1}$$

$$\geq \frac{1}{2}\sup_\psi \frac{(-\psi, \eta)_0}{\|\psi\|_1} + \frac{1}{2}\sup_v \frac{-(v, \text{div}\,\eta)_0}{\|v\|_1} \tag{5.18}$$

$$= \frac{1}{2}\|\eta\|_{-1} + \frac{1}{2}\|\text{div}\,\eta\|_{-1}.$$

This establishes the inf-sup condition for the pairing (5.16).

Similarly, we have $|(\nabla v - \psi, \eta)| \leq \|v\|_1\|\text{div}\,\eta\|_{-1} + \|\psi\|_1\|\eta\|_{-1}$. Once we establish the continuity, we have all of the hypotheses of Theorem III.4.3, and the general theory implies existence, stability, and

$$\|w\|_1 + \|\theta\|_1 + \|\gamma\|_{H^{-1}(\text{div},\Omega)} \leq c\|f\|_{-1}. \tag{5.19}$$

For convex domains, this estimate is obviously weaker than the regularity result (5.15).

It is not entirely obvious that the two pairings (5.12) and (5.15) lead to the same solution of the variational problem (5.11) for $f \in H^{-1}(\Omega)$. For the solution based on the spaces (5.12), the regularity assertion (5.15) garantees the inclusion $\gamma \in L_2(\Omega) \subset H^{-1}(\text{div}, \Omega)$, and thus that the solution is also consistent with the other pairing. Conversely, the second equation in (5.11) asserts that w has a weak derivative in H_0^1 which lies in H_0^2. However, this result requires a homogeneous right-hand side in the second equation of (5.11).

DKT Elements

Finite element computations with C^1 elements are circumvented if the normal hypothesis (and C^1 continuity) is satisfied at the nodes of a triangular partition rather than on the entire domain. We call this a *discrete Kirchhoff condition*, and the corresponding element a *discrete Kirchhoff triangle* or for short a DKT element.

Strictly speaking, DKT elements are nonconforming elements for the displacement formulation (5.9). Nevertheless, the connection with mixed methods simplifies the analysis since the consistency error can be estimated in terms of the Lagrange multiplier of the mixed formulation.

We consider two examples which require different treatments. Both involve reduced polynomials of degree 3. Given a triangle, let a_1, a_2, a_3 be its vertices, and let a_0 be its center of gravity:

$$\mathcal{P}_{3,\text{red}} := \{p \in \mathcal{P}_3; \ 6p(a_0) - \sum_{i=1}^{3} [2p(a_i) - \nabla p(a_i) \cdot (a_i - a_0)] = 0\}.$$

Here the constraint excludes the bubble function, which in other cases is usually appended to polynomials of lower degree. In comparison with standard interpolation using cubic polynomials, the interior point is missing, and only the nine points on the boundary are used. – Instead of using these nine function values, we can also use the function values and first derivatives at the three vertices; cf. Problem II.5.13.

5.3 Example. Following Batoz, Bathe, and Ho [1980], let

$$
\begin{aligned}
W_h &:= \{w \in H_0^1(\Omega); \ \nabla w \text{ is continuous at all nodes of } \mathcal{T}_h, \\
&\qquad w|_T \in \mathcal{P}_{3,\text{red}} \text{ for } T \in \mathcal{T}_h\}, \\
\Theta_h &:= \{\theta \in H_0^1(\Omega)^2; \ \theta|_T \in (\mathcal{P}_2)^2 \text{ and } \theta \cdot n \in \mathcal{P}_1(e) \\
&\qquad \text{for every edge } e \in \partial T \text{ and } T \in \mathcal{T}_h\}.
\end{aligned}
\tag{5.20}
$$

Every deflection is associated with a rotation via a discrete Kirchhoff condition:

$$
\begin{aligned}
w_h \longmapsto \theta_h(w_h): \quad \theta_h(a_i) &= \nabla w_h(a_i) \quad \text{for the vertices } a_i, \\
\theta_h(a_{ij}) \cdot t &= \nabla w_h(a_{ij}) \cdot t \text{ for the midpoint } a_{ij}.
\end{aligned}
\tag{5.21}
$$

As usual, n denotes a unit normal vector, and t is a tangential vector (to points on the edges). Moreover, $a_{ij} := \frac{1}{2}(a_i + a_j)$. The mapping $W_h \to \Theta_h$ in (5.21) is well defined since by the definition of the space Θ_h, the normal components at the midpoints of the sides are already determined by their values at the vertices:

$$\theta_h(a_{ij}) \cdot n = \frac{1}{2}[\nabla w_h(a_i) + \nabla w_h(a_j)] \cdot n.$$

Thus, $\theta_h(w_h) \in (P_2)^2$ is defined by the interpolation conditions at the six canonical points.

We note that on the edges, the tangential components of ∇w_h are always quadratic polynomials, and by construction coincide with the tangential components of $\theta_h = \theta_h(w_h)$. Thus, θ_h can only vanish in a triangle T if w_h is constant

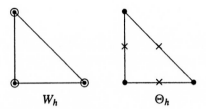

Fig. 61. The DKT element of Batoz, Bathe, and Ho (The tangential component only is fixed at the nodes marked with ×.)

on ∂T. Since the restriction of W_h to a triangle has dimension 9, the restriction of $\theta_h(W_h)$ has the same dimension as ∇W_h, i.e., 8.

Note that $\theta_h(W_h)$ is a proper subspace of Θ_h. Since $\dim(\mathcal{P}_2)^2 = 12$, the three kinematic relations $[2\,\theta(a_{ij}) - \theta(a_i) - \theta(a_j)]n = 0$ in (5.20) imply that locally, Θ_h has dimension $12 - 3 = 9$. The fact that the mapping (5.21) only gives a subspace is not a problem for either the implementation or the analysis of the elements.

The associated finite element approximation is the solution of the equation

$$a_h(w_h, v_h) = (f, v_h)_0 \quad \text{for all } v_h \in W_h$$

with the bilinear form

$$a_h(w_h, v_h) := a(\theta_h(w_h), \theta_h(v_h)). \tag{5.22}$$

For the analysis, a_h is defined on $W_h \oplus H^3(\Omega) \cap H_0^1(\Omega)$ by extending the mapping (5.21) to this space in the canonical way. In addition, following Pitkäranta [1988], we make use of the mesh-dependent norms

$$|v|_{s,h} := \left(\sum_{T \in \mathcal{T}_h} |v|_{s,T}^2\right)^{\frac{1}{2}}, \qquad \|v\|_{s,h} := \left(\sum_{T \in \mathcal{T}_h} \|v\|_{s,T}^2\right)^{\frac{1}{2}}. \tag{5.23}$$

5.4 Remark. The semi-norms

$$\begin{aligned}
&\|\nabla v_h\| \text{ and } \|\theta_h(v_h)\|_0, \\
&|\nabla v_h|_{1,h} \text{ and } |\theta_h(v_h)|_1, \\
&\|\nabla v_h\|_{1,h} \text{ and } \|\theta_h(v_h)\|_1, \quad \text{respectively,}
\end{aligned} \tag{5.24}$$

are equivalent on W_h, provided that the triangulations are shape regular.

Proof. In view of the finite dimensionality and the linearity of the mapping $\nabla v_h \mapsto \theta(v_h)$, on the reference triangle we have

$$\begin{aligned}
\|\theta_h(v_h)\|_{0,T_{\text{ref}}} &\leq c\|\nabla v_h\|_{0,T_{\text{ref}}}, \\
|\theta_h(v_h)|_{1,T_{\text{ref}}} &\leq c\|\nabla v_h\|_{1,T_{\text{ref}}}.
\end{aligned}$$

Now $\theta_h - \nabla v_h = 0$ for $v_h \in \mathcal{P}_1$. By the Bramble–Hilbert lemma, we can estimate $|\theta_h - \nabla v_h|_{1,T_{\text{ref}}}$ by $c|\nabla v_h|_{1,T_{\text{ref}}}$, and thus insert the semi-norm in the second part of (5.24). Since the mapping $\nabla v_h \mapsto \theta(v_h)$ is injective, the converse follows in the same way. Finally, the transformation theorems of Ch. II, §6 imply the estimate for the triangles T of a shape-regular triangulation. The result for the domain Ω follows by summation. $\Box$

5.5 Corollary. *The DKT element 5.3 satisfies*

$$a(\theta_h(v_h), \theta_h(v_h)) \geq c \sum_{T \in \mathcal{T}_h} \|v_h\|_{2,T}^2$$

for some constant $c > 0$ independent of h.

Proof. Since $\theta_h \in H_0^1(\Omega)^2$ and $v_h \in H_0^1(\Omega)$, it follows from Korn's inequality, Friedrichs' inequality, and the previous remark that

$$a(\theta_h(v_h), \theta_h(v_h)) \geq c\|\theta_h(v_h)\|_1^2 = c \sum_{T \in \mathcal{T}_h} \|\theta_h(v_h)\|_{1,T}^2$$

$$\geq c\|\nabla v_h\|_{1,h}^2 = c(|\nabla v_h|_{1,h}^2 + |v_h|_{1,\Omega}^2)^2$$

$$\geq c'(|\nabla v_h|_{1,h}^2 + \|v_h\|_{1,\Omega}^2) \geq c'\|v_h\|_{2,h}^2. \qquad \Box$$

Since the gradient of linear functions is exactly interpolated, the same argument implies that

$$\|\theta_h(v_h) - \nabla v_h\|_0 \leq c\,h|v_h|_{2,h} \quad \text{for all } v_h \in W_h.$$

We are now in a position to estimate the consistency error via the second lemma of Strang. Since $H^3(\Omega) \subset C^1(\Omega)$, we can directly extend the mapping θ_h in (5.21) to $W_h \oplus H^3(\Omega)$. We also have to take into consideration the fact that, in general, $\theta_h(w) \neq \nabla w$ for the solution w of (5.11). Equality holds if ∇w is linear. Hence, it follows from the Bramble–Hilbert lemma that

$$\|\theta_h(w) - \nabla w\|_0 \leq ch^2\|w\|_3.$$

In the first equation of the mixed formulation (5.11) we now set $v = v_h$ and $\psi = \theta_h(v_h)$. Together with the regularity result (5.15) for the mixed method, this implies

$$a_h(w, v_h) - (f, v_h)_0 = a(\theta_h(w), \theta_h(v_h)) - (f, v_h)_0$$

$$= [a(\theta, \theta_h(v_h)) - (f, v_h)_0] + a(\theta_h(w) - \theta, \theta_h(v_h))$$

$$= (\nabla v_h - \theta_h(v_h), \gamma)_0 + a(\theta_h(w) - \theta, \theta_h(v_h))$$

$$\leq \|\nabla v_h - \theta_h(v_h)\|_0 \|\gamma\|_0 + ch\|w\|_3\|v_h\|_{2,h}$$

$$\leq c\,h\|v_h\|_{2,h} \|f\|_{-1}.$$

Since the approximation error $\inf\{\|w - \psi_h\|_{2,h}; \ \psi_h \in W_h\}$ is also of order $\mathcal{O}(h)$, the second lemma of Strang implies

5.6 Theorem. *Let $\mathcal{T}_h$ be a family of shape-regular triangulations. Then the finite element solution using the DKT element 5.3 satisfies*

$$\|w - w_h\|_{2,h} \le ch\|f\|_{-1},$$

with a constant c independent of h.

In Example 5.3 we have used a rather large finite element space for the rotations. The amount of computation can be reduced by using smaller spaces.

5.7 Example *(Zienkiewicz triangle).* Let

$$W_h := \{w \in H_0^1(\Omega); \ \nabla w \text{ is continuous at all nodes of } \mathcal{T}_h,$$

$$w|_T \in \mathcal{P}_{3,\text{red}} \text{ for } T \in \mathcal{T}_h\}, \tag{5.25}$$

$$\Theta_h := \{\theta \in H_0^1(\Omega)^2; \ \theta|_T \in (\mathcal{P}_1)^2 \text{ for } T \in \mathcal{T}_h\} = (\mathcal{M}_{0,0}^1)^2.$$

Every deflection can be associated with a rotation by means of a discrete Kirchhoff condition:

$$w_h \longmapsto \theta_h(w_h) : \quad \theta_h(a_i) = \nabla w_h(a_i) \text{ for the vertices } a_i.$$

It is clear from a dimensionality argument that $\theta_h = 0$ can hold in a triangle, even though $\nabla w_h \ne 0$. Thus, the bilinear form a_h must be defined differently than in (5.22):

$$a_h(w_h, v_h) := \sum_{T \in \mathcal{T}_h} a(\nabla w_h, \nabla v_h)_T.$$

This element is called the *Zienkiewicz triangle*, see Zienkiewicz [1971]. It is interesting to note that both theoretical results and numerical experience show that the convergence of this nonconforming method occurs only for *three-direction meshes*, i.e. for meshes where the grid lines run only in three directions; see Lascaux and Lesaint [1975].

It is possible to dispense with the restriction to three-direction meshes by adding a penalty term, which leads to the quadratic form

$$a_h(v_h, v_h) := a(\theta_h(v_h), \theta_h(v_h)) + \sum_{T \in \mathcal{T}_h} \frac{1}{h_T^2} \int_T |\nabla v_h - \theta_h(v_h)|^2 dx. \tag{5.26}$$

It is clear from (5.7) that the penalty term was selected on the basis of the theory of the Mindlin–Reissner plate.

For simplicity, we restrict ourselves to a uniform grid. Then the various mesh sizes h_T can be replaced by a global one:

$$a_h(v_h, v_h) := a(\theta_h(v_h), \theta_h(v_h)) + \frac{1}{h^2}\|\nabla v_h - \theta_h(v_h)\|_0^2.$$

In addition to the norms (5.23), the energy norm

$$\||v, \psi\||_2 := (\|\psi\|_1^2 + \frac{1}{h^2}\|\nabla v - \psi\|_0^2)^{\frac{1}{2}}$$

enters into the analysis.

5.8 Lemma. *For all $v_h \in W_h$, $h \leq 1$,*

$$c^{-1}\|v_h\|_{2,h} \leq \||v_h, \theta_h(v_h)\||_2 \leq c\|v_h\|_{2,h}, \tag{5.27}$$

where c is a constant independent of h.

Proof. (1) By the approximation results,

$$\|\nabla v - \theta_h(v)\|_{s,T} \leq c\, h^{1-s}|v|_{2,T} \quad \text{for } s = 0, 1.$$

With $s = 0$, we have $\|\nabla v_h - \theta_h(v_h)\|_{0,\Omega} \leq c\, h\|v_h\|_{2,h}$. Thus, with $s = 1$,

$$\|\theta_h(v_h)\|_{1,T}^2 \leq (\|v_h\|_{1,T} + \|v_h - \theta_h(v_h)\|_{1,T})^2 \leq 2\|v_h\|_{1,T}^2 + 2c|v_h|_{2,T}^2.$$

Hence, $\|\theta_h(v_h)\|_{1,\Omega} \leq 2(1+c)\|v_h\|_{2,h}$, and the inequality on the right in (5.27) is proved.

(2) To prove the other inequality, we establish the stronger assertion

$$c^{-1}\|v_h\|_{2,h} \leq \||v_h, \psi_h\||_2 \quad \text{for all } v_h \in W_h, \psi_h \in \Theta_h. \tag{5.28}$$

First we use Friedrichs' inequality:

$$\frac{1}{c}\|v_h\|_1^2 \leq \|\nabla v_h\|_0^2 \leq 2\|\nabla v_h - \psi_h\|_0^2 + 2\|\psi_h\|_0^2$$
$$\leq 2h^{-2}\|\nabla v_h - \psi_h\|_0^2 + 2\|\psi_h\|_1^2 \leq 2\||v_h, \psi_h\||_2^2.$$

Similar estimates along with the usual inverse inequality lead to

$$\sum_T |v_h|_{2,T}^2 \leq \sum_T (|\nabla v_h - \psi_h|_{1,T} + |\psi_h|_{1,T})^2$$
$$\leq 2\sum_T (|\nabla v_h - \psi_h|_{1,T}^2 + |\psi_h|_{1,T}^2)$$
$$\leq 2\sum_T h^{-2}|\nabla v_h - \psi_h|_{0,T}^2 + 2\|\psi_h\|_{1,\Omega}^2 = 2\||v_h, \psi_h\||_2^2,$$

and the proof is complete. ∎

Taking account of Remark III.1.3, we can now directly carry over the method of proof of Theorem 5.6 to establish convergence. For details, see Problem 5.11.

5.9 Theorem. *Let T_h be a family of shape-regular triangulations. Then the finite element solution using the DKT element 5.7 satisfies*

$$\|w - w_h\|_{2,h} \leq c\, h\|f\|_{-1},$$

where c is a constant independent of h.

For the treatment of DKT elements by multigrid methods, see e.g. Peisker, Rust, and Stein [1990].

Problems

5.10 Generalize Theorem 5.9 to shape-regular grids, and verify (5.27) for a_h with the associated energy norm

$$|||v, \psi|||_2 := \left(\|\psi\|_1^2 + \sum_{T \in \mathcal{T}_h} \frac{1}{h_T^2} \int_T |\nabla v - \psi|^2 dx \right)^{\frac{1}{2}}.$$

5.11 To treat the consistency error $a_h(w, v_h) - (f, v_h)_0$ in Theorem 5.9, estimate the contribution of the penalty term

$$\frac{1}{h^2} \int_\Omega (\nabla w - \theta_h(w)) \cdot (\nabla v_h - \theta_h(v_h)) \, dx \quad \text{for } v_h \in W_h$$

in terms of $|w|_{3,\Omega}|v_h|_{2,h}$ with the correct power of h.

5.12 Express the dimension of the finite element spaces with DKT elements in terms of the number of triangles, nodes, and edges.

5.13 Show that the mixed method for the Timoshenko beam which corresponds to (5.11) is stable for $X := H_0^1(0, b) \times H_0^1(0, b)$ and $M := L_2(0, b)$. With which Standard Sobolev space does $H^{-1}(\text{div}, (0, b))$ coincide? [Hence, we do not need the space $H^{-1}(\text{div})$ in contrast to the investigation of plates.]

§ 6. The Mindlin–Reissner Plate

The Mindlin–Reissner model for a bending plate involves minimizing (5.7) over $(w, \theta) \in X := H_0^1(\Omega) \times H_0^1(\Omega)^2$. Here the shear term does not vanish since the normal hypothesis is not assumed. The Mindlin plate turns out to be a singular perturbation problem. This is consistent with the observation that the directions of the rotations differ from the normal directions only near the boundary; see Arnold and Falk [1989], Pitkäranta and Suri [1997].

Here t should be thought of as a small parameter. In order to avoid shear locking, we treat the plate as a mixed problem with penalty term. Now we proceed in exactly the same way as we did in going from (3.32) to (3.37). Introducing the shear term

$$\gamma := t^{-2}(\nabla w - \theta), \tag{6.1}$$

we get the following – at first purely formal – mixed problem with penalty term: Find $(w, \theta) \in X := H_0^1(\Omega) \times H_0^1(\Omega)^2$ and $\gamma \in M := L_2(\Omega)^2$ such that

$$
\begin{aligned}
a(\theta, \psi) + (\nabla v - \psi, \gamma)_0 &= (f, v) && \text{for all } (v, \psi) \in X, \\
(\nabla w - \theta, \eta)_0 - t^2(\gamma, \eta)_0 &= 0 && \text{for all } \eta \in M.
\end{aligned}
\tag{6.2}
$$

Here the bilinear form a is defined in (5.5). The equation (6.2) and the mixed formulation (5.11) for the Kirchhoff plate differ only by the penalty term $-t^2(\gamma, \eta)_0$. However, we should not overlook the fact that here the variable γ in the equation has a different meaning. In (5.11) it serves as a *Lagrange multiplier*, while by (6.1), here it is to be regarded as a *normed shear term*.

The variational problem cannot be dealt with directly using the general results for the saddle point theory in Ch. III, §4. From the theory of the Kirchhoff plate, we know that $H^{-1}(\text{div}, \Omega)$ is the natural space for the Lagrange multiplier. Theorems III.4.11 and III.4.13 are not applicable since $H^{-1}(\text{div}, \Omega) \not\subset L_2(\Omega)$, and because the bilinear form a is not elliptic on the entire space X.

Since these arguments may appear quite formal to the reader, we present another reason: For $\theta \in H_0^1(\Omega)^2$, $w \in H_0^1(\Omega)$, $\text{rot}\,\theta \in L_2(\Omega)$ and $\text{rot grad}\, w = 0 \in L_2(\Omega)$, and thus

$$\nabla w - \theta \in H_0(\text{rot}, \Omega).$$

The rotation is defined in (4.9). Hence, the proper space is not $L_2(\Omega)$, but rather

$$H_0(\text{rot}, \Omega) := \{\eta \in L_2(\Omega)^2;\ \text{rot}\,\eta \in L_2(\Omega),\ \eta \cdot \tau = 0 \text{ on } \partial\Omega\}. \tag{6.3}$$

Here $\tau = \tau(x)$ is defined (almost everywhere) on $\partial\Omega$ as the direction of the tangent in the counterclockwise direction. We endow the space (6.3) with the norm

$$\|\eta\|_{0,\mathrm{rot}} := (\|\eta\|_0^2 + \|\operatorname{rot}\eta\|_0^2)^{\frac{1}{2}}. \tag{6.4}$$

In terms of the general theory in Ch. III, §4, the bilinear form b for the mixed formulation is given by $b(w, \theta, \eta) := (\nabla w - \theta, \eta)_0$. Now we specialize η to be an element of $L_2(\Omega)^2$ of the form $\eta = \operatorname{curl} p$. Then because of the orthogonality of the rotation and the gradient,

$$b(w, \theta; \eta) = (\nabla w - \theta, \eta)_0 = 0 - (\theta, \eta)_0 \le \|\theta\|_1 \|\eta\|_{-1},$$

for $w \in H_0^1(\Omega)$, $\theta \in H_0^1(\Omega)^2$, and thus

$$\sup_{w,\theta} \frac{b(w, \theta, \eta)}{\|w\|_1 + \|\theta\|_1} \le \|\eta\|_{-1}.$$

In order to ensure that the inf-sup condition holds, we have to endow M with a norm which is weaker than the L_2-norm, that is with the one which is dual to (6.4), i.e., $\|\cdot\|_{H^{-1}(\mathrm{div},\Omega)}$; see below.

As shown in Brezzi and Fortin [1986], the analyis is simplified if we use the Helmholtz decomposition of the shear term into a gradient field and a rotational field. Using the decomposition we get expressions and estimates which involve the usual Sobolev norms.

The Helmholtz Decomposition

In the following we shall see that the space

$$H^{-1}(\mathrm{div}, \Omega) := \{\eta \in H^{-1}(\Omega)^2; \ \operatorname{div}\eta \in H^{-1}(\Omega)\}$$

with the graph norm (5.17) is the dual space of $H_0(\mathrm{rot}, \Omega)$. Clearly,

$$H_0(\mathrm{rot}, \Omega) \subset L_2(\Omega)^2 \subset H^{-1}(\mathrm{div}, \Omega).$$

As usual, we identify functions in $L_2(\Omega)/\mathbb{R}$ which differ only by a constant. The norm of an element in this space is just the L_2-norm of the representer which is normalized to have zero integral; see Problem III.6.6.

6.1 Lemma. *Assume that $\Omega \subset \mathbb{R}^2$ is simply connected. Then every function $\eta \in H^{-1}(\text{div}, \Omega)$ is uniquely decomposable in the form*

$$\eta = \nabla \psi + \text{curl } p \qquad (6.5)$$

with $\psi \in H_0^1(\Omega)$ and $p \in L_2(\Omega)/\mathbb{R}$. Moreover, the norms

$$\|\eta\|_{H^{-1}(\text{div},\Omega)} \quad \text{and} \quad (\|\psi\|_{1,\Omega}^2 + \|p\|_0^2)^{\frac{1}{2}} \qquad (6.6)$$

are equivalent, where p is the representer satisfying $\int_\Omega p\,dx = 0$.

Proof. By hypothesis, $\chi := \text{div } \eta \in H^{-1}(\Omega)$. Let $\psi \in H_0^1(\Omega)$ be the solution of the equation $\Delta \psi = \chi$. Then $\text{div}(\eta - \nabla \psi) = \text{div } \eta - \Delta \psi = 0$. By classical estimates, every divergence-free function in Ω can be represented as a rotation, i.e., $\eta - \nabla \psi = \text{curl } p$ with a suitable $p \in L_2(\Omega)/\mathbb{R}$. This establishes the decomposition.

We also observe that

$$\|\text{div } \eta\|_{-1} = \|\Delta \psi\|_{-1} = |\psi|_1, \qquad (6.7)$$

and

$$\|\eta\|_{-1} = \|\nabla \psi + \text{curl } p\|_{-1} \le \|\nabla \psi\|_{-1} + \|\text{curl } p\|_{-1} \le \|\psi\|_0 + \|p\|_0.$$

After summation, it follows that $\|\eta\|_{H^{-1}(\text{div},\Omega)}^2 \le 2\|\psi\|_1^2 + 2\|p\|_0^2$.

In view of (6.7), to complete the proof we need only show that $\|p\|_0 \le c\|\eta\|_{H^{-1}(\text{div},\Omega)}$. Note that $\int_\Omega p\,dx = 0$. It is known from the Stokes problem (see Problem III.6.7) that there exists a function $v \in H_0^1(\Omega)^2$ with

$$\text{div } v = p \quad \text{and} \quad \|v\|_1 \le c\|p\|_0. \qquad (6.8)$$

Then for $\xi = (\xi_1, \xi_2) := (-v_2, v_1)$, it clearly follows that $\xi \in H_0^1(\Omega)^2$,

$$\text{rot } \xi = p \quad \text{and} \quad \|\xi\|_1 \le c\|p\|_0.$$

Moreover, taking account of (6.7), we see that the decomposition (6.5) satisfies

$$\begin{aligned}
\|p\|_0^2 &= (p, \text{rot } \xi) = (\text{curl } p, \xi) = (\eta - \nabla \psi, \xi) \\
&\le \|\eta\|_{-1}\|\xi\|_1 + |\psi|_1\|\xi\|_0 \\
&\le c(\|\eta\|_{-1} + \|\text{div } \eta\|_{-1})\|p\|_0. \qquad \square
\end{aligned}$$

Supplement. If η lies in $L_2(\Omega)^2$ and not just in $H^{-1}(\text{div}, \Omega)$, then we even have $p \in H^1(\Omega)/\mathbb{R}$ for the second component of the Helmholtz decomposition (6.5), and $L_2(\Omega) = \nabla H_0^1(\Omega) \oplus \mathrm{curl}(H^1(\Omega)/\mathbb{R})$. Thus, p is a solution of the Neumann problem $(\mathrm{curl}\, p, \mathrm{curl}\, q)_0 = (\mathrm{div}(\eta - \nabla\psi), \mathrm{curl}\, q)_0$ for given $q \in H^1(\Omega)$.

We assert that

$$(H_0(\mathrm{rot}, \Omega))' = H^{-1}(\mathrm{div}, \Omega) \qquad (6.9)$$

but we will verify only the inclusion $H^{-1}(\mathrm{div}, \Omega) \subset (H_0(\mathrm{rot}, \Omega))'$, and leave the proof of the converse to Problem 6.11. Let $\gamma \in H_0(\mathrm{rot}, \Omega)$ and $\eta = \nabla\psi + \mathrm{curl}\, p \in H^{-1}(\mathrm{div}, \Omega)$. Using Lemma 6.1, we conclude that

$$\begin{aligned}
(\gamma, \eta)_0 &= (\gamma, \nabla\psi)_0 + (\gamma, \mathrm{curl}\, p)_0 \\
&= (\gamma, \nabla\psi)_0 + (\mathrm{rot}\,\gamma, p)_0 \\
&\le \|\gamma\|_0 \|\psi\|_1 + \|\mathrm{rot}\,\gamma\|_0 \|p\|_0.
\end{aligned}$$

Now by (6.6) and the Cauchy–Schwarz inequality for $\mathbb{R}^2$, we get

$$(\gamma, \eta)_0 \le c\|\gamma\|_{H_0(\mathrm{rot},\Omega)} \cdot \|\eta\|_{H^{-1}(\mathrm{div},\Omega)}.$$

This shows that the bilinear form $(\gamma, \eta)_0$ can be extended from a dense subset to all of $H_0(\mathrm{rot}, \Omega) \times H^{-1}(\mathrm{div}, \Omega)$, and the inclusion is established.

The Mixed Formulation with the Helmholtz Decomposition

We now return to the variational problem (6.2) for the Mindlin–Reissner plate. Following Brezzi and Fortin [1986], we now assume that the shear term has the form

$$\gamma = \nabla r + \mathrm{curl}\, p \qquad (6.10)$$

with $r \in H_0^1(\Omega)$ and $p \in H^1(\Omega)/\mathbb{R}$. We decompose the test function η in the same way as $\eta = \nabla z + \mathrm{curl}\, q$. Also note that the gradients and rotations are L_2-orthogonal; see Problem 6.10. Now we apply Green's formula to the rotation, so that (6.2) leads to the equivalent system

$$\begin{aligned}
(\nabla r, \nabla v)_0 &= (f, v)_0 && \text{for } v \in H_0^1(\Omega), \\
a(\theta, \psi) - (\mathrm{rot}\,\psi, p)_0 &= (\nabla r, \psi)_0 && \text{for } \psi \in H_0^1(\Omega)^2, \\
-(\mathrm{rot}\,\theta, q)_0 - t^2(\mathrm{curl}\, p, \mathrm{curl}\, q)_0 &= 0 && \text{for } q \in H^1(\Omega)/\mathbb{R}, \\
(\nabla w, \nabla z)_0 &= (\theta, \nabla z)_0 + t^2(f, z)_0 && \text{for } z \in H_0^1(\Omega).
\end{aligned}$$
$$(6.11)$$

The first equation is a Poisson equation which can be solved first. The second and third equations together constitute an equation of Stokes type with penalty term. The fourth equation is also a Poisson equation which is independent of the others, and can be solved afterwards.

We now show that the middle equations

$$a(\theta, \psi) - (\text{rot } \psi, p)_0 = (\nabla r, \psi)_0,$$
$$-(\text{rot } \theta, q)_0 - t^2(\text{curl } p, \text{curl } q)_0 = 0 \qquad (6.12)$$

of (6.11) are indeed of Stokes type. By Korn's inequality, the bilinear form a is H^1-elliptic. Moreover,

$$\text{rot } \psi = \text{div } \psi^\perp \quad \text{with the convention} \quad x^\perp := (x_2, -x_1)$$

for any vector in $\mathbb{R}^2$. Clearly, $\|\psi^\perp\|_{s,\Omega} = \|\psi\|_{s,\Omega}$ for $\psi \in H^s(\Omega)$. Thus, (6.12) represents a (generalized) Stokes problem for $\theta^\perp$ with singular penalty term.

6.2 Theorem. *Equations (6.12) and (6.2) describe stable variational problems on* $H_0^1(\Omega)^2 \times H^1(\Omega)/\mathbb{R}$ *w.r.t. the norm*

$$\|\theta\|_1 + \|p\|_0 + t \| \text{curl } p\|_0,$$

and on $H_0^1(\Omega) \times H_0^1(\Omega)^2 \times L_2(\Omega)^2$ *w.r.t. the norm*

$$\|w\|_1 + \|\theta\|_1 + \|\gamma\|_{H^{-1}(\text{div},\Omega)} + t \|\gamma\|_0.$$

The constants in the associated inf-sup conditions are independent of t.

Proof. The first assertion follows immediately from Theorem III.4.13 with $X := H_0^1(\Omega)^2$, $M := L_2(\Omega)/\mathbb{R}$, and $M_c := H(\text{rot}, \Omega)/\mathbb{R}$.

As mentioned above, to prove the second assertion, we cannot directly apply Theorem III.4.13. However, once we check the inf-sup condition for every component of (6.11), it follows for (6.2) with the help of the Helmholtz decompositions of $H^{-1}(\text{div}, \Omega)$ and $L_2(\Omega)^2$, respectively. The argument proceeds in exactly the same way as going from (6.2) to (6.11). $\qquad \square$

MITC Elements

The system which arises from the discretization of (6.2) can be transformed in the sense of (3.39) into displacement form (with reduction operators):

$$a(\theta_h, \theta_h) + t^{-2}\|\nabla w_h - R_h\theta_h\|_0^2 - 2(f, w_h)_0 \longrightarrow \min_{w_h,\theta_h} !, \qquad (6.13)$$

where the minimization is performed over the spaces W_h and Θ_h, respectively. Here

$$R_h : H^1(\Omega)^2 \longrightarrow \Gamma_h \qquad (6.14)$$

is a so-called *reduction operator*, i.e., a linear mapping defined on the finite element space for the shear terms which does not affect the elements in Γ_h.

If possible, the finite element calculations are performed using the displacement model (6.13), since it leads to systems of equations with positive definite matrices and with fewer unknowns. On the other hand, for the convergence analysis, it is still best to use the mixed formulations. However, there is a problem: In general, the functions in Γ_h cannot be represented in the form

$$\gamma_h = \text{grad } r_h + \text{curl } p_h,$$

where r_h and p_h again belong to finite element spaces. Thus, except for a special case treated by Arnold and Falk [1989], a modification of the Helmholtz decomposition is necessary.

The notation for the following finite element spaces is suggested by the variables in (6.11).

6.3 The Axioms of Brezzi, Bathe, and Fortin [1989]. Suppose the spaces

$$W_h \subset H_0^1(\Omega), \quad \Theta_h \subset H_0^1(\Omega)^2, \quad Q_h \subset L_2(\Omega)/\mathbb{R}, \quad \Gamma_h \subset H_0(\text{rot}, \Omega)$$

and the mapping R_h defined in (6.14) have the following properties:

(P_1) $\nabla W_h \subset \Gamma_h$, i.e. the discrete shear term $\gamma_h := t^{-2}(\nabla w_h - R_h \theta_h)$ lies in Γ_h.

(P_2) rot $\Gamma_h \subset Q_h$ – this requirement is consistent with $\gamma_h \in H(\text{ro}, \Omega)$, and thus with rot $\gamma_h \in L_2(\Omega)$.

(P_3) The pair (Θ_h, Q_h) satisfies the inf-sup condition

$$\inf_{q_h \in Q_h} \sup_{\psi_h \in \Theta_h} \frac{(\text{rot } \psi_h, q_h)}{\|\psi_h\|_1 \|q_h\|_0} =: \beta > 0,$$

where β is independent of h. – The spaces are thus suitable for the Stokes problem.

(P_4) Let P_h be the L_2-projector onto Q_h. Then

$$\text{rot } R_h \eta = P_h \text{ rot } \eta \quad \text{for all } \eta \in H_0^1(\Omega)^2,$$

i.e. the following diagram is commutative:

$$
\begin{array}{ccc}
H_0^1(\Omega)^2 & \xrightarrow{\ \text{rot}\ } & L_2(\Omega) \\
R_h \downarrow & & \downarrow P_h \\
\Gamma_h & \xrightarrow{\ \text{rot}\ } & Q_h.
\end{array}
$$

(P_5) If $\eta_h \in \Gamma_h$ and rot $\eta_h = 0$, then $\eta_h \in \nabla W_h$. This means that the sequence

$$W_h \xrightarrow{\text{grad}} \Gamma_h \xrightarrow{\text{rot}} Q_h$$

is exact.[22] – This condition corresponds to the fact that rotation-free fields are gradient fields.

We now observe that the *rotation* can be defined as a *weak derivative*, in analogy with Definition II.1.1: A function $u \in L_2(\Omega)$ lies in $H(\text{rot}, \Omega)$ and $v \in L_2(\Omega)^2$ is equal to curl u in the weak sense provided that

$$\int_\Omega v \cdot \varphi \, dx = \int_\Omega u \, \text{rot} \, \varphi \, dx \quad \text{for all } \varphi \in C_0^\infty(\Omega)^2.$$

Similarly, we now define the rotation on the finite element space Q_h in the weak sense; see Peisker and Braess [1992]. The *discrete curl operator* is used indirectly also by Brezzi, Fortin, and Stenberg [1991].

6.4 Definition. The mapping

$$\text{curl}_h : Q_h \longrightarrow \Gamma_h$$

called *discrete curl operator* is defined by

$$(\text{curl}_h \, q_h, \eta)_0 = (q_h, \text{rot} \, \eta)_0 \quad \text{for } \eta \in \Gamma_h. \tag{6.15}$$

Since $\Gamma_h \subset H_0(\text{rot}, \Omega)$, the functional $\eta \longmapsto (q_h, \text{rot} \, \eta)_0$ is well defined and continuous. Thus, $\text{curl}_h \, q_h$ is uniquely determined by (6.15).

6.5 Theorem. *Suppose properties (P_1), (P_2) and (P_5) hold. Then*

$$\Gamma_h = \nabla W_h \oplus \text{curl}_h \, Q_h$$

defines an L_2-orthogonal decomposition (which is called a discrete Helmholtz decomposition).

Proof. (1) It follows directly from Definition 6.1 and (P_1) that $\nabla W_h \oplus \text{curl}_h \, Q_h \subset \Gamma_h$.

(2) For $q_h \in Q_h$ and $w_h \in W_h$, we have $(\text{curl}_h \, q_h, \nabla w_h)_0 = (q_h, \text{rot} \, \nabla w_h)_0 = (q_h, 0)_0 = 0$. Thus, $\text{curl}_h \, q_h$ and ∇w_h are orthogonal in $L_2(\Omega)$.

---

[22] A sequence of linear mappings $A \xrightarrow{f} B \xrightarrow{g} C$ is called *exact* provided that the image of f coincides with the kernel of g.

(3) Given $\gamma_h \in \Gamma_h$, let η_h be the L_2-projection onto $\mathrm{curl}_h \, Q_h$. Then η_h is characterized by

$$(\gamma_h - \eta_h, \mathrm{curl}_h \, q_h)_0 = 0 \quad \text{for all } q_h \in Q_h.$$

By Definition 6.4, $(\mathrm{rot}(\gamma_h - \eta_h), q_h)_0 = 0$ for $q_h \in Q_h$. Since $\mathrm{rot}(\gamma_h - \eta_h) \in Q_h$, it follows that $\mathrm{rot}(\gamma_h - \eta_h) = 0$. By (P_5) we deduce that $\gamma_h - \eta_h \in \nabla W_h$, which implies that $\gamma_h \in \mathrm{curl}_h \, Q_h \oplus \nabla W_h$. $\qquad\square$

We can now follow the same arguments leading from the variational problem (6.7) to the equation (6.11) in exactly the same way as for the finite element version (6.13), see [Peisker and Braess 1992 or Brezzi and Fortin 1991]. This leads to the following problem: Find $(r_h, \theta_h, p_h, w_h) \in W_h \times \Theta_h \times Q_h \times W_h$ such that

$$
\begin{aligned}
(\nabla r_h, \nabla v_h)_0 &= (f, v_h)_0 & v_h &\in W_h, \\
a(\theta_h, \psi_h) - (p_h, \mathrm{rot}\,\psi_h)_0 &= (\nabla r_h, \psi_h)_0 & \psi_h &\in \Theta_h, \\
-(\mathrm{rot}\,\theta_h, q_h)_0 - t^2(\mathrm{curl}_h \, p_h, \mathrm{curl}_h \, q_h)_0 &= 0 & q_h &\in Q_h, \\
(\nabla w_h, \nabla z_h)_0 &= (\theta_h, \nabla z_h)_0 + t^2(f, z_h)_0 & z_h &\in W_h.
\end{aligned}
$$
$$(6.16)$$

Finite element spaces with properties (P_1)–(P_5) automatically satisfy inf-sup conditions analogous to (6.2) with constants which are independent of t and h.

We compare this also with the discrete version of (6.2): Find $(w_h, \theta_h) \in X_h$ and $\gamma_h \in M_h$ such that

$$
\begin{aligned}
a(\theta_h, \psi_h) + (\nabla v_h - \psi_h, \gamma_h)_0 &= (f, v_h) & \text{for all } (\psi_h, v_h) \in X_h, \\
(\nabla w_h - \theta_h, \eta_h)_0 - t^2(\gamma_h, \eta_h)_0 &= 0 & \text{for all } \eta_h \in M_h.
\end{aligned}
$$
$$(6.17)$$

6.6 Example. The so-called MITC7 element (element with *mixed interpolated tensorial components*) is a triangular element involving up to seven degrees of freedom per triangle and variable: The shear terms belong to a more complicated space; see Fig. 62:

$$\Gamma_h := \{\eta \in H_0(\mathrm{rot}, \Omega); \ \eta_{|T} = \begin{pmatrix} p_1 \\ p_2 \end{pmatrix} + p_3 \begin{pmatrix} y \\ -x \end{pmatrix}, \ p_1, p_2, p_3 \in \mathcal{P}_1, \ T \in \mathcal{T}_h\},$$

$$W_h := \mathcal{M}^2_{0,0}, \qquad \Theta_h := \mathcal{M}^2_{0,0} \oplus B_3, \qquad Q_h := \mathcal{M}^1/\mathbb{R}.$$

Here we have made use of the usual notations as in (4.13). Finally, we define the operator R_h by

$$
\begin{aligned}
\int_e (\eta - R_h\eta)\tau \, p_1 ds &= 0 \quad \text{for every edge } e \text{ and every } p_1 \in \mathcal{P}_1, \\
\int_T (\eta - R_h\eta)dx &= 0 \quad \text{for every } T \in \mathcal{T}_h.
\end{aligned}
$$
$$(6.18)$$

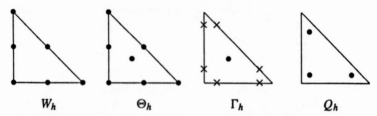

$$W_h \qquad \Theta_h \qquad \Gamma_h \qquad Q_h$$

Fig. 62. MITC7 Element (only tangential components are fixed at the points marked with ×)

We recall that elements in $H_0(\mathrm{rot}, \Omega)$ have to have continuous tangential components along the edges between the triangles. As with the Raviart–Thomas element, it is easy to check that the *tangential components* of the vector expression

$$\begin{pmatrix} y \\ -x \end{pmatrix}$$

are constant on every edge. Thus, the functions in Γ_h are linear on the edges, and so are determined by the values at two points. In particular, these points can be the sample points for a Gaussian quadrature formula which exactly integrates quadratic polynomials. Thus (in agreement with Fig. 62) the six degrees of freedom for functions in Γ_h are determined by the values on the sides. These six values along with the two components at the midpoint of the triangle determine the eight local degrees of freedom.

Therefore, the restriction operator R_h described by (6.18) can be computed from interpolation at the above six points on the sides along with two integrals over the triangle. Thus, the values of $R_h \eta$ in a triangle depend only on the values of η in the same triangle. This means that the system matrix can be assembled locally, triangle by triangle. – This would not have been the case if we had used the L_2-projector in place of R_h.

For numerical results using this element, see Bathe, Brezzi, and Cho [1989] and Bathe, Bucalem, and Brezzi [1991/92].

The Model without a Helmholtz Decomposition

The formulation (6.2) for the Mindlin plate is motivated by the similarity with the mixed formulation of the Kirchhoff plate. Arnold and Brezzi [1993] developed a clever modification which permits a development without using the Helmholtz decomposition. There is a simple treatment in terms of the theory of saddle point problems with penalty term developed in Ch. III, §4; cf. also Braess [1996]. A similar modification can also be found in the treatment of shells by Pitkäranta [1992]. We will partly follow the modification of Chapelle and Stenberg [1998].

Let $t, h < 1$. We again start with the minimization of the functional (6.7), but now combine a part of the shear term with the bending part:

$$\Pi(u) = \frac{1}{2}a_p(w, \theta; w, \theta) + \frac{t'^{-2}}{2}\int_\Omega |\nabla w - \theta|^2 dx - \int_\Omega f w \, dx, \qquad (6.19)$$

where

$$a_p(w, \theta; v, \phi) := a(\theta, \phi) + \frac{1}{h^2 + t^2}\int_\Omega (\nabla w - \theta)\cdot(\nabla v - \phi)\,dx,$$

$$\frac{1}{t^2} = \frac{1}{h^2 + t^2} + \frac{1}{t'^2} \quad \text{or} \quad t'^2 = t^2\frac{h^2}{h^2 + t^2}. \qquad (6.20)$$

Thus, we seek $(w, \theta) \in X = H_0^1(\Omega) \times H_0^1(\Omega)^2$ such that

$$a_p(w, \theta; v, \phi) + \frac{1}{t'^2}(\nabla w - \theta, \nabla v - \phi)_0 = (f, w)_0 \quad \text{for all } (v, \phi) \in X. \quad (6.21)$$

By analogy with the derivation of (6.2) from (6.7), with the introduction of (modified) shear terms $\gamma := t'^{-2}(\nabla w - \theta)$, we now get the following mixed problem with penalty term. Find $(w, \theta) \in X$ and $\gamma \in M$ such that

$$\begin{aligned} a_p(w, \theta; v, \phi) + (\nabla v - \phi, \gamma)_0 &= (f, v)_0 \quad \text{for all } (v, \phi) \in X, \\ (\nabla w - \theta, \eta)_0 - t'^2(\gamma, \eta)_0 &= 0 \qquad \text{for all } \eta \in M. \end{aligned} \qquad (6.22)$$

The essential difference compared to (6.2) is the coercivity of the enhanced form a_p.

6.7 Lemma. *There exists a constant $c := c(\Omega) > 0$ such that*

$$a_p(w, \theta; w, \theta) \geq c(\|w\|_1^2 + \|\theta\|_1^2) \quad \text{for all } w \in H_0^1(\Omega), \ \theta \in H_0^1(\Omega)^2. \quad (6.23)$$

Proof. By Korn's inequality, $a(\phi, \phi) \geq c_1\|\phi\|_1^2$. In addition,

$$\|\nabla w\|_0^2 \leq (\|\nabla w - \theta\|_0 + \|\theta\|_0)^2 \leq 2\|\nabla w - \theta\|_0^2 + 2\|\theta\|_1^2.$$

Friedrichs' inequality now implies

$$\|w\|_1^2 \leq c_2|w|_1^2 \leq 2c_2(\|\nabla w - \theta\|_0^2 + \|\theta\|_1^2),$$

and so

$$\begin{aligned} \|w\|_1^2 + \|\theta\|_1^2 &\leq (1 + 2c_2)\left(\|\nabla w - \theta\|_0^2 + \|\theta\|_1^2\right) \\ &\leq (1 + 2c_2)\left(\|\nabla w - \theta\|_0^2 + c_1^{-1}a(\theta, \theta)\right) \\ &\leq (1 + 2_2)(1 + c_1^{-1})\, a_p(w, \theta; w, \theta). \end{aligned}$$

This establishes the coercivity with the constant $c := (1 + 2c_2)^{-1}(1 + c_1^{-1})^{-1}$. $\square$

The additional term ensures that the coercivity of the quadratic form holds on more than the kernel. This is essential not only for theoretical reasons, but numerical computations have shown that the factor in front of the shear term in (6.21) has to be chosen appropriately and that it should be of the order $O(h^{-2})$. Now Theorem III.4.13 is applicable since the Brezzi condition holds by (5.18).

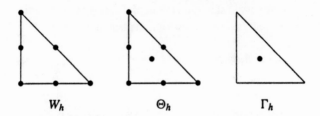

W_h Θ_h Γ_h

Fig. 63. Plate element without the Helmholtz decomposition: $W_h = \mathcal{M}^2_{0,0}$, $\Theta_h = (\mathcal{M}^2_{0,0} \oplus B_3)^2$ and $\Gamma_h = (\mathcal{M}^0)^2$.

6.8 Theorem. *The variational formulation (6.21) for the Mindlin–Reissner plate is stable w.r.t. the spaces*

$$X = H^1_0(\Omega) \times H^1_0(\Omega)^2, \quad M := H^{-1}(\mathrm{div}, \Omega), \quad M_c := L_2(\Omega)^2. \qquad (6.24)$$

In particular, we have stability w.r.t. the norm

$$\|w\|_1 + \|\theta\|_1 + \|\gamma\|_{H^{-1}(\mathrm{div},\Omega)} + t\|\gamma\|_0. \qquad (6.25)$$

Using Fortin's criterion (Lemma III.4.8), it is now easy to show that

$$W_h := \mathcal{M}^2_{0,0}, \quad \Theta_h := (\mathcal{M}^2_{0,0} \oplus B_3)^2, \quad \Gamma_h = (\mathcal{M}^0)^2$$

provides a stable combination of finite element spaces. Here we assume that the domain Ω either is convex or has a smooth boundary.

We make use of the operators $\pi^0_h : H^1_0(\Omega) \to \mathcal{M}^1_{0,0}$ and $\Pi^1_h := H^1_0(\Omega)^2 \to (\mathcal{M}^1_{0,0} \oplus B_3)^2$ appearing in the proof of Theorem III.7.2. In particular, $\int_T (\Pi_h v - v)_i dx = 0$ for every $T \in \mathcal{T}_h$ and $i = 1, 2$. Since Γ_h contains piecewise constant fields, $\int_\Omega (\Pi_h \theta - \theta) \cdot \gamma_h = 0$ for all $\gamma_h \in \Gamma_h$ and $\theta \in H^1_0(\Omega)^2$.

The transversal displacement can be treated analogously. For every edge e of the triangulation, we define a linear mapping $\pi^2_h : H^1_0(\Omega) \to \mathcal{M}^2_{0,0}$ with

$$\int_e (\pi^2_h v - v) ds = 0.$$

This requires one degree of freedom per edge, which can be the value at the midpoint of the edge. In analogy with Π_h, we set

$$\Pi^2_h v := \pi^0_h v + \pi^2_h(v - \pi^0_h v),$$

so that $\int_e (\Pi^2_h v - v) ds = 0$ for every edge e of the triangulation. Using Green's formula, we have

$$\int_T \mathrm{grad}(\Pi^2_h v - v) \cdot \gamma_h dx = \int_{\partial T} (\pi^2_h v - v) \gamma_h \cdot n ds - \int_T (\pi^2_h v - v)\, \mathrm{div}\, \gamma_h dx = 0.$$

The contour integral vanishes by construction, and the second integral also vanishes since γ_h is constant on T. The boundedness of Π_h^2 follows as for π_h^1, and so the hypotheses of Fortin's criterion are satisfied since $(\operatorname{grad}\Pi_h^2 w - \Pi_h\theta, \gamma_h)_0 = (\operatorname{grad} w - \theta, \gamma_h)_0$ for $\gamma_h \in \mathcal{M}^0$. $\qquad\qquad\qquad\square$

More recently, Chapelle and Stenberg [1998] analyzed the finite element discretization with $\Theta_h := (\mathcal{M}_{0,0}^1 \oplus B_3)^2$ keeping W_h and Γ_h as in Fig. 63. They avoided the $H^{-1}(\operatorname{div})$-norm by using mesh-dependent norms and showed stability with respect to the norm whose square is

$$\|w\|_1^2 + \|\theta\|_1^2 + \frac{1}{h^2+t^2}\|\nabla w - \theta\|_0^2 + (h^2+t^2)\,\|\gamma\|_0^2.$$

In particular, the duality argument of Aubin–Nitsche could be more easily performed with these norms.

Problems

6.9 Let $\eta \in L_2(\Omega)^2$. Show that the spaces for the components of the decomposition (6.5) can be exchanged, i.e., that we can choose $\psi \in H^1(\Omega)/\mathbb{R}$ and $p \in H_0^1(\Omega)$. To this end, decompose $\eta^\perp$ according to (6.5), and write the result for $\eta^\perp$ as a decomposition of η.

6.10 Does $\psi \in H^1(\Omega)$, $q \in H^1(\Omega)/\mathbb{R}$ suffice to establish the orthogonality relation

$$(\nabla\psi, \operatorname{curl} q)_0 = 0,$$

or is a zero boundary condition required?

6.11 Show that

$$\|\operatorname{div}\eta\|_{-1} \le \operatorname{const}\, \sup_\gamma \frac{(\gamma, \eta)_0}{\|\gamma\|_{H(\operatorname{rot},\Omega)}},$$

and thus that $\operatorname{div}\eta \in H^{-1}(\Omega)$ for $\eta \in (H_0(\operatorname{rot}, \Omega))'$. Since $H_0(\operatorname{rot}, \Omega) \supset H_0^1(\Omega)$ implies $(H_0(\operatorname{rot}, \Omega))' \subset H^{-1}(\Omega)$, this completes the proof of (6.9).

6.12 In what sense do the solutions of (5.9) and (6.2) satisfy

$$\operatorname{div}\gamma = f?$$

6.13 Let $t > 0$, and suppose $H^1(\Omega)$ is endowed with the norm

$$\||v\|| := (\|v\|_0^2 + t^2\|v\|_1^2)^{1/2}.$$

The norm of the dual space $\||u\||_{-1} := \sup_v \frac{(u,v)_0}{\||v\||}$ can be estimated easily from above by

$$\||u\||_{-1} \leq \min\{\|u\|_0, \frac{1}{t}\|u\|_{-1}\}.$$

Give an example to show that there is no corresponding estimate from below with a constant independent of t by computing the size of

$$u(x) := \sin x + n \sin n^2 x \in H^0[0, \pi] \quad (1/t \leq n \leq 2/t)$$

sufficiently exactly in each of these norms.

6.14 The finite element space W_h contains H^1 conforming elements, and thus lies in $C(\Omega)$. Show that $\nabla W_h \subset H_0(\text{rot}, \Omega)$. What property of the rotation is responsible for this?

References

Argyris, J.H. (1957): Die Matrizentheorie der Statik. Ingenieur-Archiv XXV, 174–194

Arnold, D.N. (1981): Discretization by finite elements of a model parameter dependent problem. Numer. Math. 37, 405–421

Arnold, D.N. and Brezzi, F. (1985): Mixed and nonconforming finite element methods: implementation, postprocessing and error estimates. M^2AN 19, 7–32

Arnold, D.N. and Brezzi, F. (1993): Some new elements for the Reissner-Mindlin plate model. In *Boundary Value Problems for Partial Differential Equations and Applications* (J.-L. Lions and C. Baiocchi, eds.) pp. 287–292, Masson, Paris

Arnold, D.N., Brezzi, F., and Douglas, J. (1984): PEERS: A new mixed finite element for plane elasticity. Japan J. Appl. Math. 1, 347–367

Arnold, D.N., Brezzi, F., and Fortin, M. (1984): A stable finite element for the Stokes equations. Calcolo 21, 337–344

Arnold, D.N. and Falk, R.S. (1989): A uniformly accurate finite element method for the Mindlin-Reissner plate. SIAM J. Numer. Anal. 26, 1276–1290

Arnold, D.N., Scott, L.R., and Vogelius M. (1988): Regular inversion of the divergence operator with Dirichlet boundary conditions on a polygon, Ann. Scuola Norm. Sup. Pisa CI. Sci. (4), 15, 169–192

Aubin, J.P. (1967): Behaviour of the error of the approximate solution of boundary value problems for linear elliptic operators by Galerkin's and finite difference methods. Ann. Scuola Norm. Sup. Pisa 21, 599–637

Axelsson, O. (1980): Conjugate gradient type methods for unsymmetric and inconsistent systems of linear equations. Linear Algebra Appl. 29, 1–16

Axelsson, O. and Barker, V.A. (1984): *Finite Element Solution of Boundary Value Problems: Theory and Computation.* 432 pp., Academic Press, New York – London

Babuška, I. (1971): Error bounds for finite element method. Numer. Math. 16, 322–333

Babuška, I. and Aziz, A.K. (1972): Survey lectures on the mathematical foundations of the finite element method. In *The Mathematical Foundation of the Finite Element Method with Applications to Partial Differential Equations* (A.K. Aziz, ed.) pp. 3–363, Academic Press, New York – London

Babuška, I., Duran, R., and Rodríguez, R. (1992): Analysis of the efficiency of an a posteriori error estimator for linear triangular finite elements. SIAM J. Numer. Anal. 29, 947–964

Babuška, I., Osborn, J., and Pitkäranta, J. (1980): Analysis of mixed methods using mesh dependent norms. Math. Comp. 35, 1039–1062

Babuška, I. and Pitkäranta, J. (1990): The plate paradox for hard and soft simple supported plate. SIAM J. Math. Anal. 21, 551–576

Babuška, I. and Rheinboldt, W.C. (1978a): Error estimates for adaptive finite element computations. SIAM J. Numer. Anal. 15, 736–754

Babuška, I. and Rheinboldt, W.C. (1978b): A posteriori error estimates for the finite element method. Int. J. Numer. Meth. Engrg. 12, 1597–1607

Babuška, I. and Suri, M. (1992): Locking effects in the finite element approximation of elasticity problems. Numer. Math. 62, 439–463

Bachvalov, N.S. (1966): On the convergence of a relaxation method with natural constraints on the elliptic operator. USSR Comput. Math. math. Phys. 6(5), 101–135

Bank, R.E. (1990): *PLTMG: A Software Package for Solving Elliptic Partial Differential Equations. User's Guide 6.0.* SIAM, Philadelphia

Bank, R.E. and Dupont, T. (1981): An optimal order process for solving finite element equations. Math. Comp. 36, 35–51

Bank, R.E., Dupont, T., and Yserentant, H. (1988): The hierarchical basis multigrid method. Numer. Math. 52, 427–458

Bank, R.E., Weiser, A. (1985): Some a posteriori error estimators for elliptic partial differential equations. Math. Comp. 44, 283–301

Bank, R.E., Welfert, B., and Yserentant, H. (1990): A class of iterative methods for solving saddle point problems. Numer. Math. 56, 645–666

Bathe, K.-J. (1986): *Finite-Elemente-Methoden* [Engl. translation: *Finite Element Procedures in Engineering Analysis*]. 820 pp., Springer-Verlag, Berlin – Heidelberg – New York

Bathe, K.-J., Brezzi, F., and Cho, S.W. (1989): The MITC7 and MITC9 plate bending elements. J. Computers & Structures 32, 797–814

Bathe, K.J., Bucalem, M.L., and Brezzi, F. (1991/92): Displacement and stress convergence of our MITC plate bending elements. J. Engin. Comp.

Batoz, J.-L., Bathe, K.-J., and Ho, L.W. (1980): A study of three-node triangular plate bending elements. Int. J. Num. Meth. Engrg. 15, 1771–1812

Berger, A., Scott, R., and Strang, G. (1972): Approximate boundary conditions in the finite element method. Symposia Mathematica 10, 295–313

Blum, H. (1991): Private communication

Blum, H., and Rannacher, R. (1980): On the boundary value problem of the biharmonic operator on domains with angular corners. Math. Methods Appl. Sci. 2, 556–581

Bornemann, F. and Deuflhard, P. (1996): The cascadic multigrid method for elliptic problems, Numer. Math. 75, 135–152

Braess, D. (1981): The contraction number of a multigrid method for solving the Poisson equation. Numer. Math. 37, 387–404

Braess, D. (1988): A multigrid method for the membrane problem. Comput. Mechanics 3, 321–329

Braess, D. (1996): Stability of Saddle Point Problems with Penalty. RAIRO Anal. Numér. 30, n° 6, 731–742

Braess, D. (1998): Enhanced assumed strain elements and locking in membrane problems. Comput. Methods Appl. Mech. Engrg. 165, 155–174 (1998)

Braess, D. and Blömer, C. (1990): A multigrid method for a parameter dependent problem in solid mechanics. Numer. Math. 57, 747–761

Braess D. and Dahmen W. (1999): A cascadic multigrid algorithm for the Stokes problem. Numer. Math. 82, 179–191

Braess, D., Deuflhard, P., and Lipnikov, L. (1999): A subspace cascadic multigrid method for mortar elements. Preprint SC 99-07, Konrad–Zuse–Zentrum, Berlin

Braess, D., Dryja, M., and Hackbusch, W. (1999): A multigrid method for nonconforming FE-discretisations with application to non-matching grids. Computing 63, 1–25

Braess, D. and Hackbusch, W. (1983): A new convergence proof for the multigrid method including the V-cycle. SIAM J. Numer. Anal. 20, 967–975

Braess, D., Klaas, O., Niekamp, R., Stein, E., and Wobschal, F. (1995): Error indicators for mixed finite elements in 2-dimensional linear elasticity. Comput. Methods Appl. Mech. Engrg. 127, 345–356

Braess, D. and Peisker, P. (1986): On the numerical solution of the biharmonic equation and the role of squaring matrices for preconditioning. IMA J. Numer. Anal. 6, 393–404

Braess, D. and Peisker, P. (1992): Uniform convergence of mixed interpolated elements for Reissner-Mindlin plates. M^2AN (RAIRO) 26, 557–574

Braess, D. and Sarazin, R. (1997): An efficient smoother for the Stokes problem. Appl. Num. Math. 23, 3–19

Braess, D. and Verfürth, R. (1990): Multigrid methods for nonconforming finite element methods. SIAM J. Numer. Anal. 27, 979–986

Braess, D. and Verfürth, R. (1996): A posteriori error estimators for the Raviart-Thomas element. SIAM J. Numer. Anal. 33, 2431–2444

Bramble, J.H. and Hilbert, S.R. (1970): Estimation of linear functionals on Sobolev spaces with applications to Fourier transforms and spline interpolation. SIAM J. Numer. Anal. 7, 113–124

Bramble, J.H. and Pasciak, J. (1988): A preconditioning technique for indefinite systems resulting from mixed approximations of elliptic problems. Math. Comp. 50, 1–18

Bramble, J.H., Pasciak, J., and Xu, J. (1990): Parallel multilevel preconditioners. Math. Comp. 55, 1–22

Bramble, J.H., Pasciak, J., Wang, J., and Xu, J. (1991): Convergence estimates for multigrid algorithms without regularity assumptions. Math. Comp. 57, 23–45

Brandt, A. (1977): Multi-level adaptive solutions to boundary-value problems. Math. Comp. 31, 333–390

Brandt, A. and Dinar, D. (1979): Multigrid solutions to elliptic flow problems. In *Numerical Methods for Partial Differential Equations* (S. Parter, ed.) pp. 53–147, Academic Press

Brenner, S. (1989): An optimal-order multigrid method for P1 nonconforming finite elements. Math. Comp. 52, 1–15

Brezzi, F. (1974): On the existence, uniqueness and approximation of saddle-point problems arising from Lagrangian multipliers. RAIRO Anal. Numér. 8, R-2, 129–151

Brezzi, F., Bathe, K.-J., and Fortin, M. (1989): Mixed-interpolated elements for Reissner-Mindlin plates. Int. J. Num. Meth. Eng. 28, 1787–1801

Brezzi, F., Douglas, J., Durán, R., and Fortin, M. (1987): Mixed finite elements for second order elliptic problems in three variables. Numer. Math. 51, 237–250

Brezzi, F., Douglas, J., and Marini, L.D. (1985): Two families of mixed finite elements for second order elliptic problems. Numer. Math. 47, 217–235

Brezzi, F. and Fortin, M. (1986): Numerical approximation of Mindlin-Reissner plates. Math. Comp. 47, 151–158

Brezzi, F. and Fortin, M. (1991): *Mixed and Hybrid Finite Element Methods.* 356 pp., Springer-Verlag, Berlin – Heidelberg – New York

Brezzi, F., Fortin, M., and Stenberg, R. (1991): Error analysis of mixed-interpolated elements for Reissner-Mindlin plates. Math. Models and Methods in Appl. Sci. 1, 125–151

Briggs, W.L. (1987): *A Multigrid Tutorial.* 88 pp., SIAM, Philadelphia

Ciarlet, Ph. (1978): *The Finite Element Method for Elliptic Problems.* 530 pp., North-Holland, Amsterdam

Chapelle, D. and Stenberg, R. (1998): An optimal low-order locking-free finite element method for Mindlin–Reissner plates. Math. Models and Methods in Appl. Sci. 8, 407–430

Ciarlet, Ph. (1988): *Mathematical Elasticity. Volume I: Three-Dimensional Elasticity.* 451 pp., North-Holland, Amsterdam

Ciarlet, Ph. and Lions, J.L. (eds.) (1990): *Handbook of Numerical Analysis I. Finite difference methods (Part 1),* Solution of equations in $\mathbb{R}^n$ (Part 1). North-Holland, Amsterdam, 652 pp.

Clément, P. (1975): Approximation by finite element functions using local regularization. RAIRO Anal. Numér. 9, R-2, 77–84

Courant, R. (1943): Variational methods for the solution of problems of equilibrium and vibrations. Bull. Amer. Math. Soc. 49, 1–23

Crouzeix, M. and Raviart, P.A. (1973): Conforming and nonconforming finite element methods for solving the stationary Stokes equations. RAIRO Anal. Numér. 7, R-3, 33–76

Deuflhard, P., Leinen, P., and Yserentant, H. (1989): Concepts of an adaptive hierarchical finite element code. Impact Computing Sci. Engrg. 1, 3–35

Dörfler, W. (1996): A convergent adaptive algorithm for Poisson's Equation. SIAM J. Numer. Anal. 33, 1106–1124

Duvaut, G. and Lions, J.L. (1976): *Les Inéquations en Mécanique et en Physique.* Dunod, Paris

Falk, R.S., (1991): Nonconforming finite element methods for the equations of linear elasticity. Math. Comp. 57, 529–550

Fedorenko, R.P. (1961): A relaxation method for solving elliptic difference equations. USSR Comput. Math. math. Phys. 1(5), 1092–1096

Fedorenko, R.P. (1964): The speed of convergence of one iterative process. USSR Comput. Math. math. Phys. 4(3), 227–235

Fortin, M. (1977): An analysis of the convergence of mixed finite element methods. RAIRO Anal. Numér. 11, 341–354

Fortin, M. and Glowinski, R. (eds.) (1983): *Augmented Lagrangian Methods: Applications to the Numerical Solution of Boundary-Value Problems.* 340 pp., North-Holland, Amsterdam – New York

Frehse, J. and Rannacher, R. (1978): Asymptotic L^∞-error estimates for linear finite element approximations of quasilinear boundary value problems SIAM J. Numer. Anal. 15, 418–431

Freund, Gutknecht, and Nachtigal (1993) An implementation of the look-ahead Lanczos algorithm for non-Hermitean matrices SIAM J. Sci. Stat. Comput. 14, 137–158

Freudenthal, H (1942): Simplizialzerlegungen von beschränkter Flachheit. Annals Math. 43, 580–582

Gilbarg, D. and Trudinger, N.S. (1983): *Elliptic Partial Differential Equations of Second Order.* 511 pp., Springer-Verlag, Berlin – Heidelberg – New York

Girault, V. and Raviart, P.-A. (1986): *Finite Element Methods for Navier-Stokes Equations.* 374 pp., Springer-Verlag, Berlin – Heidelberg – New York

Glowinski, R. (1984): *Numerical Methods for Nonlinear Variational Problems.* 493 pp., Springer-Verlag, Berlin – Heidelberg – New York

Golub, G.H. and van Loan, S.F. (1983): *Matrix Computations.* 476 pp., The John Hopkins University Press, Baltimore, Maryland

Griebel, M. (1994): *Multilevelmethoden als Iterationsverfahren über Erzeugendensystemen* Teubner, Stuttgart

Gustafsson, I. (1978): A class of first order factorization methods. BIT 18, 142–156

Hackbusch, W. (1976): A fast iterative method for solving Poisson's equation in a general domain. In *Numerical Treatment of Differential Equations, Proc. Oberwolfach, Juli 1976* (R. Bulirsch, R.D. Grigorieff, and J. Schröder, eds.) pp. 51–62, Springer-Verlag, Berlin – Heidelberg – New York 1977

Hackbusch, W. (1985): *Multi-Grid Methods and Applications.* 377 pp., Springer-Verlag, Berlin – Heidelberg – New York

Hackbusch, W. (1986): *Theorie und Numerik elliptischer Differentialgleichungen.* 270 pp., Teubner, Stuttgart

Hackbusch, W. (1989): On first and second order box schemes. Computing 41, 277–296

Hackbusch, W. (1991): *Iterative Lösung großer schwachbesetzter Gleichungssysteme.* 382 pp., Teubner, Stuttgart, [Engl. translation: *Iterative Solution of Large Sparse Systems of Equations.*] 429 pp. Springer-Verlag, Berlin – Heidelberg – New York 1994

Hackbusch, W. and Reusken, A. (1989): Analysis of a damped nonlinear multilevel method. Numer. Math. 55, 225–246

Hackbusch, W. and Trottenberg, U. (eds.) (1982): *Multigrid Methods.* Springer-Verlag, Berlin – Heidelberg – New York

Hadamard, J. (1932): *Le problème de Cauchy et les équations aux dérivées partielles linéaires hyperboliques.* Hermann, Paris

Hemker, P.W. (1980): The incomplete LU-decomposition as a relaxation method in multigrid algorithms. In *Boundary and Interior Layers – Computational and Asymptotic Methods* (J.J.H. Miller, ed.), Boole Press, Dublin

Hestenes, M.R. and Stiefel, E. (1952): Methods of conjugate gradients for solving linear systems. J. Res. NBS 49, 409–436

Huang, Z. (1990): A multi-grid algorithm for mixed problems with penalty. Numer. Math. 57, 227–247

Hughes, T.J.R., Ferencz, R.M., and Hallquist, J.O. (1987): Large-scale vectorized implicit calculations in solid mechanics on a CRAY-XMP48 utilizing EBE preconditioned conjugate gradients. Comput. Meth. Appl. Mech. Engrg. 62, 215–248

Johnson, C. and Pitkäranta, J. (1982): Analysis of some mixed finite element methods related to reduced integration. Math. Comp. 38, 375–400

Kadlec, J. (1964): On the regularity of the solution of the Poisson problem on a domain with boundary locally similar to the boundary of a convex open set [Russian]. Czech. Math. J. 14(89), 386–393

Kirmse, A. (1990): Private communication

Lascaux, P. and Lesaint, P. (1975): Some nonconforming finite elements for the plate bending problem. RAIRO Anal. Numér. 9, R-1, 9–53

Marsden, J.E. and Hughes, T.J.R. (1983): *Mathematical Foundations of Elasticity.* Prentice-Hall, Englewood Cliffs, New Jersey

McCormick, S.F. (1989): *Multilevel Adaptive Methods for Partial Differential Equations.* SIAM, Philadelphia

Meier, U. and Sameh, A. (1988): The behaviour of conjugate gradient algorithms on a multivector processor with a hierarchical memory. J. comp. appl. Math. 24, 13–32

Meijerink, J.A. and van der Vorst, H.A. (1977): An iterative solution method for linear systems of which the coefficient matrix is a symmetric M-matrix. Math. Comp. 31, 148–162

Nicolaides, R.A. (1977): On the ℓ^2-convergence of an algorithm for solving finite element equations. Math. Comp. 31, 892–906

Nitsche, J.A. (1968): Ein Kriterium für die Quasioptimalität des Ritzschen Verfahrens. Numer. Math. 11, 346–348

Nitsche, J.A. (1981): On Korn's second inequality. RAIRO Anal Numér. 15, 237–248, Numer. Math. 11, 346–348

Ortega, J.M. (1988): *Introduction to Parallel and Vector Solution of Linear Systems.* Plenum Press, New York

Ortega, J.M. and Voigt, R.G. (1985): Solution of partial differential equations on vector and parallel computers. SIAM Review 27, 149–240

Oswald, P. (1994): *Multilevel Finite Element Approximation.* Teubner, Stuttgart

Paige, C.C. and Saunders, M.A. (1975): Solution of sparse indefinite systems of linear equations. SIAM J. Numer. Anal. 12, 617–629

Parter, S.V. (1987): Remarks on multigrid convergence theorems. Appl. Math. Comp. 23, 103–120

Peisker, P. and Braess, D. (1992): Uniform convergence of mixed interpolated elements for Reissner-Mindlin plates. RAIRO Anal. Numér. 26, 557–574

Peisker, P., Rust, W., and Stein, E. (1990): Iterative solution methods for plate bending problems: multi-grid and preconditioned cg algorithm. SIAM J. Numer. Anal. 27, 1450–1465

Pitkäranta, J. (1988): Analysis of some low-order finite element schemes for Mindlin-Reissner and Kirchhoff plates. Numer. Math. 53, 237–254

Pitkäranta, J. (1992): The problem of membrane locking in finite element analysis of cylindrical shells. Numer. Math. 61, 523–542

Pitkäranta, J. (19998): The first locking-free plane-elastic finite element: Historia mathematica. Helsinki, Report A411

Pitkäranta, J. and Suri, M. (1996): Design principles and error analysis for reduced-shear plate-bending finite elements. Numer. Math. 75,223–266

Powell, M.J.D. (1977): Restart procedures for the conjugate gradient method. Math. Programming 12, 241–254

Prager, W. and Synge, J.L. (1949): Approximations in elasticity based on the concept of function spaces. Quart. Appl. Math. 5, 241–269

Rannacher, R. and Turek, S. (1992) Simple nonconforming quadrilateral Stokes element. Numer. Meth. for Partial Differential Equs. 8, 97–111

Raviart, P.A. and Thomas, J.M. (1977): A mixed finite element method for second order elliptic problems. In *Mathematical Aspects of Finite Element Methods* (I. Galligani and E. Magenes, eds.) pp. 292–315, Springer-Verlag, Berlin – Heidelberg – New York

Reddy, B.D. (1988): Convergence of mixed finite element approximations for the shallow arc problem. Numer. Math. 53, 687–699

Reid, J.K. (ed.) (1971): *Large Sparse Sets of Linear Equations.* Academic Press, New York

Reissner, E. (1950): On a variational theorem in elasticity. J. Math. Phys. 29, 90–95

Reusken, A. (1992): On maximum convergence of multigrid methods for two-point boundary value problems. SIAM J. Numer. Anal. 29, 1569–1578

344   References

Ries, M., Trottenberg, U., and Winter, G. (1983): A note on MGR methods. Linear Algebra Appl. 49, 1–26

Ritz, W. (1908): Über eine neue Methode zu Lösung gewisser Variationsprobleme der mathematischen Physik. J. reine angew. Math. 135, 1–61

Rivara, M.C. (1984) Algorithms for refining triangular grids suitable for adaptive and multigrid techniques. Int. J. Numer. Meth. Engrg. 20, 745–756

Rivlin, R.S. and Ericksen, J.L. (1955): Stress-deformation relations for isotropic materials. J. Rational Mech. Anal. 4, 323–425

Saad, Y. (1993): A flexible inner-outer preconditioned GMRES-algorithm SIAM J. Sci. Stat. Comput. 14, 461–469

Saad, Y. and Schultz, M.H. (1985): Conjugate gradient-like algorithms for solving nonsymmetric linear systems. Math. Comp. 44, 417–424

Saad, Y. and Schultz, M.H. (1986): GMRES: A generalized minimal residual algorithm for solving nonsymmetric linear systems. SIAM J. Sci. Stat. Comput. 7, 856–869

Schellbach (1851): Probleme der Variationsrechnung. J. reine angew. Math. 41, 293–363 (see §30 and Fig. 11)

Schwab, Ch. (1998): *p- and hp- Finite Element Methods.* Clarendon Press, Oxford

Scott, L.R. and Zhang, S. (1998): Finite element interpolation of nonsmooth functions. Math. Comp. 54, 483–493

Shaidurov, V.V. (1996): Some estimates of the rate of convergence for the cascadic conjugate-gradient method. Computers Math. Applic. 31, 161–171

Simo, J.C. and Rifai, M.S. (1990): A class of assumed strain methods and the method of incompatible modes. Int. J. Numer. Meth. Engrg. 29, 1595–1638

Stein, E. and Wriggers, P. (1997): *Finite Element Methods in Non-Linear Continuum Mechanics.* Springer-Verlag, Berlin – Heidelberg – New York

Stein, E. and Wunderlich, W. (1973): Finite-Element-Methoden als direkte Variationsverfahren der Elastostatik. In *Finite Elemente in der Statik* (K.E. Buck, D.W. Scharpf, E. Stein, and W. Wunderlich, eds.) pp. 71–125. Verlag von Wilhelm Ernst & Sohn, Berlin

Stenberg, R. (1988): A family of mixed finite elements for the elasticity problem. Numer. Math. 53, 513–538

Stoer, J. (1983): Solution of large linear systems of equations by conjugate gradient type methods. In *Mathematical Programming: The State of the Art* (A. Bachem, M. Grötschel, B. Korte, eds.) pp. 540–565, Springer-Verlag, Berlin – Heidelberg – New York

Stolarski, H. and Belytschko, T. (1987): Limitation principles for mixed finite elements based on the Hu-Washizu variational formulation. Comput. Methods Appl. Mech. Eng. 60, 195–216

Strang, G. and Fix, G.J. (1973): *An Analysis of the Finite Element Method.* 306 pp., Prentice-Hall, Englewood Cliffs, New Jersey

Suri, M., Babuška, I., and Schwab, C. (1995): Locking effects in the finite element approximation of plate models. Math. Comp. 64, 461–482

Truesdell, C. (1977/1991): *A First Course in Rational Continuum Mechanics.* Academic Press, New York – London

Turner, M.J., Clough, R.M., Martin, H.C., and Topp, L.J. (1956): Stiffness and deflection analysis of complex structures. J. Aeron. Sci. 23, 805–823, 854

van der Vorst, H.A. (1992): BiCG-STAB: A fast and smoothly converging variant of BiCG for the solution of nonsymmetric linear systems. SIAM J. Sci. Stat. Comput. 13, 631–644

Varga, R.S. (1960): Factorization and normalized iterative methods. In *Boundary Value Problems in Differential Equations* (R.E. Langer, ed.) pp. 121–142. University of Wisconsin Press, Madison

Varga, R.S. (1962): *Matrix Iterative Analysis.* 322 pp., Prentice-Hall, Englewood Cliffs, New Jersey

Verfürth, R. (1984): Error estimates for a mixed finite element approximation of the Stokes equations. RAIRO Anal. Numér. 18, 175–182

Verfürth, R. (1988): Multi-level algorithms for mixed problems II. Treatment of the Mini-Element. SIAM J. Numer. Anal. 25, 285–293

Verfürth, R. (1994): A posteriori error estimation and adaptive mesh-refinement techniques. J. Comp. Appl. Math. 50, 67–83

Verfürth, R. (1996): *A Review of A Posteriori Error Esimation and Adaptive Mesh-Refinement Techniques.* 129 pp., Wiley-Teubner, Chichester – New York – Stuttgart

Verfürth, R. (1997): A posteriori error estimates for low order finite elements.

Vogelius, M. (1983): An analysis of the p-version of the finite element method for nearly incompressible materials. Uniformly valid, optimal error estimates, Numer. Math. 41, 39–53

Wachspress, E.L. (1971): A rational basis for function approximation. J. Inst. Math. Appl. 8, 57–68

Washizu, K. (1968): *Variational Methods in Elasticity and Plasticity.* Pergamon Press, Cambridge

Werner, H. (1982): *Praktische Mathematik I.* 285 pp., Springer-Verlag, Berlin – Heidelberg – New York

Wesseling, W. (1992): *An Introduction to Multigrid Methods.* 287 pp., John Wiley & Sons, Chichester

Widlund, O. (1988): Iterative substructuring methods: Algorithms and theory for elliptic problems in the plane. In *Proceedings of the First International Symposium on Domain Decomposition Methods for Partial Differential Equations* (R. Glowinski, G.H. Golub, G.A. Meurant, and J. Periaux, eds.), SIAM, Philadelphia

Wittum, G. (1989): On the convergence of multi-grid methods with transforming smoothers. Theory with applications to the Navier-Stokes equations. Numer. Math. 57, 15–38

Wittum, G. (1989a): Private communication

Wittum, G. (1989b): On the robustness of ILU smoothing. SIAM J. Sci. Stat. Comput. 10, 699–717

Xu, J. (1992): Iterative methods by space decomposition and subspace correction. SIAM Review 34, 581–613

Yeo, S.T., and Lee, B.C. (1996): Equivalence between enhanced assumed strain method and assumed stress hybrid method based on the Hellinger–Reissner principle. Int. J. Numer. Meths. Engrg. 39, 3083–3099

Yosida, K. (1971): *Functional Analysis* (3rd edition). 475 pp., Springer-Verlag, Berlin – Heidelberg – New York

Young, D.M. (1971): *Iterative Solution of Large Linear Systems.* 570 pp., Academic Press, New York – London

Young, D.M. and Kang, C. Jea (1980): Generalized conjugate-gradient acceleration of nonsymmetrizable iterative methods. Linear Algebra Appl. 34, 159–194

Yserentant, H. (1986): On the multi-level splitting of finite element spaces. Numer. Math. 49, 379–412

Yserentant, H. (1990): Two preconditioners based on the multi-level splitting of finite element spaces. Numer. Math. 58, 163–184

Yserentant, H. (1993): Old and new convergence proofs for multigrid methods. Acta Numerica 3, 285–326

Zienkiewicz, O.C. (1971): *The Finite Element Method in Structural and Continuum Mechanics.* McGraw-Hill, London

Zienkiewicz, O.C. and Zhu, J.Z. (1987): A simple error estimator and adaptive procedure for practical engineering analysis. Int. J. Numer. Meth. Engrg. 24, 337–357

Some Additional Books on Finite Elements

Akin, J. Ed. (1986): *Finite Element Analysis for Undergraduates*. Academic Press, New York – London

Bramble, J.H. (1993): *Multigrid Methods*. John Wiley & Sons, Chichester

Brenner, S.C. and Scott, L.R. (1996): *The Mathematical Theory of Finite Elements*. Springer-Verlag, Berlin – Heidelberg – New York

Cuvelier, C., Segal, A., and Steenhoven, A.A. (1986): *Finite Element Methods and Navier-Stokes Equations*. 483 pp. D. Reidel, Dordrecht – Boston

Goering, H., Roos, H.-G. and Tobiska, L. (1988): *Finite-Element-Methode*. 193 pp. Wissenschaftliche Taschenbücher Bd. 285, Akademie-Verlag, Berlin

Hughes, T.J.R. (1987): *The Finite Element Method*. 802 pp., Prentice Hall, Englewood Cliffs, New Jersey

Johnson, C. (1987): *Numerical Solution of Partial Differential Equations by the Finite Element Method*. Cambridge University Press

Laurie, D.P. (ed.) (1983): *Numerical Solution of Partial Differential Equations: Theory, Tools and Case Studies*. 341 pp., Birkhäuser, Basel

Marsal, D. (1976): *Die numerische Lösung partieller Differentialgleichungen in Wissenschaft und Technik*. 574 pp., BI Mannheim - Wien - Zürich

Mitchell, A. and Wait, R. (1976): *The Finite Element Method in Partial Differential Equations*. John Wiley & Sons, Chichester

Quarteroni, A. and Valli, A. (1994) *Numerical Approximation of Partial Differential Equations*. Springer-Verlag, Berlin – Heidelberg – New York

Raviart, P.A. et Thomas, J.M. (1987): *Introduction à l'Analyse Numérique des Équations aux Dérivées Partielles*. Masson, Paris

Schwarz, H.R. (1984): *Methode der finiten Elemente*. 346 pp. (2. Auflage), Teubner, Stuttgart

Temam, R. (1977): *Navier-Stokes Equations. Theory and Numerical Analysis*. 500 pp., North-Holland, Amsterdam

Wachspress, E.L. (1966): *Iterative Solution of Elliptic Systems*. Prentice Hall, Englewood Cliffs, New Jersey

348

Index

Printed in the United Kingdom
by Lightning Source UK Ltd.
105064UKS00001B/124